T0214620

Theory of Distributions

Svetlin G. Georgiev

Theory of Distributions

Second Edition

 Springer

Svetlin G. Georgiev
Department of Differential Equations
Faculty of Mathematics and Informatics
University of Sofia "St. Kliment Ohridski"
Sofia, Bulgaria

ISBN 978-3-030-81264-5   ISBN 978-3-030-81265-2   (eBook)
https://doi.org/10.1007/978-3-030-81265-2

1st edition: © Springer International Publishing 2015

2nd edition: © The Editor(s) (if applicable) and The Author(s), under exclusive license to Springer Nature Switzerland AG 2021

This work is subject to copyright. All rights are solely and exclusively licensed by the Publisher, whether the whole or part of the material is concerned, specifically the rights of translation, reprinting, reuse of illustrations, recitation, broadcasting, reproduction on microfilms or in any other physical way, and transmission or information storage and retrieval, electronic adaptation, computer software, or by similar or dissimilar methodology now known or hereafter developed.

The use of general descriptive names, registered names, trademarks, service marks, etc. in this publication does not imply, even in the absence of a specific statement, that such names are exempt from the relevant protective laws and regulations and therefore free for general use.

The publisher, the authors, and the editors are safe to assume that the advice and information in this book are believed to be true and accurate at the date of publication. Neither the publisher nor the authors or the editors give a warranty, expressed or implied, with respect to the material contained herein or for any errors or omissions that may have been made. The publisher remains neutral with regard to jurisdictional claims in published maps and institutional affiliations.

This Springer imprint is published by the registered company Springer Nature Switzerland AG.
The registered company address is: Gewerbestrasse 11, 6330 Cham, Switzerland

Preface to the Second Edition

In the 5 years since the first edition of this book was published, I have received a lot of messages and letters from readers commenting on the book and suggesting how it could be improved. With the aim of this information, I have revised the first edition of the book. The changes its second edition are as follows.

In Chap. 1, a section titled "L^p Spaces" has been added. In this section, L^p spaces for $p \geq 1$ are introduced. They are deduced from the Hölder, Young, Minkowski inequalities and an interpolation inequality. In this section, certain criteria for uniform integrability of some classes of functions are given. The Riesz-Fischer theorem and the L^p dominated convergence theorem are studied. Also, the conception for separable and dual spaces is introduced and the Riesz representation theorem for bounded linear functionals on L^p spaces is proven. A new section titled "Change of Variables" has been added to Chap. 2. Here, the change of variables for some classes of distributions is investigated and a representation of the Dirac delta function is provided. In Chap. 3, a new section titled "The Local Structure of Distributions" has been added. In this section, a criterion for linear and continuous extension of a distribution with compact support is provided and that any distribution with compact support has a finite order is proven.

A new section called "Notes and References" has been introduced in all chapters. In this section, some additional materials are provided for each chapter. Some problems are provided with detailed proofs. The book's index has been updated.

The aim of the second edition is to present a clear and well-organized treatment of the concept behind the development of mathematics and solution techniques. The material of this book is presented in a highly readable, mathematically solid format. Many practical problems are illustrated, displaying the scope of the theory of distributions.

Paris, France   Svetlin G. Georgiev
February 2021

Preface to the First Edition

The theory of partial differential equations is without a doubt one of the branches of analysis in which ideas and methods of different fields of mathematics manifest themselves and are interlaced—from functional and harmonic analysis to differential geometry and topology. Because of that, the study of this topic represents a constant endeavour and requires undertaking several challenges. The main aim of this book is to explain many of the fundamental ideas underlying the theory of distributions.

The book consists of ten chapters. The first chapter deals with the well-known classical theory regarding the space $\mathscr{C}^{\infty}$, the Schwartz space and the convolution of locally integrable functions. It may also serve as an introduction to typical questions related to cones in $\mathbb{R}^{n}$. Chapter 2 collects the definitions of distributions, their order, sequences, support and singular support, and multiplication by $\mathscr{C}^{\infty}$ functions. In Chaps. 3 and 4 we introduce differentiation and homogeneous distributions. The notion of direct multiplication of distributions is developed in Chap. 5. The following two Chaps. 6 and 7, deal with specific problems about convolutions and tempered distributions. In Chaps. 8 and 9 we collected basic material and problems regarding integral transforms. Sobolev spaces are discussed in the tenth, and final, chapter.

The volume is aimed at graduate students and mathematicians seeking an accessible introduction to some aspects of the theory of distributions, and is well suited for a one-semester lecture course.

It is a pleasure to acknowledge the great help I received from Professor Mokhtar Kirane, University of La Rochelle, La Rochelle, France, who made valuable suggestions that have been incorporated in the text.

I express my gratitude in advance to anybody who will inform me about mistakes, misprints, or express criticism or other comments, by writing to the e-mail addresses svetlingeorgiev1@gmail.com, sgg2000bg@yahoo.com.

Paris, France
January 2015

Svetlin G. Georgiev

Contents

Chapter 1
Introduction

1.1 The Spaces $\mathscr{C}_0^\infty$ and $\mathscr{S}$

With $\mathbb{N}^n \cup \{0\}$ we denote the space of multi-indices $\alpha = (\alpha_1, \alpha_2, \ldots, \alpha_n)$, $\alpha_k \in \mathbb{N} \cup \{0\}$, $k = 1, 2, \ldots, n$. For $\alpha = (\alpha_1, \alpha_2, \ldots, \alpha_n)$, $\beta = (\beta_1, \beta_2, \ldots, \beta_n) \in \mathbb{N}^n \cup \{0\}$, we will write $\alpha \leq \beta$ if $\alpha_k \leq \beta_k$, $k = 1, 2, \ldots, n$. Set $D = (D_1, D_2, \ldots, D_n)$,

$$D_k = \frac{\partial}{\partial x_k}, k = 1, 2, \ldots, n, D^\alpha = \frac{\partial^{|\alpha|}}{\partial_{x_1}^{\alpha_1} \partial_{x_2}^{\alpha_2} \ldots \partial_{x_n}^{\alpha_n}}.$$

Let $X \subset \mathbb{R}^n$ be an open set. If $K \subset X$ is a compact set we shall write $K \subset\subset X$. The following conventions will also be used throughout the book: $U(x_0, R)$ is the open ball of radius R with centre at the point x_0, $S(x_0, R) = \partial U(x_0, R)$ is the sphere of radius R with centre at x_0, $\overline{U}(x_0, R) = U(x_0, R) \cup S(x_0, R)$ is the closed ball of radius R with centre at the point x_0 and $U_R = U(0, R)$, $S_R = S(0, R)$, $\overline{U}_R = \overline{U}(0, R)$.

If A and B are sets in $\mathbb{R}^n$, by $d(A, B)$ or $\mathrm{dist}(A, B)$ we shall denote the distance between the sets A and B, that is

$$d(A, B) = \mathrm{dist}(A, B) = \inf_{x \in A, y \in B} |x - y|.$$

We shall use A^ϵ to denote the ϵ-neighbourhood of a set A, i.e., $A^\epsilon = A + U_\epsilon$. If A is an open set, then A_ϵ will designate the set of points in A that are more than ϵ away from the boundary ∂A, i.e., $A_\epsilon = \{x : x \in A, \mathrm{dist}(x, \partial A) > \epsilon\}$.

We use $\mathrm{int}A$ to denote the set of interior points of the set A. With κ_A we will denote the characteristic function of A, i.e., $\kappa_A(x) = 1$ for $x \in A$ and $\kappa_A(x) = 0$ for $x \notin A$.

Definition 1.1 The set A is called convex if for any points x and y in A the segment

$$\lambda x + (1 - \lambda) y, \qquad \lambda \in [0, 1],$$

lies entirely in A.

© The Author(s), under exclusive license to Springer Nature Switzerland AG 2021
S. G. Georgiev, *Theory of Distributions*,
https://doi.org/10.1007/978-3-030-81265-2_1

We will write ch$A$ to denote the convex hull of a set A.

Definition 1.2 We call space of basic functions the space $\mathscr{C}_0^\infty(X)$ of smooth functions with compact support defined on X.

Example 1.1 The function

$$\omega(x) = \begin{cases} Ce^{-\frac{1}{1-|x|^2}}, & |x| \le 1, \\ 0, & |x| > 1, \end{cases}$$

where the constant C is chosen so that $\displaystyle\int_{\mathbb{R}^n} \omega(x)dx = 1$, belongs in $\mathscr{C}_0^\infty(\mathbb{R}^n)$.

Set $\omega_\epsilon(x) = \dfrac{1}{\epsilon^n}\omega\left(\dfrac{x}{\epsilon}\right)$, $x \in \mathbb{R}^n$.

Definition 1.3 The function ω_ϵ is called the "cap-shaped function".

We have

$$\int_{\mathbb{R}^n} \omega_\epsilon(x)dx = \frac{1}{\epsilon^n}\int_{\mathbb{R}^n} \omega\left(\frac{x}{\epsilon}\right)dx = \int_{\mathbb{R}^n} \omega(x)dx = 1.$$

Definition 1.4 We say that the sequence $\{\phi_k\}_{k=1}^\infty$ of elements of $\mathscr{C}_0^\infty(X)$ converges to the function $\phi \in \mathscr{C}_0^\infty(X)$ if there exists a compact set $K \subset X$ such that supp$\phi_k \subset K$ for every $k \in \mathbb{N}$ and $\lim_{k\to\infty} D^\alpha\phi_k(x) = D^\alpha\phi(x)$ uniformly for every multi-index $\alpha \in \mathbb{N}^n \cup \{0\}$.

Example 1.2 Take $\phi \in \mathscr{C}_0^\infty(\mathbb{R})$. The sequence $\left\{\dfrac{1}{k}\phi(x)\right\}_{k=1}^\infty$ converges to 0 in $\mathscr{C}_0^\infty(\mathbb{R})$.

Example 1.3 Take $\phi \in \mathscr{C}_0^\infty(\mathbb{R})$. Then the sequence $\left\{\dfrac{1}{k}\phi\left(\dfrac{x}{k}\right)\right\}_{k=1}^\infty$ does not converge to 0 in $\mathscr{C}_0^\infty(\mathbb{R})$.

Lemma 1.1 *For every set $X_1 \subset X$ and every $\epsilon > 0$ there exists a function $\phi_\epsilon \in \mathscr{C}^\infty(\mathbb{R}^n)$ such that $\phi_\epsilon(x) = 1$ when $x \in X_1^\epsilon$, $\phi_\epsilon(x) = 0$ when $x \in \mathbb{R}^n \backslash X_1^{3\epsilon}$ and $0 \le \phi_\epsilon(x) \le 1$ when $x \in \mathbb{R}^n$.*

Proof Let $\kappa_{X_1^{2\epsilon}}$ be the characteristic function of the set $X_1^{2\epsilon}$, i.e., $\kappa_{X_1^{2\epsilon}}(x) = 1$ for $x \in X_1^{2\epsilon}$ and $\kappa_{X_1^{2\epsilon}}(x) = 0$ for $x \notin X_1^{2\epsilon}$. Define

$$\phi_\epsilon(x) = \int_{\mathbb{R}^n} \kappa_{X_1^{2\epsilon}}(y)\omega_\epsilon(x-y)dy = \int_{X_1^{2\epsilon}} \omega_\epsilon(x-y)dy, \quad x \in \mathbb{R}^n.$$

We will prove that the function ϕ_ϵ has the required properties. Since $\omega_\epsilon \in \mathscr{C}^\infty(\mathbb{R}^n)$, $0 \leq \omega_\epsilon(x)$, $x \in \mathbb{R}^n$, $\mathrm{supp}\,\omega_\epsilon = \overline{U}_\epsilon$ and

$$\int_{\mathbb{R}^n} \omega_\epsilon(x)dx = 1,$$

we get $\phi_\epsilon \in \mathscr{C}^\infty(\mathbb{R}^n)$ and

$$0 \leq \phi_\epsilon(x)$$
$$= \int_{X_1^{2\epsilon}} \omega_\epsilon(x-y)dy$$
$$\leq \int_{\mathbb{R}^n} \omega_\epsilon(x-y)dy$$
$$= \int_{\mathbb{R}^n} \omega_\epsilon(y)dy$$
$$= 1, \quad x \in \mathbb{R}^n,$$

and

$$\phi_\epsilon(x) = \int_{\mathbb{R}^n} \kappa_{X_1^{2\epsilon}}(y)\omega_\epsilon(x-y)dy$$
$$= \int_{\overline{U}(x,\epsilon)} \kappa_{X_1^{2\epsilon}}(y)\omega_\epsilon(x-y)dy$$
$$= \begin{cases} \int_{\overline{U}(x,\epsilon)} \omega_\epsilon(x-y)dy = \int_{\overline{U}_\epsilon} \omega_\epsilon(y)dy = 1, & x \in X_1^\epsilon, \\ 0, & x \notin X_1^{3\epsilon}. \end{cases}$$

This completes the proof.

Lemma 1.2 (Expansion of the Unit Function) *Let $\phi \in \mathscr{C}_0^\infty(\mathbb{R}^n)$ and $\mathrm{supp}\,\phi$ be covered by a finite number of neighbourhoods $U(x_k, r_k)$ for some $x_k \in \mathbb{R}^n$, $r_k > 0$, $k = 1, 2, \ldots, m$, and for some $m \in \mathbb{N}$. Then there exist functions $\phi h_k \in \mathscr{C}_0^\infty(U(x_k, r_k))$, $k = 1, 2, \ldots, m$, such that $\sum_{k=1}^m h_k(x) = 1$ for x in a neighbourhood of $\mathrm{supp}\,\phi$ and $\mathrm{supp}(\phi h_k) \subset U(x_k, r_k)$, $k = 1, 2, \ldots, m$.*

Proof For $k = 1, 2, \ldots, m$, take $0 < r_k' < r_k$ so that the union $\cup_{k=1}^m U(x_k, r_k')$ covers $\mathrm{supp}\,\phi$. Be Lemma 1.1, it follows that there exist functions $\eta_k \in \mathscr{C}_0^\infty(U(x_k, r_k))$,

$k = 1, 2, \ldots, m$, so that $\eta_k(x) = 1$, $x \in U(x_k, r'_k)$, $\mathrm{supp}\eta_k \subset U(x_k, r_k)$, $0 \le \eta_k(x) \le 1$, $x \in \mathbb{R}^n$, $k = 1, 2, \ldots, m$. Then, for $x \in \cup_{k=1}^m U(x_k, r'_k)$, we set

$$h(x) = \sum_{k=1}^m \eta_k(x), \quad h_k(x) = \frac{\eta_k(x)}{h(x)}, \quad k = 1, 2, \ldots, m,$$

and for $x \in \mathbb{R}^n \setminus \left(\cup_{k=1}^m U(x_k, r_k) \right)$ we set $h(x) = h_k(x) = 0$, $k = 1, 2, \ldots, m$. Note that $\phi h_k \in \mathscr{C}_0^\infty (U(x_k, r_k))$, $k = 1, 2, \ldots, m$, and

$$h(x) \ge 1, \quad \sum_{k=1}^m h_k(x) = \sum_{k=1}^m \frac{\eta_k(x)}{h(x)} = 1, \quad x \in \cup_{k=1}^m U(x_k, r'_k).$$

This completes the proof.

Definition 1.5 We say that the sequence $\{\eta_k\}_{k=1}^\infty$ in $\mathscr{C}_0^\infty(\mathbb{R}^n)$ converges to 1 in $\mathbb{R}^n$ if

1. for every $\alpha \in \mathbb{N}^n \cup \{0\}$ there exists a constant $c_\alpha > 0$ such that $|D^\alpha \eta_k(x)| \le c_\alpha$ for every $k \in \mathbb{N}$ and every $x \in \mathbb{R}^n$,
2. for every compact set K in $\mathbb{R}^n$ there exists $N = N(K) \in \mathbb{N}$ such that $\eta_k(x) = 1$ for every $k > N$ and $x \in K$.

Example 1.4 Choose $\eta \in \mathscr{C}_0^\infty(\mathbb{R}^n)$ so that $\eta(x) = 1$ for $|x| \le 1$. Set $\eta_k(x) = \eta\left(\frac{x}{k}\right)$, $x \in \mathbb{R}^n$, $k \in \mathbb{N}$. Then the sequence $\{\eta_k\}_{k=1}^\infty$ tends to 1 in $\mathbb{R}^n$.

Definition 1.6 With $\mathscr{S}(\mathbb{R}^n)$ we denote the space of $\mathscr{C}^\infty$ functions ϕ such that

$$\sup_{x \in \mathbb{R}^n} |x|^\beta |D^\alpha \phi(x)| < \infty, \quad \forall \alpha \in \mathbb{N}^n \cup \{0\}, \quad \beta \in \mathbb{N} \cup \{0\}.$$

Here $|x| = (x_1^2 + x_2^2 + \cdots + x_n^2)^{\frac{1}{2}}$ and $x = (x_1, x_2, \ldots, x_n)$. By $\| \cdot \|_{\mathscr{S},p}$, $p \in \mathbb{N}$, we shall indicate the norm

$$\|\phi\|_{\mathscr{S},p} = \sup_{x \in \mathbb{R}^n, |\alpha| \le p} (1 + |x|^2)^{\frac{p}{2}} |D^\alpha \phi(x)|, \quad \phi \in \mathscr{S}(\mathbb{R}^n). \tag{1.1}$$

Example 1.5 Let $n = 1$ and $\phi(x) = e^{-x^2}$, $x \in \mathbb{R}$. Then $\phi \in \mathscr{S}(\mathbb{R})$.

Example 1.6 Let $n = 1$, ϕ be as in Example 1.5 and $p(x) = a_0 x^k + a_1 x^{k-1} + \cdots + a_{k-1} x + a_k$, $x \in \mathbb{R}$, where $a_j \in \mathbb{R}$, $j = 0, 1, \ldots, k$, $k \in \mathbb{N}$. Then $\phi p \in \mathscr{S}(\mathbb{R})$.

Exercise 1.1 Prove that (1.1) satisfies all axioms for a norm.

Theorem 1.1 *The space $\mathscr{C}_0^\infty(\mathbb{R}^n)$ is a proper subset of the space $\mathscr{S}(\mathbb{R}^n)$.*

Proof Let $\phi \in \mathscr{C}_0^\infty(\mathbb{R}^n)$, $\alpha \in \mathbb{N}^n \cup \{0\}$ and $\beta \in \mathbb{N} \cup \{0\}$ be arbitrarily chosen. Set $K = \text{supp}\phi$ and $\psi(x) = |x|^\beta D^\alpha \phi(x)$, $x \in \mathbb{R}^n$. Then $\text{supp}\psi = K$ and there exists a constant $K_1 > 0$ so that

$$\sup_{x \in \mathbb{R}^n} |\psi(x)| = \max_{x \in K} |\psi(x)| \le K_1 < \infty.$$

Because $\alpha \in \mathbb{N}^n \cup \{0\}$ and $\beta \in \mathbb{N} \cup \{0\}$ were arbitrarily chosen, we conclude that $\phi \in \mathscr{S}(\mathbb{R}^n)$. Now, using that $\phi \in \mathscr{C}_0^\infty(\mathbb{R}^n)$ was arbitrarily chosen and we get that it is an element of the space $\mathscr{S}(\mathbb{R}^n)$, we obtain the inclusion $\mathscr{C}_0^\infty(\mathbb{R}^n) \subseteq \mathscr{S}(\mathbb{R}^n)$. Let $\psi_1(x) = e^{-|x|^2}$, $x \in \mathbb{R}^n$. Then $\psi_1 \in \mathscr{S}(\mathbb{R}^n)$ and $\psi_1 \notin \mathscr{C}_0^\infty(\mathbb{R}^n)$. Thus, $\mathscr{C}_0^\infty(\mathbb{R}^n) \subset \mathscr{S}(\mathbb{R}^n)$. This completes the proof.

Theorem 1.2 Let $\phi, \psi \in \mathscr{S}(\mathbb{R}^n)$. Then $\phi\psi \in \mathscr{S}(\mathbb{R}^n)$.

Proof Let $\alpha \in \mathbb{N}^n \cup \{0\}$ and $\beta \in \mathbb{N} \cup \{0\}$ be arbitrarily chosen. Since $\phi, \psi \in \mathscr{S}(\mathbb{R}^n)$, we have that

$$\sup_{x \in \mathbb{R}^n} |x|^\beta |D^\gamma \phi(x)| < \infty, \qquad \sup_{x \in \mathbb{R}^n} |D^{\alpha-\gamma} \psi(x)| < \infty$$

for any $\gamma \in \mathbb{N}^n \cup \{0\}$, $\gamma \le \alpha$. Then, using Leibnitz's rule, we find

$$\sup_{x \in \mathbb{R}^n} |x|^\beta |D^\alpha(\phi\psi)(x)| = \sup_{x \in \mathbb{R}^n} |x|^\beta \left| \sum_{\gamma:\gamma \le \alpha} \binom{\alpha}{\gamma} D^\gamma \phi(x) D^{\alpha-\gamma} \psi(x) \right|$$

$$\le \sup_{x \in \mathbb{R}^n} |x|^\beta \sum_{\gamma:\gamma \le \alpha} \binom{\alpha}{\gamma} |D^\gamma \phi(x)| |D^{\alpha-\gamma} \psi(x)|$$

$$\le \sum_{\gamma:\gamma \le \alpha} \binom{\alpha}{\gamma} \left(\sup_{x \in \mathbb{R}^n} |x|^\beta |D^\gamma \phi(x)| \right) \left(\sup_{x \in \mathbb{R}^n} |D^{\alpha-\gamma} \psi(x)| \right)$$

$$< \infty,$$

i.e., $\phi\psi \in \mathscr{S}(\mathbb{R}^n)$. This completes the proof.

Definition 1.7 We say that the sequence $\{\phi_k\}_{k=1}^\infty$ of elements of $\mathscr{S}(\mathbb{R}^n)$ converges to 0 in $\mathscr{S}(\mathbb{R}^n)$, if for every $p \in \mathbb{N} \cup \{0\}$ and every $\alpha \in \mathbb{N}^n \cup \{0\}$, we have

$$\lim_{k \to \infty} (1 + |x|^2)^{\frac{p}{2}} |D^\alpha \phi_k(x)| = 0$$

uniformly.

Theorem 1.3 Let $\{\phi_k\}_{k=1}^\infty$ be a sequence of elements of $\mathscr{C}_0^\infty(\mathbb{R}^n)$ such that $\phi_k \to_{k \to \infty} 0$ in $\mathscr{C}_0^\infty(\mathbb{R}^n)$. Then $\phi_k \to_{k \to \infty} 0$ in $\mathscr{S}(\mathbb{R}^n)$.

Proof Since $\phi_k \to_{k\to\infty} 0$ in $\mathscr{C}_0^\infty(\mathbb{R}^n)$, there exists a compact set $K \subset \mathbb{R}^n$ so that $\operatorname{supp}\phi_k \subset K$ for any $k \in \mathbb{N}$ and $D^\alpha \phi_k(x) \to_{k\to\infty} 0$ uniformly for any multi-index $\alpha \in \mathbb{N}^n \cup \{0\}$. Let $K_1 > 0$ be a constant so that $1 + |x|^2 \le K_1$ for any $x \in K$. Then, for any $p \in \mathbb{N} \cup \{0\}$, we have

$$0 \le \lim_{k\to\infty} \left(1 + |x|^2\right)^{\frac{p}{2}} |D^\alpha \phi(x)| \le K_1^{\frac{p}{2}} \lim_{k\to\infty} |D^\alpha \phi_k(x)| = 0$$

uniformly, i.e., $\phi_k \to_{k\to\infty} 0$ in $\mathscr{S}(\mathbb{R}^n)$. This completes the proof.

Theorem 1.4 *The space $\mathscr{C}_0^\infty(\mathbb{R}^n)$ is dense in the space $\mathscr{S}(\mathbb{R}^n)$.*

Proof Let $\phi \in \mathscr{S}(\mathbb{R}^n)$ and $\eta \in \mathscr{C}_0^\infty(\mathbb{R}^n)$ be chosen so that $\eta(x) = 1$ for $|x| < 1$. Set $\phi_k(x) = \phi(x)\eta\left(\frac{x}{k}\right)$, $x \in \mathbb{R}^n$, $k \in \mathbb{N}$. Then $\phi_k \in \mathscr{C}_0^\infty(\mathbb{R}^n)$ and $\phi_k \to_{k\to\infty} \phi$ in $\mathscr{C}_0^\infty(\mathbb{R}^n)$. Now, applying Theorem 1.3, we conclude that $\phi_k \to_{k\to\infty} \phi$ in $\mathscr{S}(\mathbb{R}^n)$ as well. This completes the proof.

Remark 1.1 Note that if $a \in \mathscr{C}^\infty(\mathbb{R}^n)$ and $\phi \in \mathscr{S}(\mathbb{R}^n)$, it does not follow that $a\phi \in \mathscr{S}(\mathbb{R}^n)$. Take for instance $a(x) = e^{|x|^2}$, $x \in \mathbb{R}^n$, and $\phi(x) = e^{-|x|^2}$, $x \in \mathbb{R}^n$. Then $a \in \mathscr{C}^\infty(\mathbb{R}^n)$, $\phi \in \mathscr{S}(\mathbb{R}^n)$ and $a(x)\phi(x) = 1$, $x \in \mathbb{R}^n$. Consequently $a\phi \notin \mathscr{S}(\mathbb{R}^n)$.

Definition 1.8 By Θ_M we denote the space of functions $a \in \mathscr{C}^\infty(\mathbb{R}^n)$ for which there exist constants $C_a > 0$ and $m_a \in \mathbb{N}$ such that

$$|D^\alpha a(x)| \le C_a (1 + |x|)^{m_a}, \quad x \in \mathbb{R}^n,$$

for every $\alpha \in \mathbb{N}^n \cup \{0\}$. Such functions are called multipliers of $\mathscr{S}(\mathbb{R}^n)$.

Theorem 1.5 *Let $a \in \Theta_M$ and $\phi \in \mathscr{S}(\mathbb{R}^n)$. Then $a\phi \in \mathscr{S}(\mathbb{R}^n)$.*

Proof Let $\beta \in \mathbb{N} \cup \{0\}$ and $\alpha \in \mathbb{N}^n \cup \{0\}$ be arbitrarily chosen. Then

$$\sup_{x\in\mathbb{R}^n} |x|^\beta |D^\alpha(a\phi)(x)| = \sup_{x\in\mathbb{R}^n} |x|^\beta \left| \sum_{\gamma\le\alpha} \binom{\alpha}{\gamma} D^\gamma a(x) D^{\alpha-\gamma}\phi(x) \right|$$

$$\le \sup_{x\in\mathbb{R}^n} |x|^\beta \left(\sum_{\gamma\le\alpha} \binom{\alpha}{\gamma} |D^\gamma a(x)||D^{\alpha-\gamma}\phi(x)| \right)$$

$$\le \sup_{x\in\mathbb{R}^n} |x|^\beta \left(\sum_{\gamma\le\alpha} \binom{\alpha}{\gamma} C_a (1 + |x|)^{m_a} |D^{\alpha-\gamma}\phi(x)| \right)$$

$$= \sum_{\gamma\le\alpha} \binom{\alpha}{\gamma} C_a \sup_{x\in\mathbb{R}^n} \left(|x|^\beta (1 + |x|)^{m_a} |D^{\alpha-\gamma}\phi(x)| \right)$$

$$< \infty.$$

Thus, $a\phi \in \mathscr{S}(\mathbb{R}^n)$. This completes the proof.

Theorem 1.6 *Let $a \in \Theta_M$. Then the map $\phi \mapsto a\phi$ from $\mathscr{S}(\mathbb{R}^n)$ to $\mathscr{S}(\mathbb{R}^n)$ is a continuous map.*

Proof Let $\phi \in \mathscr{S}(\mathbb{R}^n)$ and $\{\phi_k\}_{k=1}^\infty$ be a sequence of elements of $\mathscr{S}(\mathbb{R}^n)$ that converges to ϕ in $\mathscr{S}(\mathbb{R}^n)$. Then, for any $\beta \in \mathbb{N} \cup \{0\}$ and any $\alpha \in \mathbb{N}^n \cup \{0\}$, we have

$$\lim_{k\to\infty} \left(1 + |x|^2\right)^{\frac{\beta}{2}} \left|D^\alpha \phi_k(x) - D^\alpha \phi(x)\right| = 0$$

uniformly. Hence, for any $\beta \in \mathbb{N} \cup \{0\}$ and $\alpha \in \mathbb{N}^n \cup \{0\}$, we get

$$\lim_{k\to\infty} \left(1 + |x|^2\right)^{\frac{\beta}{2}} \left|D^\alpha(a\phi_k)(x) - D^\alpha(a\phi)(x)\right|$$

$$= \lim_{k\to\infty} \left(1 + |x|^2\right)^{\frac{\beta}{2}} \left|\sum_{\gamma \leq \alpha} \binom{\alpha}{\gamma} D^\gamma a(x) D^{\alpha-\gamma}\phi_k(x) - \sum_{\gamma \leq \alpha} \binom{\alpha}{\gamma} D^\gamma a(x) D^{\alpha-\gamma}\phi(x)\right|$$

$$= \lim_{k\to\infty} \left(1 + |x|^2\right)^{\frac{\beta}{2}} \left|\sum_{\gamma \leq \alpha} \binom{\alpha}{\gamma} D^\gamma a(x) \left(D^{\alpha-\gamma}\phi_k(x) - D^{\alpha-\gamma}\phi(x)\right)\right|$$

$$\leq \lim_{k\to\infty} \left(1 + |x|^2\right)^{\frac{\beta}{2}} \sum_{\gamma \leq \alpha} \binom{\alpha}{\gamma} \left|D^\gamma a(x)\right| \left|D^{\alpha-\gamma}\phi_k(x) - D^{\alpha-\gamma}\phi(x)\right|$$

$$\leq \sum_{\gamma \leq \alpha} \binom{\alpha}{\gamma} C_a \lim_{k\to\infty} \left(1 + |x|^2\right)^{\frac{\beta}{2}} (1 + |x|)^{m_a} \left|D^{\alpha-\gamma}\phi_k(x) - D^{\alpha-\gamma}\phi(x)\right|$$

$$= 0$$

uniformly. Thus, $a\phi_k \to_{k\to\infty} a\phi$ in $\mathscr{S}(\mathbb{R}^n)$. This completes the proof. ∎

Exercise 1.2 Prove that the maps $\phi \mapsto D^\alpha \phi$, $\alpha \in \mathbb{N}^n \cup \{0\}$, and $\phi(x) \mapsto \phi(Ax+b)$, where A is an $n \times n$ matrix with $\det A \neq 0$, are linear and continuous maps from $\mathscr{S}(\mathbb{R}^n)$ to itself.

Definition 1.9 For $p \in \mathbb{N} \cup \{0\}$, with $\mathscr{S}_p(\mathbb{R}^n)$ we will denote the completion of the space $\mathscr{S}(\mathbb{R}^n)$ with respect to $\|\cdot\|_{\mathscr{S},p}$.

Definition 1.10 Let X and Y be two normed vector spaces with norms $\|\cdot\|_X$ and $\|\cdot\|_Y$, respectively. We say that X is compactly embedded in Y, and we write $X \hookrightarrow Y$ or $Y \hookleftarrow X$, if

1. X is continuously embedded in Y. i.e., there is a constant $C > 0$ such that $\|x\|_Y \leq C\|x\|_X$ for any $x \in X$,
2. any bounded set in X is totally bounded in Y, i.e., every sequence in such a bounded set has a subsequence that is a Cauchy sequence in the norm $\|\cdot\|_Y$.

Theorem 1.7 *The spaces* $\mathscr{S}_p(\mathbb{R}^n)$, $p \in \mathbb{N} \cup \{0\}$, *are Banach spaces fitting in a chain of continuous and compact embedding*

$$\mathscr{S}_0(\mathbb{R}^n) \hookleftarrow \mathscr{S}_1(\mathbb{R}^n) \hookleftarrow \mathscr{S}_2(\mathbb{R}^n) \hookleftarrow \cdots . \tag{1.2}$$

Proof Let $p \in \mathbb{N} \cup \{0\}$ be arbitrarily chosen. Take $\phi \in \mathscr{S}_{p+1}(\mathbb{R}^n)$ arbitrarily. Then

$$\begin{aligned}
\|\phi\|_{\mathscr{S},p+1} &= \sup_{x\in\mathbb{R}^n, |\alpha|\le p} \left(1+|x|^2\right)^{\frac{p+1}{2}} |D^\alpha \phi(x)| \\
&\ge \sup_{x\in\mathbb{R}^n, |\alpha|\le p} \left(1+|x|^2\right)^{\frac{p}{2}} |D^\alpha \phi(x)| \\
&= \|\phi\|_{\mathscr{S},p}.
\end{aligned}$$

Hence, $\phi \in \mathscr{S}_p(\mathbb{R}^n)$ and the embedding (1.2) are continuous. Now, let M be an infinitely bounded set in $\mathscr{S}_{p+1}(\mathbb{R}^n)$. Then there exists a constant $C > 0$ such that $\|\phi\|_{p+1} \le C$ for every $\phi \in M$. Hence,

$$\left| D^\alpha \phi(x) \right| \le C$$

for every $x \in \mathbb{R}^n$, $\alpha \in \mathbb{N}^n \cup \{0\}$, $|\alpha| \le p$, $\phi \in M$. Therefore

$$(1+|x|^2)^{\frac{p}{2}} |D^\alpha \phi(x)| = \frac{(1+|x|^2)^{\frac{p+1}{2}} |D^\alpha \phi(x)|}{(1+|x|^2)^{\frac{1}{2}}} \le \frac{C}{(1+|x|^2)^{\frac{1}{2}}} \to_{|x|\to\infty} 0.$$

Let $\{R_k\}_{k=1}^\infty$ be an increasing sequence of positive numbers such that

$$(1+|x|^2)^{\frac{p}{2}} |D^\alpha \phi(x)| \le \frac{1}{k} \quad \text{for} \quad |x| > R_k, |\alpha| \le p.$$

By the Arzela–Ascoli theorem, it follows that there exists a sequence $\{\phi_j^{(1)}\}_{j=1}^\infty$ of elements of M that converges in $\mathscr{S}_p(\overline{U}_{R_1})$. We may then find a sequence $\{\phi_j^{(2)}\}_{j=1}^\infty$ converging in $\mathscr{S}_p(\overline{U}_{R_2})$, and so on. The sequence $\{\phi_k^{(k)}\}_{k=1}^\infty$ converges in $\mathscr{S}_p(\mathbb{R}^n)$. This completes the proof.

Theorem 1.8 *If* $\phi \in \mathscr{C}^p(\mathbb{R}^n)$ *and* $|x|^p D^\alpha \phi(x) \to_{|x|\to\infty} 0$ *for* $\alpha \in \mathbb{N}^n \cup \{0\}$, $|\alpha| \le p$, *then* $\phi \in \mathscr{S}_p(\mathbb{R}^n)$.

Proof To prove this assertion, we choose a sequence $\{\eta_k\}_{k=1}^\infty$ of elements in $\mathscr{C}_0^\infty(\mathbb{R}^n)$ such that $\eta_k \to_{k\to\infty} 1$ in $\mathbb{R}^n$. Fix $\epsilon > 0$. Since $|x|^p |D^\alpha \phi(x)| \to_{|x|\to\infty} 0$ for any $\alpha \in \mathbb{N}^n \cup \{0\}$, $|\alpha| \le p$, it follows that there exists $R = R(\epsilon) > 0$ such that the inequality

$$(1+|x|^2)^{\frac{p}{2}} |D^\alpha \phi(x)| < \epsilon$$

holds for $|x| > R$. As $\eta_k \to_{k \to \infty} 1$ in $\mathbb{R}^n$, there exists $N \in \mathbb{N}$ such that $\eta_k(x) = 1$ for every $k > N$ and $|x| \le R + 1$. Now, define

$$\phi_{\frac{1}{k}}(x) = \int_{\mathbb{R}^n} \phi(y) \omega_{\frac{1}{k}}(x - y) dy.$$

Observe that $\{\phi_{\frac{1}{k}} \eta_k\}_{k=1}^\infty$ is a sequence in $\mathscr{C}_0^\infty(\mathbb{R}^n)$ and there exists $N_1 \in \mathbb{N}$ so that

$$\sup_{\substack{x \in \mathbb{R}^n \\ |\alpha| \le p}} \left(1 + |x|^2\right)^{\frac{p}{2}} \left| D^\alpha \left(\phi - \phi_{\frac{1}{k}} \eta_k\right)(x) \right| < \epsilon$$

for $k > N_1$, $|x| \le R + 1$. Set $N_2 = \max\{N, N_1\}$. Then, for $k > N_2$, we have

$$\|\phi - \phi_{\frac{1}{k}} \eta_k\|_{\mathscr{S}, p} = \sup_{\substack{x \in \mathbb{R}^n \\ |\alpha| \le p}} (1 + |x|^2)^{\frac{p}{2}} |D^\alpha(\phi - \phi_{\frac{1}{k}} \eta_k)(x)|$$

$$\le \sup_{\substack{|x| \le R+1 \\ |\alpha| \le p}} (1 + |x|^2)^{\frac{p}{2}} |D^\alpha(\phi - \phi_{\frac{1}{k}} \eta_k)(x)|$$

$$+ \sup_{\substack{|x| > R+1 \\ |\alpha| \le p}} (1 + |x|^2)^{\frac{p}{2}} |D^\alpha(\phi - \phi_{\frac{1}{k}} \eta_k)(x)|$$

$$\le \sup_{\substack{|x| \le R+1 \\ |\alpha| \le p}} (1 + |x|^2)^{\frac{p}{2}} |D^\alpha(\phi - \phi_{\frac{1}{k}} \eta_k)(x)|$$

$$+ \sup_{\substack{|x| > R+1 \\ |\alpha| \le p}} (1 + |x|^2)^{\frac{p}{2}} \left(|D^\alpha \phi(x)| + \sum_{\beta \le \alpha} \binom{\alpha}{\beta} |D^\beta \phi_{\frac{1}{k}}(x) D^{\alpha - \beta} \eta_k(x)| \right)$$

$$< \epsilon + \sup_{\substack{|x| > R+1 \\ |\alpha| \le p}} (1 + |x|^2)^{\frac{p}{2}} |D^\alpha \phi(x)|$$

$$+ \sup_{\substack{|x| > R+1 \\ |\alpha| \le p}} (1 + |x|^2)^{\frac{p}{2}} \sum_{\beta \le \alpha} \binom{\alpha}{\beta} |D^\beta \phi_{\frac{1}{k}}(x) D^{\alpha - \beta} \eta_k(x)|$$

$$< 2\epsilon + \sup_{\substack{|x| > R+1 \\ |\alpha| \le p}} (1 + |x|^2)^{\frac{p}{2}} |D^\alpha \phi_{\frac{1}{k}}(x)|$$

$$\le 2\epsilon + \sup_{\substack{|x| > R+1 \\ |\alpha| \le p}} (1 + |x|^2)^{\frac{p}{2}} \int_{\mathbb{R}^n} |D^\alpha \phi(x - y)| \omega_{\frac{1}{k}}(y) dy$$

$$= 2\epsilon + \sup_{\substack{|x| > R+1 \\ |\alpha| \le p}} \int_{\mathbb{R}^n} \left(1 + |x - y + y|^2\right)^{\frac{p}{2}} |D^\alpha \phi(x - y)| \omega_{\frac{1}{k}}(y) dy$$

$$\leq 2\epsilon + \sup_{\substack{|x|>R+1 \\ |\alpha|\leq p}} \int_{\mathbb{R}^n} \left(1 + (|x-y|+|y|)^2\right)^{\frac{p}{2}} |D^\alpha\phi(x-y)|\, \omega_{\frac{1}{k}}(y)dy$$

$$< 2\epsilon + 2^{\frac{p}{2}} \sup_{\substack{|x|>R+1 \\ |\alpha|\leq p}} \int_{\mathbb{R}^n} \left(1 + |x-y|^2 + |y|^2\right)^{\frac{p}{2}} |D^\alpha\phi(x-y)|\, \omega_{\frac{1}{k}}(y)dy$$

$$\leq 2\epsilon + 2^{\frac{p}{2}} \sup_{\substack{|x|>R+1 \\ |\alpha|\leq p}} \int_{\mathbb{R}^n} \left(\left(1 + |x-y|^2\right) + \left(1 + |y|^2\right)\right)^{\frac{p}{2}} |D^\alpha\phi(x-y)|\, \omega_{\frac{1}{k}}(y)dy$$

$$\leq 2\epsilon + 2^{p} \sup_{\substack{|x|>R+1 \\ |\alpha|\leq p}} \int_{\mathbb{R}^n} \left(\left(1 + |x-y|^2\right)^{\frac{p}{2}} + \left(1 + |y|^2\right)^{\frac{p}{2}}\right) |D^\alpha\phi(x-y)|\, \omega_{\frac{1}{k}}(y)dy$$

$$= 2\epsilon + 2^{p} \sup_{\substack{|x|>R+1 \\ |\alpha|\leq p}} \int_{\mathbb{R}^n} \left(1 + |x-y|^2\right)^{\frac{p}{2}} |D^\alpha\phi(x-y)|\, \omega_{\frac{1}{k}}(y)dy$$

$$+ 2^{p} \sup_{\substack{|x|>R+1 \\ |\alpha|\leq p}} \int_{\mathbb{R}^n} \left(1 + |y|^2\right)^{\frac{p}{2}} |D^\alpha\phi(x-y)|\, \omega_{\frac{1}{k}}(y)dy$$

$$< 2\epsilon + 2^{p}\epsilon \int_{\mathbb{R}^n} \omega_{\frac{1}{k}}(y)dy + 2^{p}\epsilon \int_{\mathbb{R}^n} \left(1 + |y|^2\right)^{\frac{p}{2}} \omega_{\frac{1}{k}}(y)dy$$

$$\leq 2\epsilon + 2^{p}\epsilon \int_{\mathbb{R}^n} \omega_{\frac{1}{k}}(y)dy + 2^{\frac{3p}{2}}\epsilon \int_{\mathbb{R}^n} \omega_{\frac{1}{k}}(y)dy$$

$$= \epsilon\left(2 + 2^{p} + 2^{\frac{3p}{2}}\right).$$

Since $\epsilon > 0$ was arbitrarily chosen, we conclude that $\phi_{\frac{1}{k}}\eta_k \to_{k\to\infty} \phi$ in $\mathscr{S}_p(\mathbb{R}^n)$. Now, using the fact that $\mathscr{S}_p(\mathbb{R}^n)$ is a Banach space, we conclude $\phi \in \mathscr{S}_p(\mathbb{R}^n)$. This completes the proof.

Theorem 1.9 *We have*

$$\mathscr{S}(\mathbb{R}^n) = \bigcap_{p\in\mathbb{N}\cup\{0\}} \mathscr{S}_p(\mathbb{R}^n). \tag{1.3}$$

Proof Let $\phi \in \mathscr{S}(\mathbb{R}^n)$ be arbitrarily chosen. By the definition of the space $\mathscr{S}(\mathbb{R}^n)$, it follows that for any $\alpha \in \mathbb{N}^n \cup \{0\}$ and any $p \in \mathbb{N} \cup \{0\}$, we have

$$\sup_{x\in\mathbb{R}^n} |x|^p \left|D^\alpha\phi(x)\right| < \infty.$$

Hence,

$$\sup_{x\in\mathbb{R}^n} \left(1 + |x|^2\right)^{\frac{p}{2}} \left|D^\alpha\phi(x)\right| < \infty \tag{1.4}$$

for any $\alpha \in \mathbb{N}^n \cup \{0\}$ and any $p \in \mathbb{N} \cup \{0\}$. In particular, by (1.4), we get

$$\sup_{x \in \mathbb{R}^n, |\alpha| \leq p} \left(1 + |x|^2\right)^{\frac{p}{2}} \left|D^\alpha \phi(x)\right| < \infty$$

for any $p \in \mathbb{N} \cup \{0\}$. From here, we obtain that $\phi \in \mathscr{S}_p(\mathbb{R}^n)$ for any $p \in \mathbb{N} \cup \{0\}$ and so, $\phi \in \bigcap_{p \in \mathbb{N} \cup \{0\}} \mathscr{S}_p(\mathbb{R}^n)$. Because $\phi \in \mathscr{S}(\mathbb{R}^n)$ was arbitrarily chosen and we get that it is an element of $\bigcap_{p \in \mathbb{N} \cup \{0\}} \mathscr{S}_p(\mathbb{R}^n)$, we find the following inclusion

$$\mathscr{S}(\mathbb{R}^n) \subseteq \bigcap_{p \in \mathbb{N} \cup \{0\}} \mathscr{S}_p(\mathbb{R}^n). \tag{1.5}$$

Let now, $\psi \in \bigcap_{p \in \mathbb{N} \cup \{0\}} \mathscr{S}_p(\mathbb{R}^n)$ be arbitrarily chosen. Then $\psi \in \mathscr{S}_p(\mathbb{R}^n)$ for any $p \in \mathbb{N} \cup \{0\}$. From here,

$$\sup_{x \in \mathbb{R}^n, |\alpha| \leq p} \left(1 + |x|^2\right)^{\frac{p}{2}} \left|D^\alpha \psi(x)\right| < \infty \tag{1.6}$$

for any $p \in \mathbb{N} \cup \{0\}$. Now, take $\beta \in \mathbb{N} \cup \{0\}$ and $\alpha_1 \in \mathbb{N}^n \cup \{0\}$ arbitrarily. Then there is a $\gamma \in \mathbb{N} \cup \{0\}$ so that $\gamma \geq \beta$ and $|\alpha_1| \leq \gamma$. Hence,

$$\sup_{x \in \mathbb{R}^n} |x|^\beta \left|D^{\alpha_1} \psi(x)\right| \leq \sup_{x \in \mathbb{R}^n} \left(1 + |x|^2\right)^{\frac{\gamma}{2}} \left|D^{\alpha_1} \psi(x)\right|$$

$$\leq \sup_{x \in \mathbb{R}^n, |\alpha_1| \leq \gamma} \left(1 + |x|^2\right)^{\frac{\gamma}{2}} \left|D^{\alpha_1} \psi(x)\right|$$

$$< \infty,$$

where we have used (1.6). Since $\beta \in \mathbb{N} \cup \{0\}$ and $\alpha_1 \in \mathbb{N}^n \cup \{0\}$ were arbitrarily chosen, we conclude that $\psi \in \mathscr{S}(\mathbb{R}^n)$. Because $\psi \in \bigcap_{p \in \mathbb{N} \cup \{0\}} \mathscr{S}_p(\mathbb{R}^n)$ was arbitrarily chosen and we find that it is an element of $\mathscr{S}(\mathbb{R}^n)$, we arrive at the inclusion

$$\bigcap_{p \in \mathbb{N} \cup \{0\}} \mathscr{S}_p(\mathbb{R}^n) \subseteq \mathscr{S}(\mathbb{R}^n).$$

By the last relation and by (1.5), we obtain (1.3). This completes the proof.

1.2   The L^p Spaces

1.2.1   Definition

Let E be a measurable set and $\mathscr{F}$ be the collection of all measurable extended real-valued functions on E that are finite a.e. on E. With $m(E)$ we will denote the Lebesgue measure of E.

Definition 1.11 Define two functions f and g in $\mathscr{F}$ to be equivalent, and we write $f \sim g$, provided

$$f(x) = g(x) \quad \text{for} \quad \text{almost} \quad \text{all} \quad x \in E.$$

The relation $\sim$ is an equivalent relation, that is, it is reflexive, symmetric and transitive. Therefore it induces a partition of $\mathscr{F}$ into a disjoint collection of equivalence classes, which we denote by $\mathscr{F}/\sim$. For given two functions f and g in $\mathscr{F}$, their equivalence classes $[f]$ and $[g]$ and real numbers α and β, we define $\alpha[f] + \beta[g]$ to be the equivalence class of the functions in $\mathscr{F}$ that take the value $\alpha f(x) + \beta g(x)$ at points $x \in E$ at which both f and g are finite. Note that these linear combinations are independent of the choice of the representatives of the equivalence classes. The zero element in $\mathscr{F}/\sim$ is the equivalence class of functions that vanish a.e. in E. Thus, $\mathscr{F}/\sim$ is a vector space.

Definition 1.12 For $1 \leq p < \infty$, we define $L^p(E)$ to be the collection of equivalence classes $[f]$ for which

$$\int_E |f|^p < \infty.$$

This is properly defined since if $f \sim g$, then

$$\int_E |f|^p = \int_E |g|^p.$$

Note that, if $[f], [g] \in L^p(E)$, then for any real constants α and β, we have

$$\int_E |\alpha f + \beta g|^p \leq 2^p \left(\int_E |\alpha f|^p + \int_E |\beta g|^p \right)$$

$$= |2\alpha|^p \int_E |f|^p + |2\beta|^p \int_E |g|^p < \infty,$$

i.e., $\alpha[f] + \beta[g] \in L^p(E)$.

Definition 1.13 For $1 \leq p < \infty$, we define $L_{loc}^p(E)$ to be the collection of equivalence classes $[f]$ for which

$$\int_K |f|^p < \infty$$

for any compact subsets K of E. The elements of $L_{loc}^p(E)$ will be called locally p-integrable functions or p-locally integrable functions. If $p = 1$, then we will say locally integrable functions.

Definition 1.14 We call a function $f \in \mathscr{F}$ essentially bounded provided there is some $M \geq 0$, called an essential upper bound for f, for which

$$|f(x)| \leq M \quad \text{for almost all} \quad x \in E.$$

Definition 1.15 We define $L^\infty(E)$ to be the collection of equivalence classes $[f]$ for which f is essentially bounded.

Note that $L^\infty(E)$ is properly defined since if $f \sim g$, then

$$|f(x)| = |g(x)| \leq M \quad \text{for almost all} \quad x \in E.$$

Also, if $[f], [g] \in L^\infty(E)$ and $\alpha, \beta \in \mathbb{R}$, there are nonnegative constants M_1 and M_2 such that

$$|f(x)| \leq M_1, \quad |g(x)| \leq M_2 \quad \text{for almost all} \quad x \in E.$$

Hence,

$$|\alpha f(x) + \beta g(x)| \leq |\alpha| |f(x)| + |\beta| |g(x)| \leq |\alpha| M_1 + |\beta| M_2$$

for almost all $x \in E$. Therefore $\alpha[f] + \beta[g] \in L^\infty(E)$ and hence, $L^\infty(E)$ is a vector space. For simplicity and convenience, we refer to the equivalence classes in $\mathscr{F}/\sim$ as functions and denote them by f rather than $[f]$. Thus, $f = g$ means $f - g$ vanishes a.e. on E.

Definition 1.16 For $1 \leq p < \infty$, in $L^p(E)$ we define

$$\|f\|_p = \left(\int_E |f|^p \right)^{\frac{1}{p}}.$$

For $p = \infty$, we define $\|f\|_\infty$ to be the infimum of the essential upper bounds for f.

1.2.2 The Inequalities of Hölder and Minkowski

Definition 1.17 Let $p \in [1, \infty]$. The number q is said to be conjugate of p if $\frac{1}{p} + \frac{1}{q} = 1$.

Theorem 1.10 (Young's Inequality) Let $a, b > 0$, $p, q \in (1, \infty)$, $\frac{1}{p} + \frac{1}{q} = 1$. Then

$$ab \leq \frac{1}{p}a^p + \frac{1}{q}b^q.$$

Proof Since the map $x \mapsto e^x$ is convex, we get

$$ab = e^{\log a + \log b} = e^{\frac{1}{p} \log a^p + \frac{1}{q} \log b^q}$$
$$\leq \frac{1}{p}e^{\log a^p} + \frac{1}{q}e^{\log b^q} = \frac{1}{p}a^p + \frac{1}{q}b^q.$$

This completes the proof.

Theorem 1.11 (Hölder's Inequality) Let E be a measurable set, $1 \leq p < \infty$ and q be the conjugate of p. If $f \in L^p(E)$ and $g \in L^q(E)$, then the product fg is integrable over E and

$$\int_E |fg| \leq \|f\|_p \|g\|_q. \tag{1.7}$$

Proof

1. Let $p = 1$. Then $q = \infty$ and

$$\int_E |fg| = \int_E |f||g| \leq \left(\int_E |f|\right) \|g\|_\infty = \|f\|_1 \|g\|_\infty.$$

2. Let $p \in (1, \infty)$. If $\|f\|_p = 0$ or $\|g\|_q = 0$, then the assertion is evident. Let $\|f\|_p \neq 0$ and $\|g\|_q \neq 0$. We set

$$f_1 = \frac{f}{\|f\|_p}, \quad g_1 = \frac{g}{\|g\|_q}.$$

Then

$$\|f_1\|_p = \|g_1\|_q = 1.$$

Now, using Young's inequality, we have

$$|f_1 g_1| \le \frac{1}{p}|f_1|^p + \frac{1}{q}|g_1|^q.$$

Hence,

$$\frac{1}{\|f\|_p \|g\|_q} \int_E |fg| = \int_E |f_1 g_1| \le \int_E \left(\frac{1}{p}|f_1|^p + \frac{1}{q}|g_1|^q\right)$$

$$= \int_E \frac{1}{p}|f_1|^p + \int_E \frac{1}{q}|g_1|^q = \frac{1}{p}\int_E |f_1|^p + \frac{1}{q}\int_E |g_1|^q$$

$$= \frac{1}{p}\|f_1\|_p^p + \frac{1}{q}\|g_1\|_q^q = \frac{1}{p} + \frac{1}{q} = 1.$$

From the last inequality, we get the inequality (1.7). This completes the proof.

Remark 1.2 When $p = q = 2$, the Hölder inequality is known as the Cauchy–Schwartz inequality.

Theorem 1.12 *Let E be a measurable set, $p > r$, $q > r$, $r > 0$ be real numbers such that $\dfrac{1}{p} + \dfrac{1}{q} = \dfrac{1}{r}$. If $f \in L^p(E)$, $g \in L^q(E)$, then $fg \in L^r(E)$ and*

$$\|fg\|_r \le \|f\|_p \|g\|_q. \tag{1.8}$$

Proof Observe that $\dfrac{1}{\frac{p}{r}} + \dfrac{1}{\frac{q}{r}} = 1$. Set $f_1 = |f|^r$, $g_1 = |g|^r$. Then $f_1 \in L^{\frac{p}{r}}(E)$ and $g_1 \in L^{\frac{q}{r}}(E)$. Now, applying the Hölder inequality, we arrive at

$$\int_E f_1 g_1 \le \|f_1\|_{\frac{p}{r}} \|g_1\|_{\frac{q}{r}}. \tag{1.9}$$

Since

$$\int_E f_1 g_1 = \int_E |f|^r |g|^r = \int_E |fg|^r = \|fg\|_r^r$$

and

$$\|f_1\|_{\frac{p}{r}} = \left(\int\limits_E f_1^{\frac{p}{r}} \right)^{\frac{r}{p}} = \|f\|_p^r,$$

$$\|g_1\|_{\frac{q}{r}} = \left(\int\limits_E g_1^{\frac{q}{r}} \right)^{\frac{r}{q}} = \|g\|_q^r,$$

by the inequality (1.9), we find that

$$\|fg\|_r^r \le \|f\|_p^r \|g\|_q^r,$$

whereupon we get the inequality (1.8). This completes the proof.

Theorem 1.13 *Let E be a measurable set and $p_j \ge 1$, $j = 1, 2, 3$, be real numbers such that*

$$\frac{1}{p_1} + \frac{1}{p_2} + \frac{1}{p_3} = 1.$$

If $f_j \in L^{p_j}(E)$, $j = 1, 2, 3$, then

$$\int\limits_E |f_1 f_2 f_3| \le \|f_1\|_{p_1} \|f_2\|_{p_2} \|f_3\|_{p_3}.$$

Proof Let $\dfrac{1}{q} = \dfrac{1}{p_2} + \dfrac{1}{p_3}$ and $g = f_2 f_3$. By Theorem 1.12, it follows that $g \in L^q(E)$ and

$$\|g\|_q = \|f_2 f_3\|_q \le \|f_2\|_{p_2} \|f_3\|_{p_3}.$$

Now, we apply the Hölder inequality for f_1, g and p_1, q, and we find

$$\int\limits_E |f_1 f_2 f_3| = \int\limits_E |f_1 g| \le \|f_1\|_{p_1} \|g\|_q \le \|f_1\|_{p_1} \|f_2\|_{p_2} \|f_3\|_{p_3}.$$

This completes the proof.

Remark 1.3 Let E be a measurable set and $p_j \ge 1$, $j = 1, 2, \ldots, k$, be real numbers so that $\displaystyle\sum_{j=1}^{k} \frac{1}{p_j} = 1$ for some $k \in \mathbb{N}$, $k \ge 2$. If $f_j \in L^{p_j}(E)$,

$j = 1, 2, \ldots, k$, using the principle of the mathematical induction, Theorem 1.12 and Theorem 1.13, one can get the following inequality

$$\int_E \left| \prod_{j=1}^k f_j \right| \leq \prod_{j=1}^k \|f_j\|_{p_j}.$$

Lemma 1.3 (Interpolation Inequality) *Let A be an open bounded set in $\mathbb{R}^n$, $1 \leq s \leq r \leq t \leq \infty$ and*

$$\frac{1}{r} = \frac{\theta}{s} + \frac{1-\theta}{t}, \quad 0 \leq \theta \leq 1.$$

The any $u \in L^s(A) \cap L^t(A)$ belongs in $L^r(A)$ and

$$\|u\|_r \leq \|u\|_s^\theta \|u\|_t^{1-\theta}.$$

Proof We have

$$\int_A |u|^r dx = \int_A |u|^{\theta r + (1-\theta)r} dx = \int_A |u|^{\theta r} |u|^{(1-\theta)r} dx$$

$$\leq \left(\int_A |u|^s dx \right)^{\frac{r\theta}{s}} \left(\int_A |u|^t dx \right)^{\frac{r(1-\theta)}{t}},$$

from where

$$\|u\|_r \leq \|u\|_s^\theta \|u\|_t^{1-\theta}.$$

This completes the proof.

Theorem 1.14 (Minkowski's Inequality) *Let E be a measurable set, $1 \leq p \leq \infty$. If $f, g \in L^p(E)$, then*

$$\|f + g\|_p \leq \|f\|_p + \|g\|_p.$$

Proof

1. Let $p = 1$. Then

$$\|f + g\|_1 = \int_E |f + g| \leq \int_E (|f| + |g|) = \int_E |f| + \int_E |g| = \|f\|_1 + \|g\|_1.$$

2. Let $p = \infty$. Then

$$\|f + g\|_\infty \le \|f\|_\infty + \|g\|_\infty.$$

3. Let $p \in (1, \infty)$ and q be its conjugate. If $\|f + g\|_p = 0$, then the assertion is evident. Assume that $\|f + g\|_p \ne 0$. Then, applying Hölder's inequality, we get

$$\|f + g\|_p^p = \int_E |f + g|^p = \int_E |f + g||f + g|^{p-1}$$

$$\le \int_E \left(|f||f + g|^{p-1} + |g||f + g|^{p-1} \right) = \int_E |f||f + g|^{p-1} + \int_E |g||f + g|^{p-1}$$

$$\le \left(\int_E |f|^p \right)^{\frac{1}{p}} \left(\int_E |f + g|^{(p-1)q} \right)^{\frac{1}{q}} + \left(\int_E |g|^p \right)^{\frac{1}{p}} \left(\int_E |f + g|^{(p-1)q} \right)^{\frac{1}{q}}$$

$$= \|f + g\|_p^{\frac{p}{q}} \|f\|_p + \|f + g\|_p^{\frac{p}{q}} \|g\|_p.$$

Hence,

$$\|f + g\|_p^{p - \frac{p}{q}} \le \|f\|_p + \|g\|_p$$

or

$$\|f + g\|_p \le \|f\|_p + \|g\|_p.$$

This completes the proof.

1.2.3 Some Properties

Theorem 1.15 *Let E be a measurable set and $1 < p < \infty$. Suppose that $\mathscr{F}$ is a family of functions in $L^p(E)$ that is bounded in $L^p(E)$ in the sense that there is a constant $M > 0$ such that*

$$\|f\|_p \le M \quad \text{for all} \quad f \in \mathscr{F}.$$

Then the family $\mathscr{F}$ is uniformly integrable over E.

Proof Let $\epsilon > 0$ be arbitrarily chosen. Suppose that A is a measurable subset of E of finite measure. Let also, $\dfrac{1}{p} + \dfrac{1}{q} = 1$. Define g to be identically equal to 1 on A. Because $m(A) < \infty$, we have that $g \in L^q(A)$. By Hölder's inequality, for any

$f \in \mathscr{F}$, we have

$$\int_A |f| = \int_A |f| g \le \left(\int_A |f|^p \right)^{\frac{1}{p}} \left(\int_A g^q \right)^{\frac{1}{q}}.$$

On the other hand,

$$\left(\int_A |f|^p \right)^{\frac{1}{p}} \le M, \quad \left(\int_A |g|^q \right)^{\frac{1}{q}} = (m(A))^{\frac{1}{q}}$$

for any $f \in \mathscr{F}$. Therefore

$$\int_A |f| \le (m(A))^{\frac{1}{q}} M$$

for any $f \in \mathscr{F}$. Let $\delta = \left(\dfrac{\epsilon}{M} \right)^q$. Hence, if $m(A) < \delta$, we have

$$\int_A |f| \le M (m(A))^{\frac{1}{q}} < M \delta^{\frac{1}{q}} = M \frac{\epsilon}{M} = \epsilon$$

for any $f \in \mathscr{F}$. Therefore $\mathscr{F}$ is uniformly integrable over E. This completes the proof.

Theorem 1.16 *Let E be a measurable set of finite measure and $1 \le p_1 < p_2 \le \infty$. Then*

$$L^{p_2}(E) \subseteq L^{p_1}(E). \tag{1.10}$$

Proof Let $p_2 < \infty$ and $f \in L^{p_2}(E)$ be arbitrarily chosen. Then $p = \dfrac{p_2}{p_1} > 1$. We take $q > 1$ such that $\dfrac{1}{p} + \dfrac{1}{q} = 1$. Then

$$\left(\int_E |f|^{p_1 p} \right)^{\frac{1}{p}} = \left(\int_E |f|^{p_2} \right)^{\frac{p_1}{p_2}} < \infty,$$

i.e., $|f|^{p_1} \in L^p(E)$. Let $g = \kappa_E$. Since $m(E) < \infty$, then $g \in L^q(E)$. Hence, using Hölder's inequality, we get

$$\int_E |f|^{p_1} = \int_E |f|^{p_1} g \leq \left(\int_E |f|^{p_1 p} \right)^{\frac{1}{p}} \left(\int_E g^q \right)^{\frac{1}{q}}$$

$$= \|f\|_{p_2}^{p_1} (m(E))^{\frac{1}{q}} < \infty.$$

Therefore $f \in L^{p_1}(E)$. Because $f \in L^{p_2}(E)$ was arbitrarily chosen and we get that it is an element of $L^{p_1}(E)$, we obtain the relation (1.10). Let $p_2 = \infty$ and $f \in L^\infty(E)$ be arbitrarily chosen. Then there is a positive constant M such that

$$|f(x)| \leq M$$

for almost all $x \in E$. Hence,

$$\int_E |f|^{p_1} \leq M^{p_1} m(E).$$

Then $f \in L^{p_1}(E)$. Because $f \in L^\infty(E)$ was arbitrarily chosen and we get that it is an element of $L^{p_1}(E)$, we obtain the relation (1.10). This completes the proof.

Definition 1.18 Let $\{f_n\}_{n=1}^\infty$ be a sequence of elements of $L^p(E)$ and $f \in L^p(E)$. When $\|f_n - f\|_p \to 0$, as $n \to \infty$, we will say that the sequence $\{f_n\}_{n=1}^\infty$ converges to f in $L^p(E)$.

1.2.4   The Riesz–Fischer Theorem

Definition 1.19 Let X be a normed vector space. A sequence $\{f_n\}_{n=1}^\infty$ in X is said to be a rapidly Cauchy sequence provided that there is a convergent series of positive numbers $\sum_{k=1}^\infty \epsilon_k$ for which

$$\|f_{k+1} - f_k\| \leq \epsilon_k^2$$

for all $k \in \mathbb{N}$.

Lemma 1.4 *Let X be a normed vector space and $\{f_n\}_{n=1}^\infty$ be a sequence in X such that*

$$\|f_{k+1} - f_k\| \leq a_k$$

for all $k \in \mathbb{N}$, where a_k, $k \in \mathbb{N}$, are nonnegative numbers. Then

$$\|f_{n+k} - f_n\| \leq \sum_{l=n}^{\infty} a_l \qquad (1.11)$$

for all $k, n \in \mathbb{N}$.

Proof For any $n, k \in \mathbb{N}$, we have

$$f_{n+k} - f_n = f_{n+k} - f_{n+k-1} + f_{n+k-1} - \cdots + f_{n+1} - f_n = \sum_{j=n}^{n+k-1} \left(f_{j+1} - f_j \right).$$

Hence,

$$\|f_{n+k} - f_n\| = \left\| \sum_{j=n}^{n+k-1} \left(f_{j+1} - f_j \right) \right\| \leq \sum_{j=n}^{n+k-1} \|f_{j+1} - f_j\| \leq \sum_{j=n}^{n+k-1} a_j \leq \sum_{j=n}^{\infty} a_j$$

for any $k, n \in \mathbb{N}$. This completes the proof.

Theorem 1.17 *Let X be a normed vector space. Then every rapidly Cauchy sequence in X is a Cauchy sequence in X. Furthermore, every Cauchy sequence in X has a rapidly Cauchy subsequence in X.*

Proof

1. Let $\{f_n\}_{n=1}^{\infty}$ be a rapidly Cauchy sequence in X. Then there is a convergent series $\sum_{k=1}^{\infty} \epsilon_k$ of positive numbers such that

$$\|f_{k+1} - f_k\| \leq \epsilon_k^2$$

for any $k \in \mathbb{N}$. Hence and (1.11), we obtain

$$\|f_{n+k} - f_k\| \leq \sum_{l=n}^{\infty} \epsilon_l^2 \to 0, \quad \text{as} \quad n \to \infty,$$

for any $k \in \mathbb{N}$. Here we have used that the series $\sum_{k=1}^{\infty} \epsilon_k^2$ is a convergent series. Therefore $\{f_n\}_{n=1}^{\infty}$ is a Cauchy sequence.

2. Let $\{f_n\}_{n=1}^{\infty}$ is a Cauchy sequence in X. Then there are $M_1, M_2, M_3 \in \mathbb{N}$ such that

$$\left\| f_{m_1^1} - f_{m_2^1} \right\| < \frac{1}{2} \quad \text{for any} \quad m_1^1, m_2^1 \geq M_1,$$

$$\left\| f_{m_1^2} - f_{m_2^2} \right\| < \frac{1}{2^2} \quad \text{for any} \quad m_1^2, m_2^2 \geq M_2,$$

$$\left\| f_{m_1^3} - f_{m_2^3} \right\| < \frac{1}{2^3} \quad \text{for any} \quad m_1^3, m_2^3 \geq M_3.$$

In particular, for $m_1^1, m_2^1, m_2^3 \geq \max\{M_1, M_2, M_3\}$, we have

$$\left\| f_{m_1^1} - f_{m_2^1} \right\| < \frac{1}{2},$$

$$\left\| f_{m_2^3} - f_{m_1^1} \right\| < \frac{1}{2^2},$$

$$\left\| f_{m_1^3} - f_{m_2^3} \right\| < \frac{1}{2^3}.$$

We set

$$n_1 = m_2^1, \quad n_2 = m_1^1, \quad n_3 = m_2^3$$

for $m_1^1, m_2^1, m_2^3 \geq \max\{M_1, M_2, M_3\}$. Then

$$\| f_{n_2} - f_{n_1} \| < \frac{1}{2},$$

$$\| f_{n_3} - f_{n_2} \| < \frac{1}{2^2},$$

$$\left\| f_{m_1^3} - f_{n_3} \right\| < \frac{1}{2^3}$$

for $m_1^3 \geq \max\{M_1, M_2, M_3\}$. Continuing this process, we obtain a subsequence $\{f_{n_k}\}_{k=1}^{\infty}$ of the sequence $\{f_n\}_{n=1}^{\infty}$ such that

$$\| f_{n_{k+1}} - f_{n_k} \| < \frac{1}{2^k}$$

for any $k \in \mathbb{N}$. Since the series $\sum_{k=0}^{\infty} \frac{1}{\left(\sqrt{2}\right)^k}$ is convergent, we conclude that $\{f_{n_k}\}_{k=1}^{\infty}$ is a rapidly Cauchy sequence in X. This completes the proof.

Theorem 1.18 *Let E be a measurable set and $1 \leq p \leq \infty$. Then every rapidly Cauchy sequence in $L^p(E)$ converges both with respect to the $L^p(E)$ norm and pointwise a.e. in E to a function f in $L^p(E)$.*

Proof Let $1 \leq p < \infty$. We leave the case $p = \infty$ as an exercise. Let $\{f_n\}_{n=1}^{\infty}$ be a rapidly convergent sequence in $L^p(E)$. We choose $\sum\limits_{k=1}^{\infty} \epsilon_k$ to be a convergent series of positive numbers such that

$$\| f_{k+1} - f_k \|_p \leq \epsilon_k^2 \tag{1.12}$$

for any $k \in \mathbb{N}$. Let

$$E^k = \{x \in E : |f_{k+1}(x) - f_k(x)| \geq \epsilon_k\}, \quad k \in \mathbb{N}.$$

Then

$$E^k = \{x \in E : |f_{k+1}(x) - f_k(x)|^p \geq \epsilon_k^p\}, \quad k \in \mathbb{N}.$$

Hence,

$$\int_{E_k} |f_{k+1}(x) - f_k(x)|^p \geq \epsilon_k^p m\left(E^k\right).$$

Now, using (1.12), we get

$$m\left(E^k\right) \leq \frac{1}{\epsilon_k^p} \int_E |f_{k+1} - f_k|^p = \frac{1}{\epsilon_k^p} \| f_{k+1} - f_k \|_p^p$$

$$\leq \frac{1}{\epsilon_k^p} \epsilon_k^{2p} = \epsilon_k^p.$$

Since $p \geq 1$, the series $\sum\limits_{k=1}^{\infty} \epsilon_k^p$ is convergent and

$$\sum_{k=1}^{\infty} m\left(E^k\right) \leq \sum_{k=1}^{\infty} \epsilon_k^p < \infty.$$

Hence and the Borel–Cantelli Lemma, it follows that there is $E_0 \subset E$ such that $m(E_0) = 0$ and for each $x \in E \backslash E_0$ there is an index $K(x)$ such that

$$|f_{k+1}(x) - f_k(x)| < \epsilon_k$$

for any $k \geq K(x)$. Let $x \in E \backslash E_0$. Then

$$
|f_{n+k}(x) - f_n(x)| \leq \sum_{j=n}^{n+k-1} |f_{j+1}(x) - f_j(x)| \leq \sum_{j=n}^{\infty} \epsilon_j
$$

for all $n \geq K(x)$ and for any $k \in \mathbb{N}$. Therefore the sequence of real numbers $\{f_n(x)\}_{n=1}^{\infty}$ is a Cauchy sequence in $\mathbb{R}$ for any $x \in E \backslash E_0$. Consequently it is convergent for any $x \in E \backslash E_0$ and let $\lim_{n \to \infty} f_n(x) = f(x)$ for any $x \in E \backslash E_0$. By (1.12), we obtain

$$
\int_E |f_{n+k} - f_n|^p \leq \left(\sum_{j=n}^{\infty} \epsilon_j^2 \right)^p \tag{1.13}
$$

for all $n, k \in \mathbb{N}$. Since $f_n \to f$ pointwise a.e. on E, we take the limit as $k \to \infty$ in (1.13) and using Fatou's Lemma, we get

$$
\int_E |f - f_n|^p \leq \left(\sum_{j=n}^{\infty} \epsilon_j^2 \right)^p
$$

for all $n \in \mathbb{N}$. Hence, $f \in L^p(E)$ and $f_n \to f$, as $n \to \infty$, in $L^p(E)$. This completes the proof.

Theorem 1.19 (The Riesz–Fischer Theorem) *Let E be a measurable set and $1 \leq p \leq \infty$. Then $L^p(E)$ is a Banach space. Moreover, if $f_n \to f$, as $n \to \infty$, in $L^p(E)$, a subsequence of $\{f_n\}_{n=1}^{\infty}$ converges pointwise a.e. on E to f.*

Proof Let $\{f_n\}_{n=1}^{\infty}$ be a Cauchy sequence in $L^p(E)$. By Theorem 1.17, it follows that there is a subsequence $\{f_{n_k}\}_{k=1}^{\infty}$ of the sequence $\{f_n\}_{n=1}^{\infty}$ that is a rapidly Cauchy sequence in $L^p(E)$. Hence and Theorem 1.18, it follows that $\{f_{n_k}\}_{k=1}^{\infty}$ converges both with respect to the $L^p(E)$ norm and pointwise a.e. on E. Let $f_{n_k} \to f$, as $k \to \infty$, in $L^p(E)$. We take $\epsilon > 0$ arbitrarily. Then there exists $K \in \mathbb{N}$ such that

$$
\|f_{n_k} - f\|_p < \frac{\epsilon}{2} \quad and \quad \|f_l - f_m\|_p < \frac{\epsilon}{2}
$$

for any $l, m, n_k \geq K$. Hence, for any $n, n_k \geq K$, we have

$$
\|f_n - f\|_p = \|f_n - f_{n_k} + f_{n_k} - f\|_p \leq \|f_n - f_{n_k}\|_p + \|f_{n_k} - f\|_p < \frac{\epsilon}{2} + \frac{\epsilon}{2} = \epsilon.
$$

Therefore $\{f_n\}_{n=1}^{\infty}$ is convergent in $L^p(E)$ and $L^p(E)$ is a Banach space. This completes the proof.

Definition 1.20 A real-valued function f, defined on a set A, is called convex if for each pair of points $x_1, x_2 \in A$ and for each $\lambda \in [0, 1]$, we have

$$f(\lambda x_1 + (1 - \lambda)x_2) \leq \lambda f(x_1) + (1 - \lambda)f(x_2).$$

Theorem 1.20 *Let E be a measurable set and $1 \leq p < \infty$. Suppose that $\{f_n\}_{n=1}^{\infty}$ is a sequence in $L^p(E)$ that converges pointwise a.e. on E to the function $f \in L^p(E)$. Then $f_n \to f$, as $n \to \infty$, in $L^p(E)$ if and only if*

$$\lim_{n \to \infty} \int_E |f_n|^p = \int_E |f|^p. \tag{1.14}$$

Proof

1. Let $f_n \to f$, as $n \to \infty$, in $L^p(E)$. Then, by Minkowski's inequality, we have

$$\left| \|f_n\|_p - \|f\|_p \right| \leq \|f_n - f\|_p \to 0, \quad \text{as} \quad n \to \infty.$$

Hence, (1.14) holds.

2. Assume that (1.14) holds. Let $E_0 \subset E$ be such that $m(E_0) = 0$ and $f_n(x) \to f(x)$, as $n \to \infty$, for all $x \in E \backslash E_0$. Take $\psi(t) = |t|^p$, $t \in \mathbb{R}$. Then ψ is a convex function and

$$\psi\left(\frac{a+b}{2}\right) \leq \frac{\psi(a)}{2} + \frac{\psi(b)}{2}$$

for any $a, b \in \mathbb{R}$. Hence,

$$0 \leq \frac{|a|^p + |b|^p}{2} - \left|\frac{a-b}{2}\right|^p$$

for any $a, b \in \mathbb{R}$. For each $n \in \mathbb{N}$, we define the function

$$h_n(x) = \frac{|f_n(x)|^p + |f(x)|^p}{2} - \left|\frac{f_n(x) - f(x)}{2}\right|^p$$

for any $x \in E$. We have that $h_n(x) \to |f(x)|^p$, as $n \to \infty$, for any $x \in E \backslash E_0$. Hence and Fatou's Lemma, we obtain

$$\int_E |f|^p \leq \liminf_{n \to \infty} \int_E h_n = \liminf_{n \to \infty} \left(\int_E \frac{|f_n|^p + |f|^p}{2} - \int_E \left|\frac{f_n - f}{2}\right|^p \right)$$

$$= \int_E |f|^p - \limsup_{n \to \infty} \int_E \left|\frac{f_n - f}{2}\right|^p.$$

Therefore

$$\limsup_{n\to\infty} \int\limits_E \left| \frac{f_n - f}{2} \right|^p \le 0.$$

Consequently $f_n \to f$, as $n \to \infty$, in $L^p(E)$. This completes the proof.

Theorem 1.21 *Let E be a measurable set and $1 \le p < \infty$. Suppose that $\{f_n\}_{n=1}^{\infty}$ is a sequence in $L^p(E)$ that converges pointwise a.e. on E to the function f which belongs to $L^p(E)$. Then $f_n \to f$, as $n \to \infty$, in $L^p(E)$, if and only if $\{|f_n|^p\}_{n=1}^{\infty}$ is uniformly integrable and a tight over E.*

Proof

1. Let $f_n \to f$, as $n \to \infty$, in $L^p(E)$. Hence,

$$\lim_{n\to\infty} \int\limits_E |f_n - f|^p = 0.$$

Therefore $\{|f_n - f|^p\}_{n=1}^{\infty}$ is uniformly integrable and a tight over E. Because

$$|f_n|^p \le 2^p \left(|f_n - f|^p + |f|^p \right), \quad n \in \mathbb{N},$$

we conclude that $\{|f_n|^p\}_{n=1}^{\infty}$ is uniformly integrable and a tight over E.

2. Let $\{|f_n|^p\}_{n=1}^{\infty}$ is uniformly integrable and a tight over E. Since $f_n \to f$, as $n \to \infty$, pointwise a.e. on E, we have that $|f_n|^p \to |f|^p$, as $n \to \infty$, pointwise a.e. on E. Hence and the Vitalli Convergence Theorem, we conclude that $|f|^p$ is integrable over E and

$$\lim_{n\to\infty} \int\limits_E |f_n|^p = \int\limits_E |f|^p.$$

This completes the proof.

Theorem 1.22 (The L^p Dominated Convergence Theorem) *Let $\{f_n\}_{n=1}^{\infty}$ be a sequence of measurable functions that converges pointwise a.e. on E to f. For $1 \le p < \infty$, suppose that there is a function $g \in L^p(E)$ such that $|f_n| \le g$ a.e. on E for all $n \in \mathbb{N}$. Then $f_n \to f$, as $n \to \infty$, in $L^p(E)$.*

Proof Since $f_n \to f$, as $n \to \infty$, pointwise a.e. on E, we have that $|f_n|^p \to |f|^p$, as $n \to \infty$, pointwise a.e. on E. Also, $|f_n|^p \le g^p$ a.e. on E for any $n \in \mathbb{N}$ and g^p is integrable over E. Hence and the Lebesgue Dominated Convergence Theorem, we conclude that $|f|^p$ is integrable over E and

$$\lim_{n \to \infty} \int_E |f_n|^p = \int_E |f|^p .$$

This completes the proof.

1.2.5 Separability

Theorem 1.23 *Let E be a measurable set and $1 \leq p \leq \infty$. Then the subspace of simple functions in $L^p(E)$ is dense in $L^p(E)$.*

Proof Let $g \in L^p(E)$.

1. Assume $p = \infty$. Then there is $E_0 \subset E$ such that $m(E_0) = 0$ and g is bounded on $E \backslash E_0$. It follows that there is a sequence $\{f_n\}_{n=1}^{\infty}$ of simple functions on $E \backslash E_0$ that converges uniformly on $E \backslash E_0$ to g and therefore with respect to the $L^\infty(E)$ norm. Consequently the subspace of simple functions in $L^\infty(E)$ is dense in $L^\infty(E)$.
2. Let $1 \leq p < \infty$. We have that there is a sequence of simple functions $\{f_n\}_{n=1}^{\infty}$ on E such that $f_n \to g$, as $n \to \infty$, pointwise on E and

$$|f_n| \leq |g| \quad on \quad E$$

for any $n \in \mathbb{N}$. Since $g \in L^p(E)$, we have that $f_n \in L^p(E)$ for any $n \in \mathbb{N}$. Next,

$$|f_n - g| \leq |f_n| + |g| \leq 2|g| .$$

Hence and Theorem 1.22, it follows that $f_n \to g$, as $n \to \infty$, in $L^p(E)$. This completes the proof.

Theorem 1.24 *Let $[a, b]$ be a closed bounded interval and $1 \leq p < \infty$. Then the subspace of the step functions on $[a, b]$ is dense in $L^p([a, b])$.*

Proof Let A be a measurable subset of $[a, b]$. Let also, $g = \kappa_A$. Take $\epsilon > 0$ arbitrarily. Then there is a finite disjoint collection $\{I_k\}_{k=1}^{n}$ of open intervals such that if $U = \bigcup_{k=1}^{n} I_k$, then

$$m(A \backslash U) + m(U \backslash A) < \epsilon^p .$$

Let $f = \kappa_U$. Then

$$\|\kappa_U - \kappa_A\|_p = \left(\int_E |\kappa_U - \kappa_A|^p \right)^{\frac{1}{p}} = (m(A \backslash U) + m(U \backslash A))^{\frac{1}{p}} < \epsilon.$$

Therefore the step functions are dense in the simple functions with respect to the L^p norm. Hence and Theorem 1.23, it follows that the step functions on $[a, b]$ are dense in $L^p([a, b])$. This completes the proof.

Theorem 1.25 *Let E be a measurable set and $1 \leq p < \infty$. Then $L^p(E)$ is separable.*

Proof Let $[a, b]$ be a closed bounded interval and $S([a, b])$ be the collection of the step functions on $[a, b]$. Let also, $S^1([a, b])$ be the subcollection of the collection $S([a, b])$ consisting of the step functions ψ on $[a, b]$ that take rational values and for which there is a partition $P = \{x_0, \ldots, x_n\}$ of $[a, b]$ so that ψ is a rational constant on (x_{k-1}, x_k), $1 \leq k \leq n$, and x_k, $1 \leq k \leq n - 1$, are rational numbers. Using the density of the rational numbers in the real numbers, we have that $S^1([a, b])$ is dense in $S([a, b])$ with respect to the $L^p(E)$ norm. Because the set of the rational numbers is countable, we have that $S^1([a, b])$ is a countable set. We have that

$$S^1([a, b]) \subseteq S([a, b]) \subseteq L^p([a, b]).$$

Since $S^1([a, b])$ is dense in $S([a, b])$, using Theorem 1.24, it follows that $S^1([a, b])$ is dense in $L^p([a, b])$. For each $n \in \mathbb{N}$, we define by $\mathscr{F}_n$ to be the collection of the functions that vanishes outside $[-n, n]$ and whose restrictions to $[-n, n]$ belong to $S^1([-n, n])$. Let

$$\mathscr{F} = \bigcup_{n=1}^{\infty} \mathscr{F}_n.$$

Also, we have

$$\lim_{n \to \infty} \int_{[-n,n]} |f|^p = \int_{\mathbb{R}} |f|^p$$

for all $f \in L^p(\mathbb{R})$. Using the definition of $\mathscr{F}$, we conclude that $\mathscr{F}$ is a countable collection of functions that is dense in $L^p(\mathbb{R})$. Hence, the collection of the restrictions on E of the functions in $\mathscr{F}$ is a countable dense set in $L^p(E)$. Consequently $L^p(E)$ is separable. This completes the proof.

1.2.6 Duality

Definition 1.21 For a normed vector space X, a linear functional $\mathbb{T}$ on X is said to be bounded if there is a constant $M \geq 0$ such that

$$|\mathbb{T}(f)| \leq M\|f\| \tag{1.15}$$

for all $f \in X$. The infimum of all such M will be called the norm of $\mathbb{T}$ and will be denoted by $\|\mathbb{T}\|_\star$.

Let $\mathbb{T}$ be a bounded linear functional on the normed vector space X. Then, for any $f, g \in X$, we have

$$|\mathbb{T}(f) - \mathbb{T}(g)| \leq \|f - g\|. \tag{1.16}$$

Hence, if $f_n \to f$, as $n \to \infty$, in X, i.e., $f_n, f \in X, n \in \mathbb{N}$,

$$\|f_n - f\| \to 0, \quad \text{as} \quad n \to \infty,$$

using (1.16), we get

$$\mathbb{T}(f_n) \to \mathbb{T}(f), \quad as \quad n \to \infty.$$

Proposition 1.1 *Let $\mathbb{T}$ be a bounded linear functional on the normed vector space X. Then*

$$\|\mathbb{T}\|_\star = \sup_{\|f\| \leq 1} |\mathbb{T}(f)|. \tag{1.17}$$

Proof By the inequality (1.15), we get

$$|\mathbb{T}(f)| \leq \|\mathbb{T}\|_\star \|f\|$$

for any $f \in X$. Hence,

$$\sup_{\|f\| \leq 1} |\mathbb{T}(f)| \leq \|\mathbb{T}\|_\star. \tag{1.18}$$

Let $\epsilon > 0$ be arbitrarily chosen. Then there exists $g \in X, g \neq 0$, such that

$$|\mathbb{T}(g)| \geq (\|\mathbb{T}\|_\star - \epsilon)\|g\|.$$

Hence,

$$\left|\mathbb{T}\left(\frac{g}{\|g\|}\right)\right| \geq \|\mathbb{T}\|_\star - \epsilon$$

and

$$\sup_{\|f\| \leq 1} |\mathbb{T}(f)| \geq \|\mathbb{T}\|_\star - \epsilon.$$

Because $\epsilon > 0$ was arbitrarily chosen, from the last inequality, we obtain

$$\sup_{\|f\| \leq 1} |\mathbb{T}(f)| \geq \|\mathbb{T}\|_\star.$$

Hence and (1.18), we get (1.17). This completes the proof.

Theorem 1.26 *Let E be a measurable set, $1 \leq p < \infty$, $\dfrac{1}{p} + \dfrac{1}{q} = 1$, $g \in L^q(E)$, $\|g\|_q \neq 0$. Define the functional $\mathbb{T}$ on $L^p(E)$ by*

$$\mathbb{T}(f) = \int_E gf$$

for all $f \in L^p(E)$. Then $\mathbb{T}$ is a bounded linear functional on $L^p(E)$ and $\|\mathbb{T}\|_\star = \|g\|_q$.

Proof Let $f_1, f_2 \in L^p(E)$ and $\alpha, \beta \in F$, where $F = \mathbb{R}$ or $F = \mathbb{C}$. Then

$$\mathbb{T}(\alpha f_1 + \beta f_2) = \int_E g(\alpha f_1 + \beta f_2) = \int_E (\alpha g f_1 + \beta g f_2) = \int_E \alpha g f_1 + \int_E \beta g f_2$$

$$= \alpha \int_E g f_1 + \beta \int_E g f_2 = \alpha \mathbb{T}(f_1) + \beta \mathbb{T}(f_2).$$

Therefore $\mathbb{T}$ is a linear functional on $L^p(E)$. Also, using Hölder's inequality, we have

$$|\mathbb{T}(f)| = \left| \int_E gf \right| \leq \int_E |g||f|$$

$$\leq \left(\int_E |g|^q \right)^{\frac{1}{q}} \left(\int_E |f|^p \right)^{\frac{1}{p}} = \|g\|_q \|f\|_p$$

for all $f \in L^p(E)$. Consequently $\mathbb{T}$ is a bounded linear functional on $L^p(E)$. By the last inequality, we get

$$\|\mathbb{T}\|_\star \leq \|g\|_q. \tag{1.19}$$

Let

$$g_1 = \|g\|_q^{1-q} \operatorname{sign}(g)|g|^{q-1}.$$

We have

$$\int_E |g_1|^p = \int_E \|g\|_q^{p(1-q)} |g|^{p(q-1)} = \|g\|_q^{-q} \int_E |g|^q = 1,$$

i.e., $g_1 \in L^p(E)$ and $\|g_1\|_p = 1$. Next,

$$\mathbb{T}(g_1) = \int_E g g_1 = \int_E g \|g\|_q^{1-q} \operatorname{sign}(g)|g|^{q-1} = \|g\|_q^{1-q} \int_E |g|^q = \|g\|_q.$$

Therefore

$$\|\mathbb{T}\|_\star \geq \|g\|_q.$$

From the last inequality and from (1.19), we obtain $\|\mathbb{T}\|_\star = \|g\|_q$. This completes the proof.

Theorem 1.27 *Let $\mathbb{T}$ and $\mathbb{S}$ be bounded linear functionals on a normed vector space X. If $\mathbb{T} = \mathbb{S}$ on a dense subset X_0 of X, then $\mathbb{T} = \mathbb{S}$ on X.*

Proof Let $g \in X$ be arbitrarily chosen. Then there exists a sequence $\{g_n\}_{n=1}^\infty$ of elements of X_0 such that $g_n \to g$, as $n \to \infty$, in X. Hence,

$$\mathbb{T}(g_n) \to \mathbb{T}(g), \quad \mathbb{S}(g_n) \to \mathbb{S}(g), \quad as \quad n \to \infty,$$

and

$$\mathbb{T}(g_n) = \mathbb{S}(g_n)$$

for any $n \in \mathbb{N}$. Therefore $\mathbb{T}(g) = \mathbb{S}(g)$. Because $g \in X$ was arbitrarily chosen, we obtain $\mathbb{T} = \mathbb{S}$ on X. This completes the proof.

Theorem 1.28 *Let E be a measurable set and $1 \leq p < \infty$, $\dfrac{1}{p} + \dfrac{1}{q} = 1$, g is integrable over E and there is an $M \geq 0$ such that*

$$\left| \int_E gf \right| \leq M \|f\|_p \tag{1.20}$$

for any simple function f in $L^p(E)$. Then $g \in L^q(E)$ and $\|g\|_q \leq M$.

Proof Since g is integrable over E, then it is finite a.e. on E. By excising a set of measure zero from E, we can assume that g is finite on all of E.

1. Let $p > 1$. Because $|g|$ is a nonnegative measurable function on E, there exists a sequence $\{\phi_n\}_{n=1}^{\infty}$ of measurable simple functions on E that converges pointwise on E to $|g|$ and $0 \le \phi_n \le |g|$ on E. Hence, $\{\phi_n^q\}_{n=1}^{\infty}$ is a sequence of nonnegative simple functions on E such that

$$0 \le \phi_n^q \le |g|^q \quad \text{on} \quad E$$

and $\phi_n^q \to |g|^q$, as $n \to \infty$, pointwise on E. Hence and Fatou's Lemma,

$$\int_E |g|^q \le \int_E \phi_n^q \tag{1.21}$$

for every $n \in \mathbb{N}$. Let $n \in \mathbb{N}$ be arbitrarily chosen. Then

$$\phi_n^q = \phi_n \phi_n^{q-1} \le |g|\phi_n^{q-1} = g\,\text{sign}(g)\phi_n^{q-1} \quad \text{on} \quad E.$$

Let

$$f_n = \text{sign}(g)\phi_n^{q-1} \quad \text{on} \quad E.$$

Then f_n is a simple function. Because g is integrable over E, we have that ϕ_n is integrable over E. Then ϕ_n^q is integrable over E and

$$\int_E |f_n|^p = \int_E \phi_n^q.$$

Therefore $f_n \in L^p(E)$. Next, by (1.20), we obtain

$$\int_E \phi_n^q = \int_E \phi_n \phi_n^{q-1} \le \int_E |g|\phi_n^{q-1} = \int_E g\,\text{sign}(g)\phi_n^{q-1}$$

$$= \int_E g f_n \le M \, \|f_n\|_p = M \left(\int_E \phi_n^q \right)^{\frac{1}{p}}.$$

From here,

$$\left(\int_E \phi_n^q \right)^{\frac{1}{q}} \le M,$$

i.e.,

$$\|\phi_n\|_q \leq M.$$

Hence and (1.21), we obtain

$$\|g\|_q \leq M.$$

2. Let $p = 1$. Suppose that M is not an essential upper bound for g. Then there is an $\epsilon > 0$ such that the set

$$E_\epsilon = \{x \in E : |g(x)| > M + \epsilon\}$$

has finite positive measure. Let $f = \text{sign}(g)\kappa_{E_\epsilon}$. Then, by (1.20),

$$\left|\int_E fg\right| = \left|\int_{E_\epsilon} g\,\text{sign}(g)\right| = \int_{E_\epsilon} |g| > (M + \epsilon)\, m\,(E_\epsilon). \qquad (1.22)$$

On the other hand, by (1.20),

$$\left|\int_E fg\right| \leq M\|f\|_1 = M\int_E \left|\text{sign}(g)\kappa_{E_\epsilon}\right| = Mm\,(E_\epsilon),$$

which contradicts with (1.22). Therefore M is an essential upper bound for g. This completes the proof.

Theorem 1.29 *Let $[a, b]$ be a closed bounded interval, $1 \leq p < \infty$, $\dfrac{1}{p} + \dfrac{1}{q} = 1$.*

1. Suppose that $\mathbb{T}$ is a bounded linear functional on $L^p\,([a, b])$. Then there is a function $g \in L^q\,([a, b])$ such that

$$\mathbb{T}(f) = \int_a^b gf$$

for all $f \in L^p\,([a, b])$.

Proof Let $p > 1$. We leave the case $p = 1$ for an exercise. For $x \in [a, b]$, we define

$$\Phi(x) = \mathbb{T}\left(\kappa_{[a,x)}\right).$$

For $[c, d] \subseteq [a, b]$, we have

$$\kappa_{[c,d)} = \kappa_{[a,d)} - \kappa_{[a,c)}.$$

Then

$$\Phi(d) - \Phi(c) = \mathbb{T}\left(\kappa_{[a,d)}\right) - \mathbb{T}\left(\kappa_{[a,c)}\right) = \mathbb{T}\left(\kappa_{[a,d)} - \kappa_{[a,c)}\right) = \mathbb{T}\left(\kappa_{[c,d)}\right).$$

If $\{(a_k, b_k)\}_{k=1}^{n}$ is a finite disjoint collection of intervals in (a, b), then

$$\sum_{k=1}^{n} |\Phi(b_k) - \Phi(a_k)| = \sum_{k=1}^{n} \text{sign}\left(\Phi(b_k) - \Phi(a_k)\right)\left(\Phi(b_k) - \Phi(a_k)\right)$$

$$= \sum_{k=1}^{n} \text{sign}\left(\Phi(b_k) - \Phi(a_k)\right) \mathbb{T}\left(\kappa_{[a_k,b_k)}\right)$$

$$= \sum_{k=1}^{n} \mathbb{T}\left(\text{sign}\left(\Phi(b_k) - \Phi(a_k)\right)\kappa_{[a_k,b_k)}\right)$$

$$= \mathbb{T}\left(\sum_{k=1}^{n} \text{sign}\left(\Phi(b_k) - \Phi(a_k)\right)\kappa_{[a_k,b_k)}\right).$$

Consider the simple function

$$f = \sum_{k=1}^{n} \text{sign}\left(\Phi(b_k) - \Phi(a_k)\right)\kappa_{[a_k,b_k)}.$$

Then

$$|\mathbb{T}(f)| \le \|\mathbb{T}\|_\star \|f\|_p = \|\mathbb{T}\|_\star \left(\sum_{k=1}^{n} (b_k - a_k)\right)^{\frac{1}{p}}.$$

Consequently

$$\sum_{k=1}^{n} |\Phi(b_k) - \Phi(a_k)| \le \|\mathbb{T}\|_\star \left(\sum_{k=1}^{n} (b_k - a_k)\right)^{\frac{1}{p}}$$

and Φ is absolutely continuous on $[a, b]$. Therefore Φ is differentiable almost everywhere on $[a, b]$, and if $g = \Phi'$, then g is integrable over $[a, b]$ and

$$\Phi(x) = \int_a^x g$$

for all $x \in [a, b]$. Consequently, for each $[c, d] \subseteq (a, b)$,

$$\mathbb{T}\left(\kappa_{[c,d)}\right) = \Phi(d) - \Phi(c) = \int_a^b g\kappa_{[c,d)}.$$

Since the functional $\mathbb{T}$ and the functional $f \mapsto \int_a^b gf$ are linear on the vector space of step functions in $L^p([a, b])$, it follows that

$$\mathbb{T}(f) = \int_a^b gf$$

for all step functions f in $L^p([a, b])$. There is a sequence $\{\phi_n\}_{n=1}^{\infty}$ of step functions that converges to f in $L^p([a, b])$ and also it is uniformly pointwise bounded on $[a, b]$. Hence and (1.16), we get

$$\lim_{n \to \infty} \mathbb{T}(\phi_n) = \mathbb{T}(f).$$

On the other hand, by the Lebesgue Dominated Convergence Theorem, we obtain

$$\lim_{n \to \infty} \int_a^b g\phi_n = \int_a^b gf.$$

Therefore

$$\mathbb{T}(f) = \int_a^b gf$$

for all simple functions f in $L^p([a, b])$. Since $\mathbb{T}$ is bounded,

$$\left| \int_a^b gf \right| = |\mathbb{T}(f)| \leq \|\mathbb{T}\|_* \|f\|_p$$

for all simple functions f in $L^p([a, b])$. From here and from Theorem 1.28, we have that $g \in L^q([a, b])$. By Theorem 1.26, the functional $f \mapsto \int_a^b gf$ is bounded on $L^p([a, b])$. This functional agrees with $\mathbb{T}$ on the simple functions in $L^p([a, b])$. Because the set of the simple functions in $L^p([a, b])$ is dense in $L^p([a, b])$(see Theorem 1.23), using Theorem 1.27, these two functionals agree on all of $L^p([a, b])$. This completes the proof.

Theorem 1.30 (The Riesz Representation Theorem) *Let E be a measurable set, $1 \leq p < \infty$, $\dfrac{1}{p} + \dfrac{1}{q} = 1$. For each $g \in L^p(E)$, define the bounded linear functional $\mathbb{T}_g$ on $L^p(E)$ by*

$$\mathbb{T}_g(f) = \int_E gf$$

for all $f \in L^p(E)$. Then for each bounded linear functional $\mathbb{T}$ on $L^p(E)$, there is unique function $g \in L^q(E)$ for which

$$\mathbb{T}_g = \mathbb{T} \quad and \quad \|\mathbb{T}\|_\star = \|g\|_q.$$

Proof By Theorem 1.26, it follows that $\mathbb{T}_g$ is a bounded linear functional on $L^p(E)$ and $\|\mathbb{T}_g\|_\star = \|g\|_q$ for each $g \in L^p(E)$. Also, if $g_1, g_2 \in L^q(E)$, then

$$\mathbb{T}_{g_1 - g_2}(f) = \int_E (g_1 - g_2) f = \int_E (g_1 f - g_2 f) = \int_E g_1 f - \int_E g_2 f = \mathbb{T}_{g_1}(f) - \mathbb{T}_{g_2}(f)$$

for any $f \in L^p(E)$. Therefore, if $\mathbb{T}_{g_1} = \mathbb{T}_{g_2}$, then $\mathbb{T}_{g_1 - g_2} = 0$ and hence, $\|g_1 - g_2\|_q = 0$, so that $g_1 = g_2$. Therefore, for a bounded linear functional $\mathbb{T}$ on $L^p(E)$ there is at most one $g \in L^p(E)$ for which $\mathbb{T}_g = \mathbb{T}$.

1. Suppose that $E = \mathbb{R}$. Let $\mathbb{T}$ be a bounded linear functional on $L^p(\mathbb{R})$. For any $n \in \mathbb{N}$, we define the linear functional $\mathbb{T}_n$ on $L^p([-n, n])$ by

$$\mathbb{T}_n(f) = \mathbb{T}\left(\hat{f}\right)$$

for all $f \in L^p([-n, n])$, where $\hat{f}$ is the extension of f to all of $\mathbb{R}$ that vanishes outside $[-n, n]$. Then

$$\|f\|_p = \|\hat{f}\|_p$$

and

$$|\mathbb{T}_n(f)| = \left|\mathbb{T}\left(\hat{f}\right)\right| \leq \|\mathbb{T}\|_\star \|\hat{f}\|_p = \|\mathbb{T}\|_\star \|f\|_p$$

for any $f \in L^p([-n, n])$, $n \in \mathbb{N}$. Hence,

$$\|\mathbb{T}_n\|_\star \leq \|\mathbb{T}\|_\star, \quad n \in \mathbb{N}.$$

By Theorem 1.29, it follows that there is a function $g_n \in L^q([-n, n])$, $n \in \mathbb{N}$, for which

$$\mathbb{T}_n(f) = \int_E g_n f, \quad n \in \mathbb{N},$$

for all $f \in L^p([-n, n])$, $n \in \mathbb{N}$, and

$$\|g_n\|_q = \|\mathbb{T}_n\|_\star \le \|\mathbb{T}\|_\star. \tag{1.23}$$

Note that the restriction of g_{n+1} to $[-n, n]$ agree with g_n a.e. on $[-n, n]$, $n \in \mathbb{N}$. Define g to be a measurable function on $\mathbb{R}$ that agrees with g_n a.e. on $[-n, n]$ for each $n \in \mathbb{N}$. Hence, for all $f \in L^p(\mathbb{R})$ that vanishes outside a bounded set,

$$\mathbb{T}(f) = \int_{\mathbb{R}} gf.$$

By (1.23),

$$\int_{-n}^{n} |g|^q \le \|\mathbb{T}\|_\star^q, \quad n \in \mathbb{N}.$$

Because the set of all functions of $L^p(E)$ that vanishes outside a bounded set is dense in $L^p(\mathbb{R})$, using Theorem 1.27, we conclude that $\mathbb{T}_g$ agrees with $\mathbb{T}$ on all $L^p(\mathbb{R})$.

2. Let E be a measurable set and $\mathbb{T}$ be a bounded linear functional on $L^p(E)$. Define the linear functional $\hat{\mathbb{T}}$ on $L^p(E)$ by

$$\hat{\mathbb{T}}(f) = \mathbb{T}\left(f\big|_E\right), \quad f \in L^p(\mathbb{R}).$$

Then $\hat{\mathbb{T}}$ is a bounded linear functional on $L^p(\mathbb{R})$. Hence, there is a function $\hat{g} \in L^q(\mathbb{R})$ for which

$$\hat{\mathbb{T}}(f) = \int_{\mathbb{R}} \hat{g} f$$

for any $f \in L^p(\mathbb{R})$. Define $g = \hat{g}\big|_E$. Then $\mathbb{T} = \mathbb{T}_g$. This completes the proof.

Definition 1.22 Let $1 \le p \le \infty$ and q is its conjugate. The space $L^q(\cdot)$ is called the dual space of the space $L^p(\cdot)$.

Definition 1.23 Let X be a normed vector space. A sequence $\{f_n\}_{n=1}^{\infty}$ in X is said to converge weakly in X to $f \in X$ if

$$\lim_{n\to\infty} \mathbb{T}(f_n) = \mathbb{T}(f)$$

for any linear functional $\mathbb{T}$ on X. We will write

$$f_n \rightharpoonup f \quad \text{in} \quad X.$$

Remark 1.4 If X is a normed vector space, we will write $f_n \to f$ in X if

$$\|f_n - f\| \to 0, \quad \text{as} \quad n \to \infty.$$

In this case, we will say that the sequence $\{f_n\}_{n=1}^{\infty}$ converges strongly to f in X.

Theorem 1.31 *Let X be a normed vector space, $\{f_n\}_{n=1}^{\infty}$ be a sequence in X, $f \in X$. If $f_n \to f$ in X, then $f_n \rightharpoonup f$ in X.*

Proof Since $f_n \to f$ in X, we have

$$\|f_n - f\| \to 0, \quad \text{as} \quad n \to \infty.$$

Let $\mathbb{T}$ be arbitrarily chosen linear functional on X. Then

$$|\mathbb{T}(f_n) - \mathbb{T}(f)| \le \|\mathbb{T}\|_* \|f_n - f\| \to 0, \quad \text{as} \quad n \to \infty.$$

Because $\mathbb{T}$ was arbitrarily chosen linear functional on X, we conclude that $f_n \rightharpoonup f$ in X. This completes the proof.

Theorem 1.32 *Let E be a measurable set, $1 \le p < \infty$ and q is its conjugate. Then $f_n \rightharpoonup f$ in $L^p(E)$ if and only if*

$$\lim_{n\to\infty} \int_E gf_n = \int_E gf \tag{1.24}$$

for all $g \in L^q(E)$.

Proof

1. Let $f_n \rightharpoonup f$ in $L^p(E)$. Then for every linear functional $\mathbb{T}$ we have

$$\lim_{n\to\infty} \mathbb{T}(f_n) = \mathbb{T}(f).$$

Hence, using that $h \mapsto \int_E gh$, $h \in L^p(E)$, is a linear functional for each $g \in L^q(E)$, we get (1.24).

2. Assume that (1.24) holds. Let $\mathbb{T}$ be arbitrarily chosen linear functional on $L^p(E)$. By the Riesz representation theorem, it follows that for any $h \in L^p(E)$ there is a unique $g \in L^q(E)$ such that

$$\mathbb{T}(h) = \int_E gh.$$

Hence and (1.24), we obtain

$$\lim_{n \to \infty} \mathbb{T}(f_n) = \mathbb{T}(f).$$

Because $\mathbb{T}$ was arbitrarily chosen linear functional on $L^p(E)$, we conclude that $f_n \rightharpoonup f$ in $L^p(E)$. This completes the proof.

Theorem 1.33 *Let E be a measurable set, $1 \leq p < \infty$. Then a sequence in $L^p(E)$ can converge weakly to at most one function in $L^p(E)$.*

Proof Let q be the conjugate of p. Let also, $\{f_n\}_{n=1}^{\infty}$ be a sequence in $L^p(E)$ that converges weakly to $f_1, f_2 \in L^p(E)$. Then $f_1 - f_2 \in L^p(E)$ and

$$\int_E \left| \|f_1 - f_2\|_p^{1-p} \operatorname{sign}(f_1 - f_2) |f_1 - f_2|^{p-1} \right|^q \leq \|f_1 - f_2\|_p^{1-p} \int_E |f_1 - f_2|^{q(p-1)}$$

$$= \|f_1 - f_2\|_p^{1-p} \int_E |f_1 - f_2|^p = \|f_1 - f_2\|_p,$$

i.e., $\|f_1 - f_2\|_p^{1-p} \operatorname{sign}(f_1 - f_2) |f_1 - f_2|^{p-1} \in L^q(E)$. Hence and Theorem 1.32, we get

$$\int_E \|f_1 - f_2\|_p^{1-p} \operatorname{sign}(f_1 - f_2) |f_1 - f_2|^{p-1} f_1$$

$$= \lim_{n \to \infty} \int_E \|f_1 - f_2\|_p^{1-p} \operatorname{sign}(f_1 - f_2) |f_1 - f_2|^{p-1} f_n$$

$$= \int_E \|f_1 - f_2\|_p^{1-p} \operatorname{sign}(f_1 - f_2) |f_1 - f_2|^{p-1} f_2.$$

Therefore

$$0 = \int_E \|f_1 - f_2\|_p^{1-p} \operatorname{sign}(f_1 - f_2) |f_1 - f_2|^{p-1} (f_1 - f_2)$$

$$= \|f_1 - f_2\|_p^{1-p} \int_E |f_1 - f_2|^p = \|f_1 - f_2\|_p.$$

Consequently $f_1 = f_2$. This completes the proof.

Definition 1.24 Let E be a measurable set and $1 \le p < \infty$.

1. Let $\{f_n\}_{n=1}^\infty$ be a sequence in $L^p(E)$, $f \in L^p(E)$ and $f_n \rightharpoonup f$ in $L^p(E)$. The function f will be called the weak sequential limit.
2. Let $f \in L^p(E)$. The function

$$f^\star = \|f\|_p^{1-p} \operatorname{sign}(f) |f|^{p-1}$$

will be called the conjugate function of f. Note that $f^\star \in L^q(E)$, where q is the conjugate of p.

Exercise 1.3 Let E be a measurable set, $1 \le p < \infty$, $f \in L^p(E)$ and $f^\star$ be the conjugate function of f. Prove that

$$\|f^\star\|_q = 1,$$

where q is the conjugate of p.

Remark 1.5 Let X be a linear normed space and $f_n \rightharpoonup f$ in X. Suppose that the sequence $\{\|f_n\|\}_{n=1}^\infty$ is unbounded. By taking a subsequence and relabeling, we can suppose that $\|f_n\| \ge n3^n$, $n \in \mathbb{N}$. By taking a further subsequence and relabeling, we can suppose that

$$\frac{\|f_n\|}{n3^n} \to \alpha \in [1, \infty), \quad \text{as} \quad n \to \infty.$$

Let $\mathbb{T}$ be arbitrarily chosen linear functional. Then there is a constant $M > 0$ such that

$$|\mathbb{T}(f_n)| \le M, \quad |\mathbb{T}(f)| \le M$$

for any $n \in \mathbb{N}$. Hence,

$$
\left| \mathbb{T} \left(\frac{n3^n}{\|f_n\|} f_n \right) - \mathbb{T} \left(\frac{1}{\alpha} f \right) \right|
$$

$$
= \left| \mathbb{T} \left(\frac{n3^n}{\|f_n\|} f_n \right) - \mathbb{T} \left(\frac{1}{\alpha} f_n \right) + \mathbb{T} \left(\frac{1}{\alpha} f_n \right) - \mathbb{T} \left(\frac{1}{\alpha} f \right) \right|
$$

$$
= \left| \mathbb{T} \left(\left(\frac{n3^n}{\|f_n\|} - \frac{1}{\alpha} \right) f_n \right) + \mathbb{T} \left(\frac{1}{\alpha} (f_n - f) \right) \right|
$$

$$
\leq \left| \frac{n3^n}{\|f_n\|} - \frac{1}{\alpha} \right| |\mathbb{T}(f_n)| + \frac{1}{\alpha} |\mathbb{T}(f_n) - \mathbb{T}(f)|
$$

$$
\leq \left| \frac{n3^n}{\|f_n\|} - \frac{1}{\alpha} \right| M + \frac{1}{\alpha} |\mathbb{T}(f_n) - \mathbb{T}(f)| \to 0, \quad \text{as} \quad n \to \infty.
$$

Therefore

$$
\frac{n3^n}{\|f_n\|} f_n \rightharpoonup \frac{1}{\alpha} f \quad \text{in} \quad X.
$$

Theorem 1.34 *Let E be a measurable set and $1 \leq p < \infty$. Suppose that $f_n \rightharpoonup f$ in $L^p(E)$. Then $\{f_n\}_{n=1}^{\infty}$ is bounded in $L^p(E)$ and $\|f\|_p \leq \liminf\limits_{n \to \infty} \|f_n\|_p$.*

Proof Let q be the conjugate of p and $f^\star$ be the conjugate function of f. Then $f^\star \in L^q(E)$ and using Hölder's inequality, we get

$$
\int_E f^\star f_n \leq \|f^\star\|_q \|f_n\|_p = \|f_n\|_p
$$

for any $n \in \mathbb{N}$. Since $f_n \rightharpoonup f$ in $L^p(E)$, by Theorem 1.32, we get

$$
\|f\|_p = \int_E f^\star f = \lim_{n \to \infty} \int_E f^\star f_n \leq \liminf_{n \to \infty} \|f_n\|_p.
$$

Now, we assume that the sequence $\{f_n\}_{n=1}^{\infty}$ is unbounded. Using Remark 1.5 and by taking scalar multiples, we can suppose that $\|f_n\|_p = n3^n$, $n \in \mathbb{N}$. Let $f_k^\star$ be the conjugate functions of f_k, $k \in \mathbb{N}$. Let also, $\epsilon_1 = \frac{1}{3}$ and we define

$$
\epsilon_k = \frac{1}{3^k} \quad \text{if} \quad \int_E \sum_{l=1}^{k-1} \epsilon_l f_l^\star f_k \geq 0
$$

and

$$\epsilon_k = -\frac{1}{3^k} \quad \text{if} \quad \int_E \sum_{l=1}^{k-1} \epsilon_l f_l^\star f_k \leq 0$$

for $k \in \mathbb{N}, k \geq 2$. Then

$$\left| \int_E \left(\sum_{k=1}^{n} \epsilon_k f_k^\star \right) f_n \right| = \left| \int_E \left(\sum_{k=1}^{n} \epsilon_k \|f_k\|_p^{1-p} \operatorname{sign}(f_k) |f_k|^{p-1} \right) f_n \right|$$

$$= \left| \int_E \sum_{k=1}^{n-1} \epsilon_k \|f_k\|_p^{1-p} \operatorname{sign}(f_k) |f_k|^{p-1} f_n + \int_E \epsilon_n \|f_n\|_p^{1-p} |f_n|^p \right|$$

$$= \left| \int_E \sum_{k=1}^{n-1} \epsilon_k \|f_k\|_p^{1-p} \operatorname{sign}(f_k) |f_k|^{p-1} f_n + \epsilon_n \|f_n\|_p \right| \geq \frac{1}{3^n} \|f_n\|_p = n, \quad n \in \mathbb{N}.$$

Also,

$$\|\epsilon_k f_k^\star\|_q = \frac{1}{3^k}, \quad k \in \mathbb{N}.$$

Since the series $\sum_{k=1}^{\infty} \dfrac{1}{3^k}$ is a convergent series, we have that $\sum_{k=1}^{\infty} \epsilon_k f_k^\star$ is a convergent series in $L^q(E)$ and let

$$g = \sum_{k=1}^{\infty} \epsilon_k f_k^\star.$$

For any $n \in \mathbb{N}$, we have

$$\left| \int_E g f_n \right| = \left| \int_E \left(\sum_{k=1}^{\infty} \epsilon_k f_k^\star \right) f_n \right|$$

$$= \left| \int_E \left(\sum_{k=1}^{n} \epsilon_k f_k^\star \right) f_n + \int_E \left(\sum_{k=n+1}^{\infty} \epsilon_k f_k^\star \right) f_n \right|$$

$$\geq \left| \int_E \left(\sum_{k=1}^{n} \epsilon_k f_k^\star \right) f_n \right| - \left| \int_E \left(\sum_{k=n+1}^{\infty} \epsilon_k f_k^\star \right) f_n \right|$$

$$\geq n - \left\| \sum_{k=n+1}^{\infty} \epsilon_k f_k^\star \right\|_q \| f_n \|_p$$

$$\geq n - \sum_{k=n+1}^{\infty} \left\| \epsilon_k f_k^\star \right\|_q \| f_n \|_p = n - \sum_{k=n+1}^{\infty} \frac{1}{3^k} \| f_n \|_p = n - \frac{1}{2 \, (3^n)} \| f_n \|_p = \frac{n}{2},$$

which is a contradiction, because $f_n \rightharpoonup f$ in $L^p (E)$ and $g \in L^q (E)$. This completes the proof.

Theorem 1.35 *Let E be a measurable set, $1 \leq p < \infty$ and q is its conjugate. Suppose that $f_n \rightharpoonup f$ in $L^p (E)$ and $g_n \to g$ in $L^q (E)$. Then*

$$\lim_{n \to \infty} \int_E g_n f_n = \int_E g f. \tag{1.25}$$

Proof We have

$$\int_E g_n f_n - \int_E g f = \int_E (g_n - g) \, f_n + \int_E g \, (f_n - f). \tag{1.26}$$

By Theorem 1.34, it follows that there is a constant $M > 0$ such that $\| f_n \|_p \leq M$ for any $n \in \mathbb{N}$. Then, using Hölder's inequality, we obtain

$$\left| \int_E (g_n - g) \, f_n \right| \leq \int_E |g_n - g| \, |f_n| \leq \| g_n - g \|_q \| f_n \|_p \leq M \| g_n - g \|_q \to 0, \quad \text{as} \quad n \to \infty. \tag{1.27}$$

Since $f_n \rightharpoonup f$ in $L^p (E)$ and $g \in L^q (E)$, using Theorem 1.32, we obtain

$$\int_E g \, (f_n - f) \to 0, \quad \text{as} \quad n \to \infty.$$

Hence and (1.26), (1.27), we obtain (1.25). This completes the proof.

Theorem 1.36 *Let E be a measurable set, $1 \leq p < \infty$ and q be its conjugate. Let also, $\mathscr{F} \subset L^q (E)$ and its span is dense in $L^q (E)$. Suppose that $\{ f_n \}_{n=1}^{\infty}$ is a bounded sequence in $L^p (E)$ and $f \in L^p (E)$. Then $f_n \rightharpoonup f$ in $L^p (E)$ if and only if*

$$\lim_{n \to \infty} \int_E f_n g = \int_E f g \tag{1.28}$$

for all $g \in \mathscr{F}$.

Proof

1. Let $f_n \rightharpoonup f$ in $L^p(E)$. Using Theorem 1.32, we conclude that (1.28) holds.
2. Suppose that (1.28) holds. Let $g_0 \in L^q(E)$ is arbitrarily chosen. For any $g \in L^q(E)$ and for any $n \in \mathbb{N}$, we have

$$\int_E (f_n - f) g_0 = \int_E (f_n - f)(g_0 - g) + \int_E (f_n - f) g$$

and hence, using Hölder's inequality, we obtain

$$\left| \int_E (f_n - f) g_0 \right| = \left| \int_E (f_n - f)(g_0 - g) + \int_E (f_n - f) g \right|$$

$$\leq \left| \int_E (f_n - f)(g_0 - g) \right| + \left| \int_E (f_n - f) g \right| \qquad (1.29)$$

$$\leq \|f_n - f\|_p \|g_0 - g\|_q + \left| \int_E (f_n - f) g \right|.$$

We take $\epsilon > 0$ arbitrarily. Since $\{f_n\}_{n=1}^{\infty}$ is bounded in $L^p(E)$ and the span of $\mathscr{F}$ is dense in $L^q(E)$, there is $g \in \mathscr{F}$ such that

$$\|f_n - f\|_p \|g - g_0\|_q < \frac{\epsilon}{2}$$

for any $n \in \mathbb{N}$. Hence and (1.29), we conclude that

$$\int_E f_n g_0 \to \int_E f g_0, \quad \text{as} \quad n \to \infty.$$

Because $g_0 \in L^q(E)$ was arbitrarily chosen, we obtain (1.24). Therefore $f_n \rightharpoonup f$ in $L^p(E)$. This completes the proof.

1.2.7 General L^p Spaces

Let $(X, \mathscr{M}, \mu)$ be a measure space, $\mathscr{F}$ be the collection of all measurable extended real-valued functions on X that are finite a.e. on X.

Definition 1.25 Define two functions f and g in $\mathscr{F}$ to be equivalent and we write $f \sim g$, provided

$$f(x) = g(x) \quad \text{for} \quad \text{almost} \quad \text{all} \quad x \in X.$$

The relation $\sim$ is an equivalent relation, that is, it is reflexive, symmetric and transitive. Therefore it induces a partition of $\mathscr{F}$ into a disjoint collection of equivalence classes, which we denote by $\mathscr{F}/\sim$. For given two functions f and g in $\mathscr{F}$, their equivalence classes $[f]$ and $[g]$ and real numbers α and β, we define $\alpha[f] + \beta[g]$ to be the equivalence class of the functions in $\mathscr{F}$ that take the value $\alpha f(x) + \beta g(x)$ at points $x \in X$ at which both f and g are finite. Note that these linear combinations are independent of the choice of the representatives of the equivalence classes. The zero element in $\mathscr{F}/\sim$ is the equivalence class of functions that vanish a.e. in X. Thus, $\mathscr{F}/\sim$ is a vector space.

Definition 1.26 For $1 \leq p < \infty$, we define $L^p(X, \mu)$ to be the collection of equivalence classes $[f]$ for which

$$\int_X |f|^p d\mu < \infty.$$

This is properly defined since if $f \sim g$, then

$$\int_X |f|^p d\mu = \int_X |g|^p d\mu.$$

Note that, if $[f], [g] \in L^p(X, \mu)$, then for any real constants α and β, we have

$$\int_X |\alpha f + \beta g|^p d\mu \leq 2^p \left(\int_X |\alpha f|^p d\mu + \int_X |\beta g|^p d\mu \right)$$

$$= |2\alpha|^p \int_X |f|^p d\mu + |2\beta|^p \int_X |g|^p d\mu < \infty,$$

i.e., $\alpha[f] + \beta[g] \in L^p(X, \mu)$.

Definition 1.27 We call a function $f \in \mathscr{F}$ essentially bounded provided there is some $M \geq 0$, called an essential upper bound for f, for which

$$|f(x)| \leq M \quad \text{for} \quad \text{almost} \quad \text{all} \quad x \in X.$$

Definition 1.28 We define $L^\infty(X, \mu)$ to be the collection of equivalence classes $[f]$ for which f is essentially bounded.

Note that $L^\infty(X, \mu)$ is properly defined since if $f \sim g$, then

$$|f(x)| = |g(x)| \leq M \quad \text{for} \quad \text{almost} \quad \text{all} \quad x \in X.$$

Also, if $[f], [g] \in L^\infty(X, \mu)$ and $\alpha, \beta \in \mathbb{R}$, there are nonnegative constants M_1 and M_2 such that

$$|f(x)| \leq M_1, \quad |g(x)| \leq M_2 \quad \text{for almost all} \quad x \in X.$$

Hence,

$$|\alpha f(x) + \beta g(x)| \leq |\alpha| |f(x)| + |\beta| |g(x)| \leq |\alpha| M_1 + |\beta| M_2$$

for almost all $x \in X$. Therefore $\alpha[f] + \beta[g] \in L^\infty(X, \mu)$ and hence, $L^\infty(X, \mu)$ is a vector space. For simplicity and convenience, we refer to the equivalence classes in $\mathscr{F} / \sim$ as functions and denote them by f rather than $[f]$. Thus, $f = g$ means $f - g$ vanishes a.e. on X.

Definition 1.29 For $1 \leq p < \infty$, in $L^p(X, \mu)$ we define

$$\|f\|_p = \left(\int_X |f|^p d\mu \right)^{\frac{1}{p}}.$$

For $p = \infty$, we define $\|f\|_\infty$ to be the infimum of the essential upper bounds for f.

Remark 1.6 Note that the idea for the proofs of next assertions in this section is the same as the idea for the proof of the assertions in the previous sections in this chapter. Therefore we leave the proofs of the next assertions in this section.

Theorem 1.37 *Let $(X, \mathscr{M}, \mu)$ be a measure space, $1 \leq p < \infty$ and q be the conjugate of p. If $f \in L^p(X, \mu)$, $g \in L^q(X, \mu)$, then $fg \in L^1(X, \mu)$ and*

$$\int_X |fg| d\mu \leq \|f\|_p \|g\|_q.$$

Moreover, if $f \neq 0$, the function $f^\star = \|f\|_p^{1-p} \text{sign}(f) |f|^{p-1} \in L^q(X, \mu)$,

$$\int_X ff^\star d\mu = \|f\|_p \quad \text{and} \quad \|f^\star\|_q = 1.$$

Theorem 1.38 *Let $(X, \mathscr{M}, \mu)$ be a finite measure space and $1 \leq p_1 < p_2$. Then $L^{p_2}(X, \mu) \subseteq L^{p_1}(X, \mu)$ and*

$$\|f\|_{p_1} \leq c\|f\|_{p_2} \quad \text{for} \quad f \in L^{p_2}(X, \mu),$$

where

$$c = \begin{cases} (\mu(X))^{\frac{p_2-p_1}{p_1 p_2}} & if \quad p_2 < \infty \\ \mu(X) & if \quad p_2 = \infty. \end{cases}$$

Theorem 1.39 *Let $(X, \mathcal{M}, \mu)$ be a measure space and $1 \le p \le \infty$. If $\{f_n\}_{n=1}^{\infty}$ is a bounded sequence of functions in $L^p(X, \mu)$, then $\{f_n\}_{n=1}^{\infty}$ is uniformly integrable over X.*

Theorem 1.40 *Let $(X, \mathcal{M}, \mu)$ be a measure space and $1 \le p \le \infty$. Then every rapidly Cauchy sequence in $L^p(X, \mu)$ converges to a function in $L^p(X, \mu)$, both with respect to the $L^p(X, \mu)$ norm and pointwise a.e. in X.*

Theorem 1.41 (The Riesz–Ficher Theorem) *Let $(X, \mathcal{M}, \mu)$ be a measure space and $1 \le p \le \infty$. Then $L^p(X, \mu)$ is a Banach space. Moreover, if a sequence in $L^p(X, \mu)$ converges in $L^p(X, \mu)$ to a function $f \in L^p(X, \mu)$, then a subsequence converges pointwise a.e. on X to f.*

Theorem 1.42 *Let $(X, \mathcal{M}, \mu)$ be a measure space and $1 \le p < \infty$. Then the subspace of simple functions on X that vanish outside a set of finite measure is dense in $L^p(X, \mu)$.*

Theorem 1.43 (The Vitalli L^p Convergence Theorem) *Let $(X, \mathcal{M}, \mu)$ be a measure space and $1 \le p < \infty$. Suppose that $\{f_n\}_{n=1}^{\infty}$ is a sequence in $L^p(X, \mu)$ that converges pointwise a.e. to $f \in L^p(X, \mu)$. Then $f_n \to f$ in $L^p(X, \mu)$, as $n \to \infty$, if and only if $\{|f_n|^p\}_{n=1}^{\infty}$ is uniformly integrable and tight.*

For $1 \le p < \infty$, let $f \in L^q(X, \mu)$, where q is the conjugate of p. Define the linear functional $\mathbb{T}_f : L^p(X, \mu) \mapsto R$ by

$$\mathbb{T}_f(g) = \int_X fg d\mu, \quad g \in L^p(X, \mu). \tag{1.30}$$

Theorem 1.44 (The Riesz Representation Theorem for the Dual Space of the Space $L^p(X, \mu)$) *Let $(X, \mathcal{M}, \mu)$ be a σ-finite measure space, $1 \le p < \infty$ and q be the conjugate to p. For $f \in L^q(X, \mu)$, define $\mathbb{T}_f$ by (1.30). Then $\mathbb{T}_f$ is an isometric isomorphism of $L^q(X, \mu)$ onto the space of the linear functionals on $L^p(X, \mu)$.*

Remark 1.7 Let $1 \le p \le \infty$. When it is clear from the context what measure is used, μ is omitted and one just writes $L^p(X)$.

1.3   The Convolution of Locally Integrable Functions

Definition 1.30 Suppose that f and g are locally integrable functions on $\mathbb{R}^n$. If the integral

$$\int_{\mathbb{R}^n} f(x-y)g(y)dy = \int_{\mathbb{R}^n} f(y)g(x-y)dy$$

exists for almost every $x \in \mathbb{R}^n$ and defines a locally integrable function in $\mathbb{R}^n$, it is called the convolution of the functions f and g, written

$$f * g(x) = \int_{\mathbb{R}^n} f(x-y)g(y)dy = \int_{\mathbb{R}^n} f(y)g(x-y)dy.$$

We will consider two cases in which the convolution $f * g$ does exist.

Theorem 1.45 *Let f and g be locally integrable functions on $\mathbb{R}^n$ with* $\operatorname{supp} f$, $\operatorname{supp} g \subset A$, *where A is a compact set in $\mathbb{R}^n$. Then $f \star g$ exists in $\mathbb{R}^n$.*

Proof Note that

$$\int_{\mathbb{R}^n} f(x-y)g(y)dy = \int_{A} f(x-y)g(y)dy.$$

As f and g are locally integrable in $\mathbb{R}^n$ and $\operatorname{supp} f$, $\operatorname{supp} g \subset A$, also fg is locally integrable on $\mathbb{R}^n$. Therefore the integral $\int_{A} f(x-y)g(y)dy$ exists. Now, we will check that this integral defines a locally integrable function on $\mathbb{R}^n$. Indeed,

$$\int_{\mathbb{R}^n} |f * g(x)|dx = \int_{\mathbb{R}^n} \left| \int_{A} f(x-y)g(y)dy \right| dx$$

$$\leq \int_{\mathbb{R}^n} \int_{A} |f(x-y)||g(y)|dydx$$

$$= \int_{A} \int_{\mathbb{R}^n} |f(x-y)|dx|g(y)|dy$$

$$= \int_{A} \int_{\mathbb{R}^n} |f(z)|dz|g(y)|dy$$

$$= \int_{\mathbb{R}^n} |f(z)|dz \int_A |g(y)|dy$$

$$= \int_A |f(z)|dz \int_A |g(y)|dy$$

$$< \infty,$$

showing that the convolution $f * g$ exists. This completes the proof.

Theorem 1.46 Let $p \geq 1$, $q \geq 1$ and $\dfrac{1}{p} + \dfrac{1}{q} \geq 1$. Let also, $f \in L^p(\mathbb{R}^n)$ and $g \in L^q(\mathbb{R}^n)$. Then the convolution $f * g$ exists in $\mathbb{R}^n$, $f * g \in L^r(\mathbb{R}^n)$ and

$$\|f * g\|_r \leq \|f\|_p \|g\|_q, \tag{1.31}$$

where

$$\frac{1}{r} = \frac{1}{p} + \frac{1}{q} - 1.$$

Definition 1.31 The inequality (1.31) will be called the Young convolution inequality.

Proof Let us choose $\alpha \geq 0$, $\beta \geq 0$, $s \geq 1$, $t \geq 1$ in the following way:

$$\frac{1}{r} + \frac{1}{s} + \frac{1}{t} = 1, \quad \alpha r = p = (1 - \alpha)s, \quad \beta r = q = (1 - \beta)t.$$

Then

$$p + \frac{pr}{s} = p\left(1 + \frac{r}{s}\right) = p\left(1 + \frac{1 - \alpha}{\alpha}\right) = \frac{p}{\alpha} = r, \tag{1.32}$$

$$q + \frac{qr}{t} = q\left(1 + \frac{r}{t}\right) = q\left(1 + \frac{1 - \beta}{\beta}\right) = \frac{q}{\beta} = r. \tag{1.33}$$

We apply Hölder's inequality with $\dfrac{1}{r} + \dfrac{1}{s} + \dfrac{1}{t} = 1$ and $|f|^\alpha |g|^\beta$, $|f|^{1-\alpha}$, $|g|^{1-\beta}$. We get

$$\int_{\mathbb{R}^n} |f * g(x)|^r dx = \int_{\mathbb{R}^n} \left| \int_{\mathbb{R}^n} f(x - y)g(y)dy \right|^r dx$$

$$\leq \int_{\mathbb{R}^n} \left(\int_{\mathbb{R}^n} |f(y)||g(x - y)|dy \right)^r dx$$

$$= \int_{\mathbb{R}^n} \left(\int_{\mathbb{R}^n} |f(y)|^\alpha |g(x-y)|^\beta |f(y)|^{1-\alpha} |g(x-y)|^{1-\beta} dy \right)^r dx$$

$$\leq \int_{\mathbb{R}^n} \left(\int_{\mathbb{R}^n} |f(y)|^{\alpha r} |g(x-y)|^{\beta r} dy \right) \left(\int_{\mathbb{R}^n} |f(y)|^{(1-\alpha)s} dy \right)^{\frac{r}{s}}$$

$$\times \left(\int_{\mathbb{R}^n} |g(x-y)|^{(1-\beta)t} dy \right)^{\frac{r}{t}} dx$$

$$= \int_{\mathbb{R}^n} \left(\int_{\mathbb{R}^n} |f(y)|^p |g(x-y)|^q dy \right) \left(\int_{\mathbb{R}^n} |f(y)|^p dy \right)^{\frac{r}{s}} \left(\int_{\mathbb{R}^n} |g(x-y)|^q dy \right)^{\frac{r}{t}} dx$$

$$= \|f\|_p^{\frac{rp}{s}} \int_{\mathbb{R}^n} \left(\int_{\mathbb{R}^n} |f(x-y)|^p |g(y)|^q dy \right) \left(\int_{\mathbb{R}^n} |g(z)|^q dz \right)^{\frac{r}{t}} dx$$

$$= \|f\|_p^{\frac{rp}{s}} \|g\|_q^{\frac{rq}{t}} \int_{\mathbb{R}^n} |g(y)|^q \int_{\mathbb{R}^n} |f(x-y)|^p dx dy$$

$$= \|f\|_p^{\frac{rp}{s}} \|g\|_q^{\frac{rq}{t}} \|f\|_p^p \int_{\mathbb{R}^n} |g(y)|^q dy$$

$$= \|f\|_p^{\frac{rp}{s}+p} \|g\|_q^{\frac{rq}{t}} \|g\|_q^q$$

$$= \|f\|_p^{\frac{rp}{s}+p} \|g\|_q^{\frac{rq}{t}+q}.$$

Hence, using (1.32) and (1.33), we obtain

$$\int_{\mathbb{R}^n} |f*g(x)|^r dx \leq \|f\|_p^r \|g\|_q^r,$$

i.e.,

$$\|f*g\|_r^r \leq \|f\|_p^r \|g\|_q^r.$$

Therefore

$$\|f*g\|_r \leq \|f\|_p \|g\|_q < \infty. \tag{1.34}$$

Let K be a compact set in $\mathbb{R}^n$. Hölder's inequality for $\dfrac{1}{m} + \dfrac{1}{r} = 1$ tells

$$\int\limits_K |f * g(x)| dx \le \left(\int\limits_K 1^m dx\right)^{\frac{1}{m}} \left(\int\limits_K |f * g(x)|^r dx\right)^{\frac{1}{r}} = \left(\mu(K)\right)^{\frac{1}{m}} \|f * g\|_r < \infty,$$

where $\mu(K) = \int\limits_K dx$ is the measure of K. Consequently the convolution $f * g$ is well defined. This completes the proof.

Exercise 1.4 Take $f, g \in \mathscr{C}_0^\infty(\mathbb{R}^n)$ and prove that $f * g$ is well defined.

Example 1.7 Let $f(x) = e^{-x^2}$, $g(x) = 1$, $x \in \mathbb{R}$. Then

$$f * g(x) = \int\limits_{-\infty}^{\infty} e^{-(x-y)^2} dy = -\int\limits_{-\infty}^{\infty} e^{-(x-y)^2} d(x-y) = \int\limits_{-\infty}^{\infty} e^{-z^2} dz = \sqrt{\pi}.$$

Below are some properties of the convolution where we assume that all terms exist:

1. $f * g = g * f$,
2. $f * (g + h) = f * g + f * h$,
3. $a(f * g) = (af) * g = f * (ag)$ for every $a \in \mathbb{C}$,
4. $f * (g * h) = (f * g) * h$,
5. $\overline{f * g} = \overline{f} * \overline{g}$.

Definition 1.32 Let X be an open set in $\mathbb{R}^n$. If $f \in L^1_{loc}(X)$ we extend it as zero outside of X and define the function

$$f_\epsilon = f * \omega_\epsilon,$$

which will be called the regularization of f or the mean function of f.

In fact, we have

$$f_\epsilon(x) = f * \omega_\epsilon(x) = \frac{1}{\epsilon^n} \int\limits_{\mathbb{R}^n} f(y)\omega\left(\frac{x-y}{\epsilon}\right) dy = \int\limits_{\mathbb{R}^n} f(x - \epsilon z)\omega(z) dz$$

$$= \int\limits_{|z| \le 1} f(x - \epsilon z)\omega(z) dz, \quad x \in \mathbb{R}^n.$$

Theorem 1.47 Let X be an open set in $\mathbb{R}^n$ and $f \in L^p(X)$, $1 \le p \le \infty$. Then $f \in \mathscr{C}^\infty(X)$ and $\|f_\epsilon\|_p \le \|f\|_p$.

Proof By the definition for a mean function, we get that $f_\epsilon \in \mathscr{C}^\infty(X)$. Take $1 \leq p < \infty$ and $q \geq 1$ so that $\dfrac{1}{p} + \dfrac{1}{q} = 1$. Then

$$\|f_\epsilon\|_p^p = \int_X |f_\epsilon(x)|^p dx$$

$$= \int_X \left| \int_{\mathbb{R}^n} f(y)\omega_\epsilon(x - y)dy \right|^p dx$$

$$\leq \int_X \left(\int_X |f(y)|\omega_\epsilon(x - y)dy \right)^p dx$$

$$= \int_X \left(\int_X |f(y)| \, (\omega_\epsilon(x - y))^{\frac{1}{p}} \, (\omega_\epsilon(x - y))^{\frac{1}{q}} \, dy \right)^p dx$$

$$\leq \int_X \left(\int_X |f(y)|^p \omega_\epsilon(x - y)dy \right) \left(\int_{\mathbb{R}^n} \omega_\epsilon(x - y)dy \right)^{\frac{p}{q}} dx$$

$$\leq \int_{\mathbb{R}^n} \left(\int_X |f(y)|^p \omega_\epsilon(x - y)dy \right) dx$$

$$= \int_X |f(y)|^p \left(\int_{\mathbb{R}^n} \omega_\epsilon(x - y)dx \right) dy$$

$$= \int_X |f(y)|^p dy$$

$$= \|f\|_p^p.$$

Let $p = \infty$. Then

$$\|f_\epsilon\|_\infty \leq \|f\|_\infty \int_{\mathbb{R}^n} \omega_\epsilon(x - y)dy$$

$$= \|f\|_\infty.$$

This completes the proof.

Theorem 1.48 *Let X be an open set in $\mathbb{R}^n$. If $f \in \mathscr{C}_0(X)$, then $f_\epsilon \to_{\epsilon \to 0} f$ uniformly in X.*

Proof Let $\operatorname{supp} f = K \subset X$. Set

$$K_1 = \left\{ x \in \mathbb{R}^n : \operatorname{dist}(x, K) \leq \frac{1}{2}\operatorname{dist}(K, \partial X) \right\}.$$

We have that f is uniformly continuous on K_1. Take $\eta > 0$ arbitrarily. Then there exists an $\epsilon_0 > 0$ so that from $0 < \epsilon < \epsilon_0$ and $|z| \leq 1$, we have

$$|f(x - \epsilon z) - f(x)| < \eta$$

for any $x \in K$. Therefore

$$|f_\epsilon(x) - f(x)| = \left| \int_{|z| \leq 1} f(x - \epsilon z)\omega(z)dz - f(x) \int_{|z| \leq 1} \omega(z)dz \right|$$

$$= \left| \int_{|z| \leq 1} (f(x - \epsilon z) - f(x))\omega(z)dz \right|$$

$$\leq \int_{|z| \leq 1} |f(x - \epsilon z) - f(x)|\omega(z)dz$$

$$< \eta \int_{|z| \leq 1} \omega(z)dz$$

$$= \eta, \quad x \in K.$$

This completes the proof.

Theorem 1.49 *Let X be an open set in $\mathbb{R}^n$, $f \in L^1(X)$ is finite in X and $\operatorname{supp} f \subseteq K$, where K is a compact subset of X. Then $f_\epsilon \in \mathscr{C}_0^\infty(X)$ and*

$$|f_\epsilon(x) - f(x)| \leq \max_{|x-y| \leq \epsilon} |f(x) - f(y)|, \quad x \in X,$$

for any $0 < \epsilon < \operatorname{dist}(K, \partial X)$.

Proof Let $0 < \epsilon < \text{dist}(K, \partial X)$. By the definition of f_ϵ, it follows that $f_\epsilon \in \mathscr{C}^\infty(X)$. Next,

$$
\begin{aligned}
|f_\epsilon(x) - f(x)| &= \left| \int_{\mathbb{R}^n} f(y)\omega_\epsilon(x-y)dy - f(x) \int_{\mathbb{R}^n} \omega_\epsilon(x-y)dy \right| \\
&= \left| \int_{\mathbb{R}^n} (f(y) - f(x))\omega_\epsilon(x-y)dy \right| \\
&\leq \int_{\mathbb{R}^n} |f(y) - f(x)|\omega_\epsilon(x-y)dy \\
&= \int_{|x-y|\leq\epsilon} |f(y) - f(x)|\omega_\epsilon(x-y)dy \\
&\leq \left(\max_{|x-y|\leq\epsilon} |f(y) - f(x)| \right) \int_{|x-y|\leq\epsilon} \omega_\epsilon(x-y)dy \\
&\leq \left(\max_{|x-y|\leq\epsilon} |f(y) - f(x)| \right) \int_{\mathbb{R}^n} \omega_\epsilon(x-y)dy \\
&= \max_{|x-y|\leq\epsilon} |f(y) - f(x)|, \quad x \in X.
\end{aligned}
$$

This completes the proof.

Theorem 1.50 *Let X be an open set in $\mathbb{R}^n$, $f \in L^p(X)$, $p \geq 1$. Then $f_\epsilon \to_{\epsilon \to 0} f$ in $L^p(X)$.*

Proof Take $\eta > 0$ arbitrarily. Then there exists a function $g \in \mathscr{C}_0(X)$ so that $\|f - g\|_p < \dfrac{\eta}{3}$. By Theorem 1.49, it follows that there exists an $\epsilon_0 > 0$ so that $\|f_\epsilon - g\|_p < \dfrac{\eta}{3}$ and $\|f_\epsilon - g_\epsilon\| < \dfrac{\eta}{3}$ for any $0 < \epsilon < \epsilon_0$. Now, applying the triangle inequality, we arrive at

$$
\begin{aligned}
\|f_\epsilon - f\|_p &= \|f_\epsilon - g_\epsilon + g_\epsilon - g + g - f\|_p \\
&\leq \|f_\epsilon - g_\epsilon\|_p + \|g_\epsilon - g\|_p + \|g - f\|_p \\
&< \frac{\eta}{3} + \frac{\eta}{3} + \frac{\eta}{3} \\
&= \eta, \quad 0 < \epsilon < \epsilon_0.
\end{aligned}
$$

This completes the proof.

Corollary 1.1 *Let X be an open set in $\mathbb{R}^n$. Then the space $\mathscr{C}_0^\infty(X)$ is dense in $L^p(X)$.*

Proof Let $f \in L^p(X)$ be arbitrarily chosen. Take $\eta > 0$ arbitrarily. Let $g \in \mathscr{C}_0(X)$ be chosen so that $\|f - g\|_p < \dfrac{\eta}{2}$. Since $g_\epsilon \in \mathscr{C}_0^\infty(X)$, there is an $\epsilon_0 > 0$ so that $\|g_\epsilon - g\|_p < \dfrac{\eta}{2}$ for any $0 < \epsilon < \epsilon_0$. Hence,

$$
\begin{aligned}
\|f - g_\epsilon\|_p &= \|f - g + g - g_\epsilon\|_p \\
&\leq \|f - g\|_p + \|g - g_\epsilon\|_p \\
&< \frac{\eta}{2} + \frac{\eta}{2} \\
&= \eta, \quad 0 < \epsilon < \epsilon_0.
\end{aligned}
$$

This completes the proof.

1.4 Cones in $\mathbb{R}^n$

Definition 1.33 A cone in $\mathbb{R}^n$ with vertex at 0 is a set Γ with the property that if $x \in \Gamma$, then $\lambda x \in \Gamma$ for every $\lambda > 0$. The symbol $\mathrm{pr}\,\Gamma$ will denote the intersection of Γ with the unit sphere centred at 0. A cone Γ' is called compact in the cone Γ if $\overline{\mathrm{pr}\,\Gamma'} \subset \mathrm{pr}\,\Gamma$, in which case we write $\Gamma' \subset\subset \Gamma$. The cone

$$
\Gamma^* = \{\xi \in \mathbb{R}^n : (\xi, x) \geq 0 \quad \forall x \in \Gamma\},
$$

where $(\cdot, \cdot)$ is the standard inner product in $\mathbb{R}^n$, is called the conjugate cone of Γ.

Exercise 1.5 Prove that Γ^* is a closed, convex cone with vertex at 0.

Exercise 1.6 Prove $(\Gamma^*)^* = \overline{\mathrm{ch}\Gamma}$.

Definition 1.34 A cone Γ is said to be acute if for any $e \in \mathrm{pr}\,\mathrm{int}\Gamma^*$ the set $\{x : 0 \leq (e, x) \leq 1, x \in \overline{\mathrm{ch}\Gamma}\}$ is bounded in $\mathbb{R}^n$.

Example 1.8 Let $\{e_1, e_2, \ldots, e_n\}$ be a basis in $\mathbb{R}^n$. Then

$$
\Gamma = \{x \in \mathbb{R}^n : (e_k, x) > 0, k = 1, 2, \ldots, n\}
$$

is an acute cone.

Exercise 1.7 Let Γ be an acute cone. Prove that $\overline{\mathrm{ch}\Gamma}$ does not contain an integral straight line.

Exercise 1.8 Suppose $\overline{\mathrm{ch}\Gamma}$ does not contain a straight line. Prove that $\mathrm{int}\Gamma^* \neq \emptyset$.

Exercise 1.9 Let $\mathrm{int}\varGamma^* \neq \emptyset$. Prove that for every $C' \subset\subset \mathrm{int}\varGamma^*$ there exists a constant $\sigma > 0$ such that

$$(\xi, x) \geq \sigma |\xi| |x|$$

for every $\xi \in C'$ and every $x \in \overline{\mathrm{ch}\varGamma}$.

Definition 1.35 The function

$$\mu_\varGamma(\xi) = - \inf_{x \in \mathrm{pr}\varGamma} (\xi, x)$$

is called the indicator of the cone $\varGamma$.

Exercise 1.10 Prove that $\mu_\varGamma(\xi)$ is a convex function.

Definition 1.36 Let $\varGamma \subset \mathbb{R}^n$ be a closed, convex, acute cone. Set $C = \mathrm{int}\varGamma^*$. A smooth $(n-1)$-dimensional surface S without an edge is said to be C-like if each straight line $x = x_0 + te$, $-\infty < t < \infty$, $e \in \mathrm{pr}\varGamma$, intersects it in a unique point.

Every C-like surface S cuts $\mathbb{R}^n$ in two unbounded regions S_+ and S_- such that

1. S_+ lies "above" S,
2. S_- lies "below" S,
3. $\mathbb{R}^n = S_+ \cup S \cup S_-$.

Exercise 1.11 Let $\varGamma$ be a closed, convex, acute cone and suppose S is a C-like surface. Prove that

$$\overline{S_+} = S + \varGamma.$$

1.5 Advanced Practical Problems

Problem 1.1 Let $X_1 \subset \mathbb{R}^n$, $X_2 \subset \mathbb{R}^m$ be open sets. Prove that for every $\phi \in \mathscr{C}_0^\infty(X_1 \times X_2)$ there exist sequences $\{\phi_k\}_{k=1}^\infty \subset \mathscr{C}_0^\infty(X_1 \times X_2)$ and $\{N_k\}_{k=1}^\infty \subset \mathbb{N} \cup \{0\}$ such that

$$\phi_k(x, y) = \sum_{i=1}^{N_k} \phi_{ik}(x)\psi_{ik}(y), \quad (x, y) \in X_1 \times X_2,$$

and $\phi_k \to_{k \to \infty} \phi$ in $\mathscr{C}_0^\infty(X_1 \times X_2)$.

Solution Let $\mathrm{supp}\phi \subset\subset \tilde{X}_1 \times \tilde{X}_2 \subset\subset X_1' \times X_2' \subset\subset X_1 \times X_2$. By the Weierstrass theorem, it follows that there exists a sequence of polynomials P_k, $k \in \mathbb{N}$, such that

$$\left| D^\alpha P_k(x, y) - D^\alpha \phi(x, y) \right| < \frac{1}{k}, \quad |\alpha| \leq k, \quad (x, y) \in \overline{X_1'} \times \overline{X_2'}.$$

Choose functions $\xi \in \mathscr{C}_0^\infty(X_1')$ and $\eta \in \mathscr{C}_0^\infty(X_2')$ so that $\xi(x) = 1$ for $x \in \tilde{X}_1$ and $\eta(y) = 1$ for $y \in \tilde{X}_2$. Define

$$\phi_k(x, y) = \xi(x)\eta(y)P_k(x, y), \quad k \in \mathbb{N}, \quad (x, y) \in \overline{X_1'} \times \overline{X_2'}.$$

We have $\mathrm{supp}\phi_k \subset X_1' \times X_2' \subset\subset X_1 \times X_2, k \in \mathbb{N}$, and

$$\left| D^\alpha \phi(x, y) - D^\alpha \phi_k(x, y) \right| \le \begin{cases} \dfrac{1}{k} & \text{if } (x, y) \in \tilde{X}_1 \times \tilde{X}_2, \\ \dfrac{c_\alpha}{k} & \text{if } (x, y) \in X_1' \times X_2' \end{cases}$$

for $|\alpha| \le k$. Here the constants c_α are obtained by using, for $\beta \le \alpha$, $\sup\limits_{x \in X_1} \left| D^\beta \xi(x) \right|$ and $\sup\limits_{y \in X_2} \left| D^\beta \eta(y) \right|$. Therefore $\phi_k \to_{k\to\infty} \phi$ in $\mathscr{C}_0^\infty(X_1 \times X_2)$.

Problem 1.2 Prove that for every function $\phi_1 \in \mathscr{C}_0^\infty(\mathbb{R})$ there exists a function $\phi_2 \in \mathscr{C}_0^\infty(\mathbb{R})$ such that $\phi_1(x) = \phi_2'(x)$, for every $x \in \mathbb{R}$, if and only if

$$\int_{-\infty}^\infty \phi_1(x)dx = 0.$$

Solution

1. Let $\phi_1 \in \mathscr{C}_0^\infty(\mathbb{R})$ and $\int_{-\infty}^\infty \phi_1(x)dx = 0$. We consider the function

$$\phi_2(x) = \int_{-\infty}^x \phi_1(s)ds, \quad x \in \mathbb{R}.$$

Since $\phi_1 \in \mathscr{C}_0^\infty(\mathbb{R})$, it follows that $\phi_2 \in \mathscr{C}^\infty(\mathbb{R})$. If $\mathrm{supp}\phi_1 \subset [a, b] \subset \mathbb{R}$, $a < b$, then $\phi_2(x) = 0$ for $x < a$. Therefore $\mathrm{supp}\phi_2 \subset [a, \infty)$. Since $\phi_2(\infty) = \int_{-\infty}^\infty \phi_1(x)dx = 0$, there exists a $c > a$ such that $\mathrm{supp}\phi_2 \subset [a, c]$.

2. Let $\phi_1, \phi_2 \in \mathscr{C}_0^\infty(\mathbb{R})$ and

$$\phi_1(x) = \phi_2'(x) \quad \text{for} \quad x \in \mathbb{R}.$$

Integrating from $-\infty$ to x, we find

$$\int_{-\infty}^x \phi_1(s)ds = \phi_2(x). \tag{1.35}$$

Since $\phi_2 \in \mathscr{C}_0^\infty(\mathbb{R})$, we have $\phi_2(\infty) = 0$. Hence, using (1.35), we obtain

$$\int\limits_{-\infty}^{\infty} \phi_1(s)ds = 0.$$

Problem 1.3 Prove that for every $\phi \in \mathscr{C}_0^\infty(\mathbb{R})$ there exists a function $\phi_1 \in \mathscr{C}_0^\infty(\mathbb{R})$ such that

$$\phi(x) = \phi_0(x)\int\limits_{-\infty}^{\infty} \phi(s)ds + \phi_1'(x), \quad x \in \mathbb{R},$$

where $\phi_0 \in \mathscr{C}_0^\infty(\mathbb{R})$, if and only if $\int\limits_{-\infty}^{\infty} \phi_0(s)ds = 1$.

Solution

1. Let $\phi_0 \in \mathscr{C}_0^\infty(\mathbb{R})$ and $\int\limits_{-\infty}^{\infty} \phi_0(s)ds = 1$. Consider the function

$$\phi_1(x) = \int\limits_{-\infty}^{x} \phi(s)ds - \int\limits_{-\infty}^{x} \phi_0(s)ds \int\limits_{-\infty}^{\infty} \phi(s)ds. \tag{1.36}$$

Since $\phi, \phi_0 \in \mathscr{C}_0^\infty(\mathbb{R})$, it follows $\phi_1 \in \mathscr{C}^\infty(\mathbb{R})$. Let $\operatorname{supp}\phi, \operatorname{supp}\phi_0 \subset [a, b] \subset \mathbb{R}$, $a < b$. Then $\phi_1(x) = 0$ for $x < a$. Therefore $\operatorname{supp}\phi_1 \subset [a, \infty)$. From (1.36), for $x = \infty$, we have

$$\phi_1(\infty) = \int\limits_{-\infty}^{\infty} \phi(s)ds - \int\limits_{-\infty}^{\infty} \phi_0(s)ds \int\limits_{-\infty}^{\infty} \phi(s)ds = \int\limits_{-\infty}^{\infty} \phi(s)ds - \int\limits_{-\infty}^{\infty} \phi(s)ds = 0.$$

Consequently there exists a constant $c > 0$ such that $\operatorname{supp}\phi_1 \subset [a, c]$, and therefore $\phi_1 \in \mathscr{C}_0^\infty(\mathbb{R})$.
2. Let $\phi, \phi_0, \phi_1 \in \mathscr{C}_0^\infty(\mathbb{R})$ and

$$\phi_1'(x) = \phi(x) - \phi_0(x)\int\limits_{-\infty}^{\infty} \phi(s)ds, \quad x \in \mathbb{R}. \tag{1.37}$$

We integrate the Eq. (1.37) from $-\infty$ to ∞ and get

$$\int_{-\infty}^{\infty} \phi_1'(x)dx = \int_{-\infty}^{\infty} \phi(x)dx - \int_{-\infty}^{\infty} \phi_0(x)dx \int_{-\infty}^{\infty} \phi(x)dx,$$

that is,

$$0 = \int_{-\infty}^{\infty} \phi(x)dx\left(1 - \int_{-\infty}^{\infty} \phi_0(x)dx\right)$$

for every $\phi \in \mathscr{C}_0^\infty(\mathbb{R})$. In particular, the last equation is valid for every $\phi \in \mathscr{C}_0^\infty(\mathbb{R})$ for which $\int_{-\infty}^{\infty} \phi(x)dx = 1$. For such ϕ, we obtain

$$\int_{-\infty}^{\infty} \phi_0(x)dx = 1.$$

Problem 1.4 Prove that for every $\phi \in \mathscr{C}_0^\infty(\mathbb{R})$ there exists a function $\phi_2 \in \mathscr{C}_0^\infty(\mathbb{R})$ such that

$$\phi(x) = \phi_1'(x)\int_{-\infty}^{\infty}\int_{-\infty}^{s} \phi(\tau)d\tau ds + \phi_2''(x), \quad x \in \mathbb{R},$$

where $\phi_1 \in \mathscr{C}_0^\infty(\mathbb{R})$, if and only if $\int_{-\infty}^{\infty} \phi_1(x)dx = 1$.

Solution

1. Let $\phi, \phi_1 \in \mathscr{C}_0^\infty(\mathbb{R})$ and $\int_{-\infty}^{\infty} \phi_1(x)dx = 1$. Consider the function

$$\phi_2(x) = -\left(\int_{-\infty}^{x} \phi_1(s)ds\right)\left(\int_{-\infty}^{\infty}\int_{-\infty}^{s} \phi(\tau)d\tau ds\right) + \int_{-\infty}^{x}\int_{-\infty}^{y} \phi(s)dsdy, \quad x \in \mathbb{R}.$$

Since $\phi, \phi_1 \in \mathscr{C}_0^\infty(\mathbb{R})$, we conclude that $\phi_2 \in \mathscr{C}^\infty(\mathbb{R})$. Let $\mathrm{supp}\phi$, $\mathrm{supp}\phi_1 \subset [a, b] \subset \mathbb{R}$, $a < b$. Then, for $x < a$, by the definition of the function ϕ_2, we find that $\phi_2(x) = 0$. Thus, $\mathrm{supp}\phi_2 \subset [a, \infty)$. Next,

$$\phi_2(\infty) = -\left(\int_{-\infty}^{\infty} \phi_1(s)ds\right)\left(\int_{-\infty}^{\infty}\int_{-\infty}^{s} \phi(\tau)d\tau ds\right) + \int_{-\infty}^{\infty}\int_{-\infty}^{y} \phi(s)dsdy$$

$$= -\int_{-\infty}^{\infty}\int_{-\infty}^{y} \phi(s)dsdy + \int_{-\infty}^{\infty}\int_{-\infty}^{y} \phi(s)dsdy$$

$$= 0.$$

Therefore there exists a $c > 0$ so that $\mathrm{supp}\phi_2 \subset [a, c]$ and $\phi_2 \in \mathscr{C}_0^\infty(\mathbb{R})$.

2. Let $\phi, \phi_1, \phi_2 \in \mathscr{C}_0^\infty(\mathbb{R})$ and

$$\phi(x) = \phi_1'(x)\int_{-\infty}^{\infty}\int_{-\infty}^{s} \phi(\tau)d\tau ds + \phi_2''(x), \quad x \in \mathbb{R}.$$

We integrate the last equation from $-\infty$ to x and we get

$$\int_{-\infty}^{x} \phi(s)ds = \left(\int_{-\infty}^{x} \phi_1'(s)ds\right)\left(\int_{-\infty}^{\infty}\int_{-\infty}^{s} \phi(\tau)d\tau ds\right) + \int_{-\infty}^{x} \phi_2''(s)ds$$

$$= \phi_1(x)\int_{-\infty}^{\infty}\int_{-\infty}^{s} \phi(\tau)d\tau ds + \phi_2'(x), \quad x \in \mathbb{R},$$

which we integrate from $-\infty$ to x and we find

$$\int_{-\infty}^{x}\int_{-\infty}^{y} \phi(s)dsdy = \left(\int_{-\infty}^{x} \phi_1(s)ds\right)\left(\int_{-\infty}^{\infty}\int_{-\infty}^{s} \phi(\tau)d\tau ds\right) + \int_{-\infty}^{x} \phi_2'(s)ds$$

$$= \left(\int_{-\infty}^{x} \phi_1(s)ds\right)\left(\int_{-\infty}^{\infty}\int_{-\infty}^{s} \phi(\tau)d\tau ds\right) + \phi_2(x), \quad x \in \mathbb{R}^n.$$

In particular, for $x = \infty$, by the last equation and using that $\phi_2(\infty) = 0$, we get

$$\int\limits_{-\infty}^{\infty}\int\limits_{-\infty}^{y} \phi(s)ds\,dy = \left(\int\limits_{-\infty}^{\infty} \phi_1(s)ds\right)\left(\int\limits_{-\infty}^{\infty}\int\limits_{-\infty}^{s} \phi(\tau)d\tau\,ds\right) + \phi_2(\infty)$$

$$= \left(\int\limits_{-\infty}^{\infty} \phi_1(s)ds\right)\left(\int\limits_{-\infty}^{\infty}\int\limits_{-\infty}^{s} \phi(\tau)d\tau\,ds\right)$$

or

$$\left(1 - \int\limits_{-\infty}^{\infty} \phi_1(s)ds\right)\int\limits_{-\infty}^{\infty}\int\limits_{-\infty}^{s} \phi(\tau)d\tau\,ds = 0.$$

Since the last equation holds for any $\phi \in \mathscr{C}_0^\infty(\mathbb{R})$, it holds for those $\phi \in \mathscr{C}_0^\infty(\mathbb{R})$ for which

$$\int\limits_{-\infty}^{\infty}\int\limits_{-\infty}^{s} \phi(\tau)d\tau\,ds = 1.$$

For such ϕ, we obtain $\displaystyle\int\limits_{-\infty}^{\infty} \phi_1(s)ds = 1$.

Problem 1.5 Prove that for every $\phi \in \mathscr{C}_0^\infty(\mathbb{R})$ there exists a function $\phi_2 \in \mathscr{C}_0^\infty(\mathbb{R})$ such that

$$\phi(x) = \phi_1(x)\int\limits_{-\infty}^{\infty}\int\limits_{-\infty}^{s} \phi(\tau)d\tau\,ds + \phi_2''(x), \quad x \in \mathbb{R},$$

where $\phi_1 \in \mathscr{C}_0^\infty(\mathbb{R})$, if and only if $\displaystyle\int\limits_{-\infty}^{\infty}\int\limits_{-\infty}^{s} \phi_1(\tau)d\tau\,ds = 1$.

Hint For $\phi, \phi_1 \in \mathscr{C}_0^\infty(\mathbb{R})$, consider the function

$$\phi_2(x) = \int\limits_{-\infty}^{x}\int\limits_{-\infty}^{y} \phi(s)ds\,dy - \left(\int\limits_{-\infty}^{x}\int\limits_{-\infty}^{y} \phi_1(s)ds\,dy\right)\left(\int\limits_{-\infty}^{\infty}\int\limits_{-\infty}^{s} \phi(\tau)d\tau\,ds\right), \quad x \in \mathbb{R}.$$

Problem 1.6 Prove that for every $\phi \in \mathscr{C}_0^\infty(\mathbb{R})$ there exists $\phi_3 \in \mathscr{C}_0^\infty(\mathbb{R})$ such that

$$\phi(x) = \phi_1(x) \int_{-\infty}^{\infty}\int_{-\infty}^{s} \phi(\tau)d\tau ds + \phi_2'(x) \int_{-\infty}^{\infty}\int_{-\infty}^{s} \phi(\tau)d\tau ds + \phi_3''(x), \quad x \in \mathbb{R},$$

where $\phi_1, \phi_2 \in \mathscr{C}_0^\infty(\mathbb{R})$, if and only if

$$\int_{-\infty}^{\infty}\int_{-\infty}^{s} \phi_1(\tau)d\tau ds + \int_{-\infty}^{\infty} \phi_2(x)dx = 1.$$

Solution

1. Let $\phi, \phi_1, \phi_2 \in \mathscr{C}_0^\infty(\mathbb{R})$ and

$$\int_{-\infty}^{\infty}\int_{-\infty}^{s} \phi_1(\tau)d\tau ds + \int_{-\infty}^{\infty} \phi_2(x)dx = 1.$$

Define the function

$$\phi_3(x) = \int_{-\infty}^{x}\int_{-\infty}^{y} \phi(s)dsdy - \left(\int_{-\infty}^{x}\int_{-\infty}^{y} \phi_1(s)dsdy \right)\left(\int_{-\infty}^{\infty}\int_{-\infty}^{s} \phi(\tau)d\tau ds \right)$$
$$- \left(\int_{-\infty}^{x} \phi_2(s)ds \right)\left(\int_{-\infty}^{\infty}\int_{-\infty}^{s} \phi(\tau)d\tau ds \right), \quad x \in \mathbb{R}.$$

Since $\phi, \phi_1, \phi_2 \in \mathscr{C}_0^\infty(\mathbb{R})$, by the definition of the function ϕ_3, it follows that $\phi_3 \in \mathscr{C}^\infty(\mathbb{R})$. Let $\mathrm{supp}\phi$, $\mathrm{supp}\phi_1$, $\mathrm{supp}\phi_2 \subset [a, b]$. Then, for $x < a$, by the definition of ϕ_3, we find $\phi_3(x) = 0$. Consequently $\mathrm{supp}\phi_3 \subset [a, \infty)$. Note that

$$\phi_3(\infty) = \int_{-\infty}^{\infty}\int_{-\infty}^{y} \phi(s)dsdy - \left(\int_{-\infty}^{\infty}\int_{-\infty}^{y} \phi_1(s)dsdy \right)\left(\int_{-\infty}^{\infty}\int_{-\infty}^{s} \phi(\tau)d\tau ds \right)$$
$$- \left(\int_{-\infty}^{\infty} \phi_2(s)ds \right)\left(\int_{-\infty}^{\infty}\int_{-\infty}^{s} \phi(\tau)d\tau ds \right)$$
$$= \int_{-\infty}^{\infty}\int_{-\infty}^{y} \phi(s)dsdy$$

$$-\int_{-\infty}^{\infty}\int_{-\infty}^{y}\phi(s)dsdy\left(\int_{-\infty}^{\infty}\int_{-\infty}^{y}\phi_1(s)dsdy+\int_{-\infty}^{\infty}\phi_2(s)ds\right)$$

$$=\int_{-\infty}^{\infty}\int_{-\infty}^{y}\phi(s)dsdy-\int_{-\infty}^{\infty}\int_{-\infty}^{y}\phi(s)dsdy$$

$$=0.$$

From the last relation, we conclude that there exists a $c > 0$ so that supp$\phi_3 \subset [a, c]$ and $\phi_3 \in \mathscr{C}_0^{\infty}(\mathbb{R})$.

2. Let $\phi, \phi_1, \phi_2, \phi_3 \in \mathscr{C}_0^{\infty}(\mathbb{R})$ be such that

$$\phi(x) = \phi_1(x)\int_{-\infty}^{\infty}\int_{-\infty}^{s}\phi(\tau)d\tau ds + \phi_2'(x)\int_{-\infty}^{\infty}\int_{-\infty}^{s}\phi(\tau)d\tau ds + \phi_3''(x), \quad x \in \mathbb{R}.$$

We integrate the last equation from $-\infty$ to x and we get

$$\int_{-\infty}^{x}\phi(s)ds = \left(\int_{-\infty}^{x}\phi_1(s)ds\right)\left(\int_{-\infty}^{\infty}\int_{-\infty}^{s}\phi(\tau)d\tau ds\right) + \left(\int_{-\infty}^{x}\phi_2'(s)ds\right)\left(\int_{-\infty}^{\infty}\int_{-\infty}^{s}\phi(\tau)d\tau ds\right)$$

$$+ \int_{-\infty}^{x}\phi_3''(s)ds$$

$$= \left(\int_{-\infty}^{x}\phi_1(s)ds\right)\left(\int_{-\infty}^{\infty}\int_{-\infty}^{s}\phi(\tau)d\tau ds\right) + \phi_2(x)\int_{-\infty}^{\infty}\int_{-\infty}^{s}\phi(\tau)d\tau ds$$

$$+ \phi_3'(x), \quad x \in \mathbb{R},$$

which we integrate from $-\infty$ to x and we find

$$\int_{-\infty}^{x}\int_{-\infty}^{y}\phi(s)dsdy = \left(\int_{-\infty}^{x}\int_{-\infty}^{y}\phi_1(s)dsdy\right)\left(\int_{-\infty}^{\infty}\int_{-\infty}^{s}\phi(\tau)d\tau ds\right)$$

$$+ \left(\int_{-\infty}^{x}\phi_2(s)ds\right)\left(\int_{-\infty}^{\infty}\int_{-\infty}^{s}\phi(\tau)d\tau ds\right)$$

$$+ \int_{-\infty}^{x}\phi_3'(s)ds$$

$$= \left(\int_{-\infty}^{x} \int_{-\infty}^{y} \phi_1(s)dsdy \right) \left(\int_{-\infty}^{\infty} \int_{-\infty}^{s} \phi(\tau)d\tau ds \right)$$

$$+ \left(\int_{-\infty}^{x} \phi_2(s)ds \right) \left(\int_{-\infty}^{\infty} \int_{-\infty}^{s} \phi(\tau)d\tau ds \right) + \phi_3(x), \quad x \in \mathbb{R}.$$

By the last equation, for $x = \infty$, we find

$$\left(\int_{-\infty}^{\infty} \int_{-\infty}^{y} \phi(s)dsdy \right) \left(1 - \int_{-\infty}^{\infty} \int_{-\infty}^{y} \phi_1(s)dsdy - \int_{-\infty}^{\infty} \phi_2(s)ds \right) = 0.$$

Take $\phi \in \mathscr{C}_0^{\infty}(\mathbb{R})$ so that $\int_{-\infty}^{\infty} \int_{-\infty}^{y} \phi(s)dsdy = 1$. For such ϕ, we arrive at

$$\int_{-\infty}^{\infty} \int_{-\infty}^{y} \phi_1(s)dsdy + \int_{-\infty}^{\infty} \phi_2(s)ds = 1.$$

Problem 1.7 Prove that for every $\phi \in \mathscr{C}_0^{\infty}(\mathbb{R})$ there exists $\phi_3 \in \mathscr{C}_0^{\infty}(\mathbb{R})$ such that

$$\phi(x) = \phi_1(x) \int_{-\infty}^{\infty} \int_{-\infty}^{s} \phi(\tau)d\tau ds + \phi_2'(x) \int_{-\infty}^{\infty} \int_{-\infty}^{s} \phi(\tau)d\tau ds + \phi_3''(x), \quad x \in \mathbb{R},$$

where $\phi_1, \phi_2 \in \mathscr{C}_0^{\infty}(\mathbb{R})$, if and only if

$$\int_{-\infty}^{\infty} \int_{-\infty}^{s} \phi_1(\tau)d\tau ds + \int_{-\infty}^{\infty} \phi_2(x)dx = 1.$$

Hint Use the function

$$\phi_3(x) = \int_{-\infty}^{x} \int_{-\infty}^{y} \phi(s)dsdy - \left(\int_{-\infty}^{x} \int_{-\infty}^{y} \phi_1(s)dsdy \right) \left(\int_{-\infty}^{\infty} \int_{-\infty}^{s} \phi(\tau)d\tau ds \right)$$

$$- \left(\int_{-\infty}^{x} \phi_2(s)ds \right) \left(\int_{-\infty}^{\infty} \int_{-\infty}^{s} \phi(\tau)d\tau ds \right), \quad x \in \mathbb{R}.$$

Problem 1.8 Prove that for every $\phi \in \mathscr{C}_0^\infty(\mathbb{R})$ there exists $\phi_4 \in \mathscr{C}_0^\infty(\mathbb{R})$ such that

$$\phi(x) = \phi_4'''(x) + \int_{-\infty}^{\infty} \int_{-\infty}^{s_1} \int_{-\infty}^{s_2} \phi(\tau)d\tau ds_2 ds_1 \left(\phi_1(x) + \phi_2'(x) + \phi_3''(x)\right), \quad x \in \mathbb{R},$$

where $\phi_1, \phi_2, \phi_3 \in \mathscr{C}_0^\infty(\mathbb{R})$, if and only if

$$\int_{-\infty}^{\infty} \int_{-\infty}^{s_1} \int_{-\infty}^{s_2} \phi_1(\tau)d\tau ds_2 ds_1 + \int_{-\infty}^{\infty} \int_{-\infty}^{s} \phi_2(\tau)d\tau ds + \int_{-\infty}^{\infty} \phi_3(s)ds = 1.$$

Hint Use the function

$$\phi_4(x) = \int_{-\infty}^{x} \int_{-\infty}^{x_1} \int_{-\infty}^{x_2} \phi(s)ds dx_2 dx_1$$

$$- \left(\int_{-\infty}^{x} \int_{-\infty}^{x_1} \int_{-\infty}^{x_2} \phi_1(s)ds dx_2 dx_1 + \int_{-\infty}^{x} \int_{-\infty}^{x_1} \phi_2(s)ds dx_1 + \int_{-\infty}^{x} \phi_3(s)ds \right)$$

$$\times \left(\int_{-\infty}^{\infty} \int_{-\infty}^{s_1} \int_{-\infty}^{s_2} \phi(\tau)d\tau ds_2 ds_1 \right), \quad x \in \mathbb{R}.$$

Problem 1.9 Prove that for every $\phi \in \mathscr{C}_0^\infty(\mathbb{R})$ there exists $\phi_3 \in \mathscr{C}_0^\infty(\mathbb{R})$ such that

$$\phi(x) = \phi_3'''(x) + \int_{-\infty}^{\infty} \int_{-\infty}^{s_1} \int_{-\infty}^{s_2} \phi(\tau)d\tau ds_2 ds_1 \left(\phi_1(x) + \phi_2'(x)\right), \quad x \in \mathbb{R},$$

where $\phi_1, \phi_2 \in \mathscr{C}_0^\infty(\mathbb{R})$, if and only if

$$\int_{-\infty}^{\infty} \int_{-\infty}^{s_1} \int_{-\infty}^{s_2} \phi_1(\tau)d\tau ds_2 ds_1 + \int_{-\infty}^{\infty} \int_{-\infty}^{s} \phi_2(\tau)d\tau ds = 1.$$

Hint Use the function

$$\phi_3(x) = \int_{-\infty}^{x} \int_{-\infty}^{x_1} \int_{-\infty}^{x_2} \phi(s)\,ds\,dx_2\,dx_1$$

$$- \left(\int_{-\infty}^{x} \int_{-\infty}^{x_1} \int_{-\infty}^{x_2} \phi_1(s)\,ds\,dx_2\,dx_1 + \int_{-\infty}^{x} \int_{-\infty}^{x_1} \phi_2(s)\,ds\,dx_1 \right)$$

$$\times \left(\int_{-\infty}^{\infty} \int_{-\infty}^{s_1} \int_{-\infty}^{s_2} \phi(\tau)\,d\tau\,ds_2\,ds_1 \right), \qquad x \in \mathbb{R}.$$

Problem 1.10 Prove that for every $\phi \in \mathscr{C}_0^\infty(\mathbb{R})$ there exists $\phi_3 \in \mathscr{C}_0^\infty(\mathbb{R})$ such that

$$\phi(x) = \phi_3'''(x) + \int_{-\infty}^{\infty} \int_{-\infty}^{s_1} \int_{-\infty}^{s_2} \phi(\tau)\,d\tau\,ds_2\,ds_1\big(\phi_1(x) + \phi_2''(x)\big), \qquad x \in \mathbb{R},$$

where $\phi_1, \phi_2 \in \mathscr{C}_0^\infty(\mathbb{R})$, if and only if

$$\int_{-\infty}^{\infty} \int_{-\infty}^{s_1} \int_{-\infty}^{s_2} \phi_1(\tau)\,d\tau\,ds_2\,ds_1 + \int_{-\infty}^{\infty} \phi_2(\tau)\,d\tau = 1.$$

Hint Use the function

$$\phi_3(x) = \int_{-\infty}^{x} \int_{-\infty}^{x_1} \int_{-\infty}^{x_2} \phi(s)\,ds\,dx_2\,dx_1 - \left(\int_{-\infty}^{x} \int_{-\infty}^{x_1} \int_{-\infty}^{x_2} \phi_1(s)\,ds\,dx_2\,dx_1 + \int_{-\infty}^{x} \phi_2(s)\,ds \right)$$

$$\times \left(\int_{-\infty}^{\infty} \int_{-\infty}^{s_1} \int_{-\infty}^{s_2} \phi(\tau)\,d\tau\,ds_2\,ds_1 \right), \qquad x \in \mathbb{R}.$$

Problem 1.11 Prove that for every $\phi \in \mathscr{C}_0^\infty(\mathbb{R})$ there exists $\phi_3 \in \mathscr{C}_0^\infty(\mathbb{R})$ such that

$$\phi(x) = \phi_3'''(x) + \int_{-\infty}^{\infty} \int_{-\infty}^{s_1} \int_{-\infty}^{s_2} \phi(\tau)\,d\tau\,ds_2\,ds_1\big(\phi_1'(x) + \phi_2''(x)\big), \qquad x \in \mathbb{R},$$

where $\phi_1, \phi_2 \in \mathscr{C}_0^\infty(\mathbb{R})$, if and only if

$$\int\limits_{-\infty}^{\infty} \int\limits_{-\infty}^{s_1} \phi_1(\tau)d\tau ds_1 + \int\limits_{-\infty}^{\infty} \phi_2(\tau)d\tau = 1.$$

Hint Use the function

$$\phi_3(x) = \int\limits_{-\infty}^{x} \int\limits_{-\infty}^{x_1} \int\limits_{-\infty}^{x_2} \phi(s)ds dx_2 dx_1 - \left(\int\limits_{-\infty}^{x} \int\limits_{-\infty}^{x_1} \phi_1(s)ds dx_1 + \int\limits_{-\infty}^{x} \phi_2(s)ds \right)$$

$$\times \left(\int\limits_{-\infty}^{\infty} \int\limits_{-\infty}^{s_1} \int\limits_{-\infty}^{s_2} \phi(\tau)d\tau ds_2 ds_1 \right), \quad x \in \mathbb{R}.$$

Problem 1.12 Prove that for every $\phi \in \mathscr{C}_0^\infty(\mathbb{R})$ there exists $\phi_2 \in \mathscr{C}_0^\infty(\mathbb{R})$ such that

$$\phi(x) = \phi_2'''(x) + \phi_1(x) \int\limits_{-\infty}^{\infty} \int\limits_{-\infty}^{s_1} \int\limits_{-\infty}^{s_2} \phi(\tau)d\tau ds_2 ds_1, \quad x \in \mathbb{R},$$

where $\phi_1 \in \mathscr{C}_0^\infty(\mathbb{R})$, if and only if $\displaystyle\int\limits_{-\infty}^{\infty} \int\limits_{-\infty}^{s_1} \int\limits_{-\infty}^{s_2} \phi_1(\tau)d\tau ds_2 ds_1 = 1.$

Hint Use the function

$$\phi_2(x) = \int\limits_{-\infty}^{x} \int\limits_{-\infty}^{x_1} \int\limits_{-\infty}^{x_2} \phi(s)ds dx_2 dx_1 - \left(\int\limits_{-\infty}^{x} \int\limits_{-\infty}^{x_1} \int\limits_{-\infty}^{x_2} \phi_1(s)ds dx_2 dx_1 \right)$$

$$\times \left(\int\limits_{-\infty}^{\infty} \int\limits_{-\infty}^{x_1} \int\limits_{-\infty}^{x_2} \phi(s)ds dx_2 dx_1 \right), \quad x \in \mathbb{R}.$$

Problem 1.13 Prove that for every $\phi \in \mathscr{C}_0^\infty(\mathbb{R})$ there exists $\phi_2 \in \mathscr{C}_0^\infty(\mathbb{R})$ such that

$$\phi(x) = \phi_2'''(x) + \phi_1'(x) \int\limits_{-\infty}^{\infty} \int\limits_{-\infty}^{s_1} \int\limits_{-\infty}^{s_2} \phi(\tau)d\tau ds_2 ds_1, \quad x \in \mathbb{R},$$

where $\phi_1 \in \mathscr{C}_0^\infty(\mathbb{R})$, if and only if $\displaystyle\int\limits_{-\infty}^{\infty} \int\limits_{-\infty}^{s_1} \phi_1(\tau)d\tau ds_1 = 1.$

Hint Use the function

$$\phi_2(x) = \int\limits_{-\infty}^{x} \int\limits_{-\infty}^{x_1} \int\limits_{-\infty}^{x_2} \phi(s)ds dx_2 dx_1 - \left(\int\limits_{-\infty}^{x} \int\limits_{-\infty}^{x_1} \phi_1(s)ds dx_1\right)$$

$$\times \left(\int\limits_{-\infty}^{\infty} \int\limits_{-\infty}^{x_1} \int\limits_{-\infty}^{x_2} \phi(s)ds dx_2 dx_1\right), \quad x \in \mathbb{R}.$$

Problem 1.14 Prove that for every $\phi \in \mathscr{C}_0^\infty(\mathbb{R})$ there exists $\phi_2 \in \mathscr{C}_0^\infty(\mathbb{R})$ such that

$$\phi(x) = \phi_2'''(x) + \phi_1''(x) \int\limits_{-\infty}^{\infty} \int\limits_{-\infty}^{s_1} \int\limits_{-\infty}^{s_2} \phi(\tau)d\tau ds_2 ds_1, \quad x \in \mathbb{R},$$

where $\phi_1 \in \mathscr{C}_0^\infty(\mathbb{R})$, if and only if $\int\limits_{-\infty}^{\infty} \phi_1(\tau)d\tau = 1$.

Hint Use the function

$$\phi_2(x) = \int\limits_{-\infty}^{x} \int\limits_{-\infty}^{x_1} \int\limits_{-\infty}^{x_2} \phi(s)ds dx_2 dx_1 - \left(\int\limits_{-\infty}^{x} \phi_1(s)ds\right)\left(\int\limits_{-\infty}^{\infty} \int\limits_{-\infty}^{x_1} \int\limits_{-\infty}^{x_2} \phi(s)ds dx_2 dx_1\right),$$

$x \in \mathbb{R}$.

Problem 1.15 Let I be an open interval in $\mathbb{R}$, V is a Banach space with a norm $||\cdot||$, $f : I \to V$ is a smooth map. Prove

1. $||f(y) - f(x)|| \leq |y - x| \sup\limits_{t \in [0,1]} ||f'(x + t(y - x))||$, $x, y \in I$,
2. $||f(y) - f(x) - v(y - x)|| \leq |y - x| \sup\limits_{t \in [0,1]} ||f'(x + t(y - x)) - v||$, $v \in V$.

1. *Solution* Let $M = \sup\limits_{t \in [0,1]} ||f'(x + t(y - x))||$. We define the set

$$E = \left\{t : 0 \leq t \leq 1, \quad ||f(x + t(y - x)) - f(x)|| \leq Mt|y - x|\right\}.$$

Since f is a continuous function, E is a closed subset of the interval $[0, 1]$. On the other hand,

$$||f(x + 0 \cdot (y - x)) - f(x)|| = ||f(x) - f(x)|| \leq M \cdot 0 \cdot |y - x|,$$

so 0 belongs to E. From this, we conclude that E is compact and it has a maximal element s. We suppose that $s < 1$. Then we can find $t > s$ such that $t - s$ is sufficiently small. Hence,

$$\|f(x + t(y - x)) - f(x)\|$$
$$= \|f(x + t(y - x)) - f(x + s(y - x)) + f(x + s(y - x)) - f(x)\|$$
$$\leq \|f(x + t(y - x)) - f(x + s(y - x))\| + \|f(x + s(y - x)) - f(x)\|$$
$$= \|(t - s)(y - x)f'(x + \xi(y - x))\| + \|f(x + s(y - x)) - f(x)\|$$
$$\leq M(t - s)|y - x| + Ms|y - x| = Mt|y - x|,$$

where ξ is between t and s. This contradicts with the assumption that s is maximal. Therefore $s = 1$. For $t = 1$, we obtain

$$\|f(y) - f(x)\| \leq \sup_{t \in [0,1]} \|f'(x + t(y - x))\| |y - x|.$$

2. **Hint** Use the function $g(x) = f(x) - xv$ and part 1.

Problem 1.16 Let I be an open interval in $\mathbb{R}$, V is a Banach space with a norm $\|\cdot\|$, $f : I \rightarrow V$ is a continuous map that is differentiable on $I \backslash F$, where F is a closed subset of I and $f \equiv 0$ on F. Prove that if $x \in F$ and $f'(y) \rightarrow_{y \rightarrow x} 0$, $y \in I \backslash F$, then $f'(x)$ exists and it is zero for every $x \in I$.

Solution Let $y \in F$. Then $f(y) - f(x) = 0$. From this, $f'(x)$ exists for every $x \in F$ and $f'(x) = 0$ for every $x \in F$.
Now, take $y \notin F$, $x \leq y$, and let z be the point in $F \cap [x, y]$ closest to y. From the previous problem, we have

$$\|f(y) - f(x)\| = \|f(y) - f(z) + f(z) - f(x)\| = \|f(y) - f(z)\|$$

and

$$\|f(y) - f(z)\| \leq |y - z| \sup_{t \in [0,1]} \|f'(z + t(y - z))\|.$$

The last inequality implies

$$\|f(y) - f(x)\| = o(|y - x|)$$

when $y \rightarrow x$. Therefore

$$\lim_{h \rightarrow 0} \left\| \frac{f(x + h) - f(x)}{h} \right\| = 0, \quad \forall y \in I \backslash F.$$

Consequently $f'(x)$ exists for every $x \in I$ and actually $f'(x) = 0$ for every $x \in I$.

Problem 1.17 Let P be a polynomial and define

$$f(x) = \begin{cases} P\left(\dfrac{1}{x}\right)e^{-\frac{1}{x}} & \text{for} \quad x > 0, \\ 0 & \text{for} \quad x \leq 0. \end{cases}$$

Prove that

1. f is a differentiable function on $\mathbb{R}$ and $f'(0) = 0$,
2. $f \in \mathscr{C}^{\infty}(\mathbb{R})$.

Problem 1.18 Prove that there exists a function $\phi \in \mathscr{C}_0^{\infty}(\mathbb{R}^n)$ for which $\phi(0) > 0$.

Hint Use the function

$$\phi(x) = \begin{cases} e^{-\frac{1}{1-|x|^2}} & \text{for} \quad |x| < 1, \\ 0 & \text{for} \quad |x| \geq 1 \end{cases}$$

and the previous problem.

Problem 1.19 Let $R > 0$ and $f \in \mathscr{C}^k(U_R)$, $k \geq 1$. Prove

1. $f(x) - f(0) = \sum_j x_j f_j(x)$, $f_j \in \mathscr{C}^{k-1}(U_R)$, $x \in U_R$,

2. $\partial^{\alpha} f_j(0) = \partial^{\alpha} \partial_j f(0) \dfrac{1}{1+|\alpha|}$ for every multi-index α such that $|\alpha| \leq k$,

3. $\sup\limits_{U_R} \left|\partial^{\alpha} f_j\right| \leq \sup\limits_{U_R} \left|\partial^{\alpha} \partial_j f\right|$, $|\alpha| \leq k$.

Solution We will prove the assertions for $k = 1$, as for $k > 1$ one can use induction.

1. Setting $f_j(x) = \displaystyle\int_0^1 \dfrac{\partial}{\partial x_j} f(tx) dt$, $x \in U_R$, we get

$$\sum_j x_j f_j(x) = \int_0^1 \sum_j x_j \dfrac{\partial}{\partial x_j} f(tx) dt = \int_0^1 df(tx) = f(x) - f(0), \quad x \in U_R.$$

We note that $f_j \in \mathscr{C}(U_R)$.
2. From the definition of the functions f_j, it follows that

$$f_j(0) = \dfrac{\partial}{\partial x_j} f_j(0), \qquad \dfrac{\partial}{\partial x_i} f_j(0) = \dfrac{1}{2} \dfrac{\partial}{\partial x_i} \dfrac{\partial}{\partial x_j} f_j(0).$$

3. The definition of the functions f_j implies

$$|f_j(x)| \leq \int_0^1 \left| \frac{\partial}{\partial x_j} f(tx) \right| dt \leq \int_0^1 \sup_{|x| < R} \left| \frac{\partial}{\partial x_j} f(tx) \right| dt \leq \sup_{x \in U_R} \left| \frac{\partial}{\partial x_j} f(x) \right|.$$

From this, $\displaystyle \sup_{x \in U_R} |f_j(x)| \leq \sup_{x \in U_R} \left| \frac{\partial}{\partial x_j} f(x) \right|.$

Problem 1.20 Let X be an open subset of $\mathbb{R}^n$ and $f, g \in \mathscr{C}(X)$ are maps satisfying

$$\int_X f \phi \, dx = \int_X g \phi \, dx$$

for every $\phi \in \mathscr{C}_0^\infty(X)$. Prove that $f \equiv g$ on X.

Solution We have

$$\int_X (f(x) - g(x)) \phi(x) dx = 0 \qquad (1.38)$$

for every $\phi \in \mathscr{C}_0^\infty(X)$. Set $h(x) = f(x) - g(x)$, $x \in X$, and suppose that there exists $a \in X$ for which $h(a) > 0$. Since h is continuous on X, there exists a neighbourhood $U(a) \subset X$ of the point a such that $h(x) > 0$ for every $x \in U(a)$. We may choose the function $\phi \in \mathscr{C}_0^\infty(X)$ such that $\phi > 0$ on $U(a)$. Then

$$\int_X h(x) \phi(x) dx \geq \int_{U(a)} h(x) \phi(x) dx > 0,$$

which contradicts with (1.38). Consequently $f \equiv g$ on X.

Problem 1.21 Let $X \subset \mathbb{R}^n$, $U_1 \subseteq X$, $f, g \in L^1_{\text{loc}}(X)$ with

$$\int_X f \phi \, dx = \int_X g \phi \, dx$$

for every $\phi \in \mathscr{C}_0^\infty(X)$. Prove that $f \equiv g$ almost everywhere on X.

Solution Let $h(x) = f(x) - g(x)$, $x \in X$. We have

$$\int_X h(x) \phi(x) dx = 0 \qquad (1.39)$$

for every $\phi \in \mathscr{C}_0^\infty(X)$. Now, we choose a function $\phi \in \mathscr{C}_0^\infty(X)$ so that $\mathrm{supp}\phi = U_1$ and $\int_X \phi(x)dx = 1$. Then

$$\frac{1}{t^n} \int_X \phi\left(\frac{x-y}{t}\right)dy = 1, \quad x \in X.$$

Therefore

$$h(x) = h(x)\frac{1}{t^n} \int_X \phi\left(\frac{x-y}{t}\right)dy$$

$$= \frac{1}{t^n} \int_X (h(x) - h(y))\,\phi\left(\frac{x-y}{t}\right)dy + \frac{1}{t^n} \int_X h(y)\phi\left(\frac{x-y}{t}\right)dy, \quad x \in X.$$

We take t small enough so that $\left|\frac{x-y}{t}\right| < 1$, $x, y \in X$. For this t, using (1.39), we have

$$\int_X h(y)\phi\left(\frac{x-y}{t}\right)dy = 0, \quad x \in X.$$

Consequently,

$$h(x) = \frac{1}{t^n} \int_{|x-y|<t} (h(x) - h(y))\,\phi\left(\frac{x-y}{t}\right)dy, \quad x \in X,$$

whence

$$h(x) = \lim_{t \to 0} \frac{1}{t^n} \int_{|x-y|<t} (h(x) - h(y))\,\phi\left(\frac{x-y}{t}\right)dy = 0, \quad x \in X,$$

i.e., $f \equiv g$ almost everywhere on X.

Problem 1.22 Let $f \in L^1([a, b])$ and define

$$\|f\| = \int_a^b x^2|f(x)|dx.$$

Prove that this is a norm in $L^1([a, b])$.

Problem 1.23 For $f \in L^\infty([a, b])$, prove that

$$\|f\|_\infty = \min\{M : m\{x \in [a, b] : |f(x)| > M\} = 0\}.$$

Problem 1.24 Let

$$f(x) = \frac{x^{-\frac{1}{2}}}{1 + \log x}, \quad x > 1.$$

Prove that $f \in L^p((1, \infty))$ if and only if $p = 2$.

Problem 1.25 Let $f(x) = \log\left(\frac{1}{x}\right)$, $x \in (0, 1]$, $1 \le p < \infty$. Prove that $f \in L^p((0, 1])$ and $f \notin L^\infty((0, 1])$.

Problem 1.26 Let E be a measurable set, $1 \le p < \infty$ and q is the conjugate of p, $f \in L^p(E)$. Prove that $f \equiv 0$ if and only if

$$\int_E fg = 0$$

for any $g \in L^q(E)$.

Problem 1.27 Let E be a measurable set of finite measure, $1 \le p_1 < p_2 \le \infty$. Prove that if $f_n \to f$ strongly in $L^{p_2}(E)$, then $f_n \to f$ strongly in $L^{p_1}(E)$.

Problem 1.28 Let E be a measurable set, $1 \le p < \infty$, q is the conjugate of p, S is dense in $L^q(E)$. Prove that if $g \in L^p(E)$ and $\int_E fg = 0$ for any $f \in S$, then $g = 0$.

Problem 1.29 Let E be a measurable set, $1 \le p < \infty$. Prove that the functions in $L^p(E)$ that vanish outside a bounded set are dense in $L^p(E)$.

Problem 1.30 Let $[a, b]$ be a closed bounded interval and $f_n \rightharpoonup f$ in $C([a, b])$. Prove that $\{f_n\}_{n \in \mathbb{N}}$ converges pointwise on $[a, b]$ to f.

Problem 1.31 Let $[a, b]$ be a closed bounded interval and $f_n \rightharpoonup f$ in $L^\infty([a, b])$. Prove that

$$\lim_{n \to \infty} \int_a^x f_n = \int_a^x f$$

for any $x \in [a, b]$.

For any $f \in \mathscr{S}(\mathbb{R}^n)$, $\alpha, \beta \in \mathbb{N}^n \cup \{0\}$ and $l, k, m \in \mathbb{N} \cup \{0\}$, we set

$$q_{\alpha,\beta}(f) = \sup_{x \in \mathbb{R}^n} \left| x^{\alpha} D^{\beta} f(x) \right|,$$

$$q_{l,\beta}(f) = \sup_{|\alpha| \leq l} \sup_{x \in \mathbb{R}^n} \left| x^{\alpha} D^{\beta} f(x) \right|,$$

$$q_{\alpha,\beta}^*(f) = \int_{\mathbb{R}^n} \left| x^{\alpha} D^{\beta} f(x) \right| dx,$$

$$|f|_{k,m} = \sup_{|\beta| \leq m} \sup_{x \in \mathbb{R}^n} \left| \left(1 + |x|^2 \right)^k D^{\beta} f(x) \right|.$$

Problem 1.32 Prove that the following assertions are equivalent

1. $f \in \mathscr{S}(\mathbb{R}^n)$,
2. $q_{\alpha,\beta}^*(f) < \infty$ for any $\alpha, \beta \in \mathbb{N}^n \cup \{0\}$,
3. $|f|_{k,m} < \infty$ for any $k, m \in \mathbb{N} \cup \{0\}$,
4. $q_{l,\beta}(f) < \infty$ for any $l \in \mathbb{N} \cup \{0\}$, $\forall \beta \in \mathbb{N}^n \cup \{0\}$,
5. $q_{\alpha,\beta}(f) < \infty$ for any $\alpha, \beta \in \mathbb{N}^n \cup \{0\}$.

Problem 1.33 Prove that $\mathscr{S}(\mathbb{R}^n)$ embeds continuously in every space $L^p(\mathbb{R}^n)$, $p \geq 1$.

Solution Let $f \in \mathscr{S}(\mathbb{R}^n)$. Then

$$\int_{\mathbb{R}^n} |f(x)|^p dx = \int_{\mathbb{R}^n} \frac{(1 + |x|^2)^{np} |f(x)|^p}{(1 + |x|^2)^{np}} dx \leq |f|_{n,0}^p \int_{\mathbb{R}^n} \frac{dx}{(1 + |x|^2)^{np}} \leq c_1 |f|_{n,0}^p,$$

where c_1 is a constant. Let $\{f_m\}_{m=1}^{\infty}$ be a sequence in $\mathscr{S}(\mathbb{R}^n)$ such that

$$|f_m - f|_{n,0} \to_{m \to \infty} 0.$$

We obtain

$$\|f_m - f\|_{L^p} \leq C |f_m - f|_{n,0} \to_{m \to \infty} 0, \quad C = const.$$

Consequently $f_m \to_{m \to \infty} f$ in $L^p(\mathbb{R}^n)$.

Problem 1.34 Let $u \in \mathscr{C}_0^j(\mathbb{R}^n)$. Prove

1. $u * v \in \mathscr{C}^j(\mathbb{R}^n)$ for $v \in L_{loc}^1(\mathbb{R}^n)$,
2. $u * v \in \mathscr{C}^{j+k}(\mathbb{R}^n)$ for $v \in \mathscr{C}^k(\mathbb{R}^n)$.

1. *Solution* Since $u \in \mathscr{C}_0^j(\mathbb{R}^n)$, there exists a compact set $K \subset \mathbb{R}^n$ such that $\mathrm{supp}\, u \subset K$. On the other hand,

$$(u * v)(x) = \int_{\mathbb{R}^n} u(y)v(x - y)dy = \int_{x-y\in K} u(x - y)v(y)dy, \quad x \in \mathbb{R}^n.$$

From here, using that $v \in L^1_{\mathrm{loc}}(\mathbb{R}^n)$, we get

$$|(u * v)(x)| \leq \int_{x-y\in K} |u(x - y)||v(y)|dy \leq c_1 \int_{x-y\in K} |v(y)|dy \leq C, \quad c_1, C = \mathrm{const},$$

$x \in \mathbb{R}^n$. Consequently the convolution $u * v$ exists. Since $D^l u \in \mathscr{C}_0(\mathbb{R}^n)$ and $D^l(u * v) = D^l u * v$, as above, we conclude that $D^l(u * v)$ exists for every $l = 0, 1, 2, \ldots, j$. If $\{v_n\}_{n=1}^{\infty}$ is a sequence of elements of $L^1_{\mathrm{loc}}(\mathbb{R}^n)$ converging to v in $L^1_{\mathrm{loc}}(\mathbb{R}^n)$, then

$$\left| D^l(u * v_n)(x) - D^l(u * v)(x) \right| = \left| D^l u * (v_n - v)(x) \right| \leq c_2 ||v_n - v||_{L^1_{\mathrm{loc}}(\mathbb{R}^n)} \to_{n\to\infty} 0,$$

$x \in \mathbb{R}^n, l = 0, 1, 2, \ldots, j$, where $c_2 > 0$ is a constant. So,

$$D^l(u * v_n) \to_{n\to\infty} D^l(u * v), \quad l = 0, 1, 2, \ldots, j.$$

Consequently $u * v \in C^j(\mathbb{R}^n)$.

2. **Hint** Use that $D^{l+m}(u * v) = D^l u * D^m v$ for $l = 0, 1, \ldots, j, m = 0, 1, \ldots, k$.

Problem 1.35 Let Γ be a cone in $\mathbb{R}^n$ and $\sigma > 0$ a constant such that

$$(\xi, x) \geq \sigma |\xi| |x| \quad \forall \xi \in C', \forall x \in \overline{\mathrm{ch}\Gamma}$$

for every $C' \subset\subset \mathrm{int}\Gamma^*$. Prove that Γ is an acute cone.

Problem 1.36 Let Γ be a cone in $\mathbb{R}^n$ for which there exists a plane of support for $\overline{\mathrm{ch}\Gamma}$ that has a unique common point 0 with $\overline{\mathrm{ch}\Gamma}$. Prove that Γ is an acute cone.

Problem 1.37 Let Γ be a convex cone. Prove $\Gamma = \Gamma + \Gamma$.

Problem 1.38 Let Γ be a cone in $\mathbb{R}^n$. Show

$$\mu_\Gamma(\xi) \leq \mu_{\mathrm{ch}\Gamma}(\xi).$$

Problem 1.39 Let Γ be a convex cone in $\mathbb{R}^n$. Prove that for every $a \geq 0$

$$\{\xi : \mu_\Gamma(\xi) \leq a\} = \Gamma^* + \overline{U_a}.$$

Problem 1.40 Let Γ be a closed, convex, acute cone and S a C-like surface. Prove that for every $R > 0$ there exists a constant $R'(R) > 0$ such that

$$T_R = \{(x, y) : x \in S, y \in \Gamma, |x + y| \le R\} \subset U_{R'} \subset \mathbb{R}^{2n}.$$

1.6 Notes and References

In this chapter we introduce the spaces $\mathscr{C}_0^\infty$, $\mathscr{S}$, L^p, $1 \le p \le \infty$, and they are deducted some of their properties. They are formulated and proved some variants of the Hölder and Minkowski inequalities. In the chapter is defined the convolution of locally integrable functions and they are proved some of its basic properties. They are introduced some basic facts for cones in $\mathbb{R}^n$. Additional materials can be found in [3–11, 15, 17–19, 21–28, 31–36] and references therein.

Chapter 2
Generalities on Distributions

2.1 Definitions

Let X be an open set in $\mathbb{R}^n$, $n \in \mathbb{N}$ a fixed integer.

Definition 2.1 Every linear continuous map $u : \mathscr{C}_0^\infty(X) \mapsto \mathbb{C}$ is called a distribution or generalized function. In other words, a distribution is a linear map $u : \mathscr{C}_0^\infty(X) \mapsto \mathbb{C}$ such that $u(\phi_n) \to_{n\to\infty} u(\phi)$ for every sequence $\{\phi_n\}_{n=1}^\infty$ in $\mathscr{C}_0^\infty(X)$ converging to $\phi \in \mathscr{C}_0^\infty(X)$, as $n \to \infty$. The space of distributions on X will be denoted by $\mathscr{D}'(X)$. We will write $u(\phi)$ or (u, ϕ) for the value of the functional (generalized function, distribution) $u \in \mathscr{D}'(X)$ on the element $\phi \in \mathscr{C}_0^\infty(X)$.

Example 2.1 Suppose $0 \in X$ and take the map $u : \mathscr{C}_0^\infty(X) \mapsto \mathbb{C}$ defined as follows

$$u(\phi) = \phi(0) \quad \text{for} \quad \phi \in \mathscr{C}_0^\infty(X).$$

Let $\phi_1, \phi_2 \in \mathscr{C}_0^\infty(X)$ and $\alpha_1, \alpha_2 \in \mathbb{C}$. As

$$u(\phi_1) = \phi_1(0), \quad u(\phi_2) = \phi_2(0),$$
$$u(\alpha_1\phi_1 + \alpha_2\phi_2) = (\alpha_1\phi_1 + \alpha_2\phi_2)(0) = \alpha_1\phi_1(0) + \alpha_2\phi_2(0) = \alpha_1 u(\phi_1) + \alpha_2 u(\phi_2),$$

i.e., $u : \mathscr{C}_0^\infty(X) \mapsto \mathbb{C}$ is linear. Let $\{\phi_n\}_{n=1}^\infty$ be a sequence in $\mathscr{C}_0^\infty(X)$ for which $\phi_n \to_{n\to\infty} \phi$ in $\mathscr{C}_0^\infty(X)$. Then there exists a compact set $K \subset X$ such that $\mathrm{supp}\phi_n \subset K$ for every $n \in \mathbb{N}$ and $D^\alpha \phi_n \to D^\alpha \phi$ uniformly in X for every multi-index $\alpha \in \mathbb{N} \cup \{0\}$. In particular, $\phi_n(0) \to_{n\to\infty} \phi(0)$, and therefore $u(\phi_n) \to_{n\to\infty} u(\phi)$. Consequently the linear map $u : \mathscr{C}_0^\infty(X) \mapsto \mathbb{C}$ is continuous. In other words, it is a distribution on X.

© The Author(s), under exclusive license to Springer Nature Switzerland AG 2021
S. G. Georgiev, *Theory of Distributions*,
https://doi.org/10.1007/978-3-030-81265-2_2

Exercise 2.1 Let $0 \in X$. For each multi-index α prove that the map $u : \mathscr{C}_0^\infty(X) \mapsto \mathbb{C}$, defined by

$$u(\phi) = D^\alpha \phi(0) \quad \text{for} \quad \phi \in \mathscr{C}_0^\infty(X),$$

is a distribution.

Exercise 2.2 Denote by δ_a or $\delta(x - a)$, $a \in \mathbb{C}^n$, Dirac's "delta" function at the point a:

$$\delta_a(\phi) = \phi(a) \quad \text{for} \quad \phi \in \mathscr{C}_0^\infty(X).$$

Prove that δ_a is a distribution.

Exercise 2.3 Prove that the map $1 : \mathscr{C}_0^\infty(X) \mapsto \mathbb{C}$, defined by

$$1(\phi) = \int_X \phi(x)dx \quad \text{for} \quad \phi \in \mathscr{C}_0^\infty(X),$$

is a distribution.

Exercise 2.4 For $u \in L_{\text{loc}}^p(X)$, $p \geq 1$, we define $u : \mathscr{C}_0^\infty(X) \mapsto \mathbb{C}$ by

$$u(\phi) = \int_X u(x)\phi(x)dx.$$

Prove that u is a distribution.

Exercise 2.5 Let $P\dfrac{1}{x} : \mathscr{C}_0^\infty(X) \mapsto \mathbb{C}$ be the map defined by

$$P\frac{1}{x}(\phi) = P.V. \int_X \frac{\phi(x) - \phi(0)}{x}dx \quad \text{for} \quad \phi \in \mathscr{C}_0^\infty(X).$$

Prove that $P\dfrac{1}{x} \in \mathscr{D}'(X)$.

Definition 2.2 The distributions $u, v \in \mathscr{D}'(X)$ are said to be equal if

$$u(\phi) = v(\phi)$$

for any $\phi \in \mathscr{C}_0^\infty(X)$.

Definition 2.3 The linear combination $\lambda u + \mu v$ of the distributions $u, v \in \mathscr{D}'(X)$ is the functional acting by the rule

$$(\lambda u + \mu v)(\phi) = \lambda u(\phi) + \mu v(\phi), \qquad \phi \in \mathscr{C}_0^\infty(X).$$

This makes the set $\mathscr{D}'(X)$ a vector space.

Definition 2.4 Let $u \in \mathscr{D}'(X)$. We define a distribution $\bar{u} \in \mathscr{D}'(X)$, called the complex conjugate of u, by

$$\bar{u}(\phi) = \overline{u(\bar{\phi})}, \qquad \phi \in \mathscr{C}_0^\infty(X).$$

The distributions

$$\text{Re}(u) = \frac{u + \bar{u}}{2}, \qquad \text{Im}(u) = \frac{u - \bar{u}}{2i}$$

are called the real and imaginary parts of u, respectively. Equivalently,

$$u = \text{Re}(u) + i\,\text{Im}(u), \qquad \bar{u} = \text{Re}(u) - i\,\text{Im}(u).$$

If $\text{Im}(u) = 0$, then the distribution u is said to be a real distribution.

Exercise 2.6 Prove that the delta function is a real distribution.

Here are elementary properties of the distributions. If $u_1, u_2 \in \mathscr{D}'(X)$, then

1. $u_1 \pm u_2 \in \mathscr{D}'(X)$,
2. $\alpha u_1 \in \mathscr{D}'(X)$ for $\forall \alpha \in \mathbb{C}$.

These properties follow from the definition, so their proof is omitted.

Theorem 2.1 $u \in \mathscr{D}'(X)$ *if and only if for every compact subset K of X there exist constants C and k so that the inequality*

$$|u(\phi)| \leq C \sum_{|\alpha| \leq k} \sup_{x \in K} \left| D^\alpha \phi(x) \right| \tag{2.1}$$

holds for every $\phi \in \mathscr{C}_0^\infty(K)$.

Proof Let $u \in \mathscr{D}'(X)$. We will prove that the inequality (2.1) holds. Actually, we suppose that there exists a compact set K in X so that

$$|u(\phi_n)| > n \sum_{\alpha \in \mathbb{N}^n \cup \{0\}} \sup_{x \in K} \left| D^\alpha \phi_n(x) \right| \tag{2.2}$$

holds for $\phi_n \in \mathscr{C}_0^\infty(K)$. We set

$$\psi_n(x) = \frac{\phi_n(x)}{n \sum\limits_{\alpha \in \mathbb{N}^n \cup \{0\}} \sup\limits_{x \in K} \left| D^\alpha \phi_n(x) \right|}, \quad x \in K.$$

From (2.2), we obtain

$$|u(\psi_n)| > 1. \tag{2.3}$$

By the definition of ψ_n, it follows that $\psi_n \to_{n\to\infty} 0$ in $\mathscr{C}_0^\infty(X)$. Since $u : \mathscr{C}_0^\infty(X) \mapsto \mathbb{C}$ is continuous, we have $u(\psi_n) \to_{n\to\infty} 0$, which contradicts (2.3). Let now, $u : \mathscr{C}_0^\infty(X) \mapsto \mathbb{C}$ is a linear map such that for every compact set K in X there exist constants $C > 0$ and $k \in \mathbb{N} \cup \{0\}$ for which (2.1) holds. We will prove that u is a distribution on X. To show this, we will prove that $u : \mathscr{C}_0^\infty(X) \mapsto \mathbb{C}$ is continuous at 0. Let $\{\phi_n\}_{n=1}^\infty$ be a sequence in $\mathscr{C}_0^\infty(X)$ with $\phi_n \to_{n\to\infty} 0$ in $\mathscr{C}_0^\infty(X)$. Then

$$\sup_{x \in K} \left| D^\alpha \phi_n(x) \right| \to_{n\to\infty} 0$$

for every $|\alpha| \leq k$. Hence with (2.1), we conclude

$$u(\phi_n) \to_{n\to\infty} 0.$$

This completes the proof.

Exercise 2.7 The function H defined by

$$H(x) = \begin{cases} 1 & \text{for} \quad x \geq 0, \\ 0 & \text{for} \quad x < 0, \end{cases} \quad x \in \mathbb{R},$$

is called the Heaviside function. We define

$$H(\phi) = \int\limits_{-\infty}^{\infty} H(x)\phi(x)dx,$$

$\phi \in \mathscr{C}_0^\infty(\mathbb{R})$. Using the inequality (2.1), prove that $H \in \mathscr{D}'(\mathbb{R})$.

Theorem 2.2 *A linear map* $u : \mathscr{C}_0^\infty(X) \mapsto \mathbb{C}$ *is a distribution if and only if there exist functions* $\rho_\alpha \in \mathscr{C}(X)$ *such that*

$$|u(\phi)| \leq \sum_\alpha \sup_K \left| \rho_\alpha D^\alpha \phi \right|, \quad \forall \phi \in \mathscr{C}_0^\infty(K), \tag{2.4}$$

for every compact set $K \subset X$, *and only a finite number of* ρ_α *vanish identically.*

Proof

1. Let u be a linear map from $\mathscr{C}_0^\infty(X)$ to $\mathbb{C}$ and $\rho_\alpha \in \mathscr{C}(X)$ be such that the inequality (2.4) holds for every $\phi \in \mathscr{C}_0^\infty(X)$ and every compact set K. Since $\rho_\alpha \in \mathscr{C}(X)$, there exists a constant C such that

$$\sup_K |\rho_\alpha| \leq C.$$

From this and (2.4), it follows that

$$|u(\phi)| \leq C \sum_\alpha \sup_K \left| D^\alpha \phi \right|.$$

As only finitely many ρ_α vanish identically, there is a constant k such that

$$|u(\phi)| \leq C \sum_{|\alpha| \leq k} \sup_K \left| D^\alpha \phi \right|,$$

i.e., $u \in \mathscr{D}'(X)$.

2. Let $u \in \mathscr{D}'(X)$ and $\{K_j\}$ be compact subsets of X such that any compact subset of X is contained in some K_j. Take maps $\chi_j \in \mathscr{C}_0^\infty(X)$ with $\chi_j \equiv 1$ on K_j and define

$$\psi_j = \chi_j - \chi_{j-1}, \quad j > 1,$$
$$\psi_1 = \chi_1.$$

Any $\phi \in \mathscr{C}_0^\infty(X)$ satisfies

$$\phi = \sum_{j=1}^\infty \psi_j \phi. \tag{2.5}$$

Moreover,

$$\psi_j \neq 0 \quad \text{on} \quad K_j \backslash K_{j-1} \quad \text{for} \quad j > 1,$$
$$\psi_1 \neq 0 \quad \text{on} \quad K_1.$$

Consequently

$$\operatorname{supp}(\psi_j \phi) \subset \operatorname{supp}\psi_j.$$

As $\psi_j \phi$ has compact support, for every compact set K there are constants C and k_j such that

$$|u(\psi_j \phi)| \leq C \sum_{|\alpha| \leq k_j} \sup_K \left| D^\alpha (\psi_j \phi) \right|.$$

From this and (2.5), we obtain

$$|u(\phi)| = \left| \sum_{j=1}^{\infty} u(\psi_j \phi) \right|$$

$$\leq \sum_{j=1}^{\infty} |u(\psi_j \phi)|$$

$$\leq C \sum_{j=1}^{\infty} \sum_{|\alpha| \leq k_j} \sup_K \left| D^{\alpha}(\psi_j \phi) \right|$$

$$\leq C \sum_{j=1}^{\infty} \sum_{|\alpha| \leq k_j} \sum_{\beta \leq \alpha} \binom{\alpha}{\beta} \sup_K \left| D^{\beta} \psi_j \right| \sup_K \left| D^{\alpha-\beta} \phi \right|.$$

If we set

$$\rho_{\beta} = \sum_{j=1}^{\infty} \sum_{|\alpha| \leq k_j} \binom{\alpha}{\beta} D^{\beta} \psi_j,$$

we obtain

$$|u(\phi)| \leq C \sum_{\beta \leq \alpha} \sup_K \left| \rho_{\beta} D^{\alpha-\beta} \phi \right|.$$

This completes the proof.

Theorem 2.3 *Let $\{X_i\}_{i \in I}$ be an open sets of $\mathbb{R}^n$, $X = \cup_{i \in I} X_i$ and suppose that $u_i \in \mathscr{D}'(X_i)$ satisfy $u_i = u_j$ on $X_i \cap X_j$. Then there exists a unique distribution $u \in \mathscr{D}'(X)$ such that $u_{|X_i} = u_i$ for every $i \in I$.*

Proof Take $\phi_i \in \mathscr{C}_0^{\infty}(X_i)$. Define $\phi = \sum_i \phi_i$ and

$$u(\phi) = \sum_i u_i(\phi_i). \tag{2.6}$$

We claim that the definition (2.6) does not depend on the choice of the sequence $\{\phi_i\}$. For this purpose, it is enough to prove that $\sum_i \phi_i = 0$ implies $u\left(\sum_i \phi_i \right) = 0$.

Set $K = \bigcup_i \mathrm{supp}\phi_i$. Then K is a compact set. There exist functions $\psi_k \in \mathscr{C}_0^{\infty}(X_k)$ such that $0 \leq \psi_k \leq 1$ and $\sum_k \psi_k = 1$ on K. Note that only a finite number of the above summands are different from zero. Moreover, $\psi_k \phi_i \in \mathscr{C}_0^\infty (X_k \bigcap X_i)$ and $u_k(\psi_k \phi_i) = u_i(\psi_k \phi_i)$. Therefore

$$\sum_i u_i(\phi_i) = \sum_i u_i \left(\sum_k \psi_k \phi_i \right) = \sum_i \sum_k u_i(\psi_k \phi_i) = \sum_i \sum_k u_k(\psi_k \phi_i)$$

$$= \sum_k \sum_i u_k(\psi_k \phi_i) = \sum_k u_k \left(\psi_k \sum_i \phi_i \right) = \sum_k u_k(0) = 0.$$

Consequently the definition (2.6) is consistent. Let $\phi \in \mathscr{C}_0^\infty(K)$. Then $\phi = \sum_k \phi \psi_k$ and

$$\left| u(\phi) \right| = \left| \sum_i u_i(\psi_i \phi) \right| \le \sum_i \left| u_i(\phi \psi_i) \right|$$

$$\le \sum_i C_i \sum_{|\alpha| \le k} \sup_K \left| D^\alpha(\phi \psi_i) \right| \le \sum_i \tilde{C}_i \sum_{|\alpha| \le k} \sup_K \left| D^\alpha \phi \right|$$

$$\le C \sum_{|\alpha| \le k} \sup_K \left| D^\alpha \phi \right|,$$

showing u is a distribution. Here C_i, $\tilde{C}_i$ and C are nonnegative constants. We also have $u = u_i$ on X_i. Now, we will prove the uniqueness of u. Suppose there are two distributions u and $\tilde{u}$ with the previous properties. We conclude $u_{|X_i} = u_i$, $\tilde{u}_{|X_i} = u_i$, so $(u - \tilde{u})_{|X_i} = 0$ for any i. Since X is open in $\mathbb{R}^n$, it follows that $u \equiv \tilde{u}$ on X, proving uniqueness. This completes the proof.

Example 2.2 Let $f(x) = e^{\frac{1}{x}}$, $x \in \mathbb{R}\backslash\{0\}$. Suppose that $f \in \mathscr{D}'(\mathbb{R}\backslash\{0\})$. Pick $\phi_0 \in \mathscr{C}_0^\infty(\mathbb{R}\backslash\{0\})$ such that $\phi_0(x) \ge 0$ for every $x \ne 0$, $\phi_0(x) = 0$ for $x < 1$ and $x > 2$, and

$$\int_{-\infty}^{\infty} \phi_0(x)dx = 1.$$

Define the sequence $\{\phi_k\}_{k=1}^\infty$ by $\phi_k(x) = e^{-\frac{k}{2}} k \phi_0(kx)$, $x \in \mathbb{R}\backslash\{0\}$, $k \in \mathbb{N}$. It satisfies $\phi_k \to_{k \to \infty} 0$ in $\mathscr{C}_0^\infty(\mathbb{R}\backslash\{0\})$, so $f(\phi_k) \to_{k \to \infty} 0$. On the other hand,

$$f(\phi_k) = \int_{-\infty}^{\infty} e^{\frac{1}{x}} \phi_k(x)dx$$

$$= \int_1^2 e^{k\left(\frac{1}{y} - \frac{1}{2}\right)} \phi_0(y)dy$$

$$\geq \int_1^{\frac{3}{2}} e^{k\left(\frac{1}{y}-\frac{1}{2}\right)} \phi_0(y)\,dy$$

$$\geq e^{\frac{k}{6}} \int_1^{\frac{3}{2}} \phi_0(y)\,dy.$$

By this and the definition of ϕ_0, we conclude $\lim_{k\to\infty} f(\phi_k) = \infty$, which is a contradiction. Therefore $f \notin \mathscr{D}'(\mathbb{R}\backslash\{0\})$.

2.2 Order of a Distribution

Definition 2.5 If the inequality (2.1) holds for some integer k independent of the compact set $K \subset X$, the distribution u is said to be of finite order. The smallest such k is called the order of the distribution u.

The space of distributions on X of finite order is denoted by $\mathscr{D}'_F(X)$, and the space of distributions of order $\leq k$ is denoted by $\mathscr{D}'^k(X)$. Then

$$\mathscr{D}'_F(X) = \bigcup_k \mathscr{D}'^k(X).$$

Example 2.3 Dirac's "delta" function is a distribution of order 0.

Example 2.4 Let $u \in \mathscr{D}'(X)$ is defined by

$$u(\phi) = \sum_{k=1}^{\infty} \phi^{(k)}(k), \quad \phi \in \mathscr{C}_0^{\infty}(X).$$

Then the order of u is infinite.

Exercise 2.8 Prove that $P\dfrac{1}{x}$ has order 1 on $\mathbb{R}$.

Exercise 2.9 Prove that $P\dfrac{1}{x}$ is of order 0 on $\mathbb{R}\backslash\{0\}$.

Theorem 2.4 *Any $u \in \mathscr{D}'^k(X)$ can be extended in a unique way to a linear map on $\mathscr{C}_0^k(X)$ so that the inequality (2.1) holds for every $\phi \in \mathscr{C}_0^k(X)$.*

Proof Since the space $\mathscr{C}_0^\infty(X)$ is everywhere dense in $\mathscr{C}_0^k(X)$, for every $\phi \in \mathscr{C}_0^k(X)$ there exists a sequence $\{\phi_n\}_{n=1}^\infty$ in $\mathscr{C}_0^\infty(X)$ for which $\phi_n \to_{n\to\infty} \phi$ in $\mathscr{C}_0^k(X)$. Hence,

$$|u(\phi_n) - u(\phi_l)| \le C \sum_{|\alpha|\le k} \sup_K \left|D^\alpha\phi_n - D^\alpha\phi_l\right| \to_{n,l\to\infty} 0.$$

Therefore $\{u(\phi_n)\}_{n=1}^\infty$ is a Cauchy sequence in $\mathbb{R}$, and as such it converges to, say,

$$u(\phi) = \lim_{n\to\infty} u(\phi_n). \tag{2.7}$$

The claim is that (2.7) is consistent. In fact, let $\{\phi_n\}_{n=1}^\infty$, $\{\psi_n\}_{n=1}^\infty$ be two sequences in $\mathscr{C}_0^\infty(X)$ for which

$$\lim_{n\to\infty} \phi_n = \lim_{n\to\infty} \psi_n = \phi$$

in $\mathscr{C}_0^k(X)$. Then $u(\phi) = \lim_{n\to\infty} u(\gamma_n) = \lim_{n\to\infty} u(\phi_n) = \lim_{n\to\infty} u(\psi_n)$, where $\{\gamma_n\}_{n=1}^\infty = \{\phi_n\}_{n=1}^\infty \cup \{\psi_n\}_{n=1}^\infty$. For the sequence $\{\gamma_n\}_{n=1}^\infty$ we have

$$\left|u(\gamma_n)\right| \le C \sum_{|\alpha|\le k} \sup_K \left|D^\alpha\gamma_n\right|,$$

so

$$\left|u(\phi)\right| \le C \sum_{|\alpha|\le k} \sup_K \left|D^\alpha\phi\right|$$

when $n \to \infty$. This completes the proof.

2.3 Change of Variables

Let X_1 be an open set in $\mathbb{R}^n$. Take $f \in L^1_{loc}(X_1)$ and suppose that $x = Ay + b$ is a nonsingular transformation from X to X_1, i.e., A is a $n \times n$ matrix with $\det A \ne 0$ and $b \in \mathbb{R}^n$. Then, for any $\phi \in \mathscr{C}_0^\infty(X)$, we have

$$\int_X f(Ay + b)\phi(y)dy = \frac{1}{|\det A|} \int_{X_1} f(x)\phi\left(A^{-1}(x - b)\right) dx.$$

This equality motivates us to give the following definition.

Definition 2.6 For any $u \in \mathscr{D}'(X)$, define

$$(u(Ay+b), \phi(y)) = \left(u(x), \frac{\phi\left(A^{-1}(x-b)\right)}{|\det A|} \right), \quad x \in X, \quad y \in X_1. \qquad (2.8)$$

Note that the map $\phi(x) \mapsto \phi\left(A^{-1}(x-b)\right)$, $x \in X$, is a linear and continuous map from $\mathscr{C}_0^\infty(X)$ to $\mathscr{C}_0^\infty(X_1)$. Therefore the functional $u(Ay+b)$, $y \in X_1$, defined by (2.8), is well defined and it belongs to $\mathscr{D}'(X_1)$. In particular, for $u \in \mathscr{D}'(X)$ and $a \in \mathbb{C}^n$, $|a| \neq 0$, $b \in \mathbb{C}$, $b \neq 0$, we have following distributions

1. $u(\phi)(x+a) = u(\phi(x-a))(x)$, $\quad \phi \in \mathscr{C}_0^\infty(X)$, $\quad x \in X$,
2. $u(\phi)(bx) = \dfrac{1}{|b|^n} u\left(\phi\left(\dfrac{x}{b}\right)\right)(x)$, $\quad \phi \in \mathscr{C}_0^\infty(X)$, $\quad x \in X$.

Example 2.5 For $\phi \in \mathscr{C}_0^\infty(\mathbb{R})$, we have

$$\delta(\phi)(-x) = \delta(\phi)(x),$$

$$\delta(\phi)(x+1-2i) = \delta(\phi(x-1+2i))(x) = \phi(-1+2i),$$

$$\delta(\phi)(2ix) = \frac{1}{2}\delta\left(\phi\left(\frac{x}{2i}\right)\right)(x) = \frac{1}{2}\phi(0), \quad x \in \mathbb{R}.$$

Exercise 2.10 Compute $\delta(\phi)(2x+3i)$, $x \in \mathbb{R}$, for $\phi \in \mathscr{C}_0^\infty(\mathbb{R})$.

Answer $\dfrac{1}{2}\phi\left(-\dfrac{3i}{2}\right)$.

Definition 2.7 For $a \in \mathscr{C}^1(\mathbb{R})$, define

$$\delta(a(x)) = \lim_{\epsilon \to 0} \omega_\epsilon(a(x)), \quad x \in \mathbb{R}.$$

Theorem 2.5 *If $a \in \mathscr{C}^1(\mathbb{R})$ has isolated simple zeros $x_1, x_2, \ldots$, then*

$$\delta(a(x)) = \sum_k \frac{\delta(x-x_k)}{|a'(x_k)|}, \quad x \in \mathbb{R}.$$

Proof It is enough to prove the assertion on a neighbourhood of the simple zero x_k. Since x_k is an isolated simple zero of a, there exists $\epsilon_k > 0$ such that $a(x) \neq 0$ for every $x \in (x_k - \epsilon_k, x_k + \epsilon_k)$, $x \neq x_k$, $a(x_k) = 0$. As

$$\Big(\delta(a(x)), \phi(x) \Big) = \int\limits_{x_k - \epsilon_k}^{x_k + \epsilon_k} \delta(a(x)) \phi(x) dx$$

$$= \lim_{\epsilon \to 0} \int\limits_{x_k - \epsilon_k}^{x_k + \epsilon_k} \omega_\epsilon(a(x)) \phi(x) dx$$

$$= \lim_{\epsilon \to 0} \int\limits_{a(x_k - \epsilon_k)}^{a(x_k + \epsilon_k)} \omega_\epsilon(y) \frac{\phi(a^{-1}(y))}{|a'(a^{-1}(y))|} dy$$

$$= \lim_{\epsilon \to 0} \int\limits_{a(x_k - \epsilon_k)}^{a(x_k + \epsilon_k)} \omega_\epsilon(y) \frac{\phi(a^{-1}(a(x)))}{|a'(a^{-1}(a(x)))|} dy$$

$$= \frac{\phi(x_k)}{|a'(x_k)|}$$

$$= \left(\frac{\delta(x - x_k)}{|a'(x_k)|}, \phi(x) \right), \quad x \in (x_k - \epsilon_k, x_k + \epsilon_k),$$

for $\phi \in \mathscr{C}_0^\infty(x_k - \epsilon_k, x_k + \epsilon_k)$. Next, let $(c_1, d_1) \subset \mathbb{R}$ does not contain a single zero of x_k and $\phi \in \mathscr{C}_0^\infty((c_1, d_1))$. Then

$$(\delta(a(x)), \phi(x)) = \lim_{\epsilon \to 0} \int_{c_1}^{d_1} \omega_\epsilon(a(x)) \phi(x) dx$$

$$= 0, \quad x \in (c_1, d_1).$$

This completes the proof.

Example 2.6 Let us consider $\delta(\cos x)$, $x \in \mathbb{R}$. Here $a(x) = \cos x$, $x \in \mathbb{R}$, and its isolated zeros are $x_k = \dfrac{(2k + 1)\pi}{2}$, $k \in \mathbb{Z}$. We notice that $|a'(x_k)| = 1$ for $k \in \mathbb{Z}$. So,

$$\delta(\cos x) = \sum_k \delta\left(x - \frac{(2k + 1)\pi}{2} \right), \quad x \in \mathbb{R}.$$

Exercise 2.11 Compute $\delta(x^4 - 1)$, $x \in \mathbb{R}$.

Answer $\dfrac{\delta(x - 1) + \delta(x + 1)}{4}$, $x \in \mathbb{R}$.

2.4 Sequences and Series

Definition 2.8 We say that the sequence $\left\{u_n\right\}_{n=1}^{\infty}$ of elements of $\mathscr{D}'(X)$ tends to the distribution $u \in \mathscr{D}'(X)$ if

$$\lim_{n\to\infty} u_n(\phi) = u(\phi) \quad \text{for} \quad \text{any} \quad \phi \in \mathscr{C}_0^{\infty}(X).$$

If so we write

$$\lim_{n\to\infty} u_n = u \quad \text{or} \quad u_n \to_{n\to\infty} u.$$

Theorem 2.6 *If $\{u_n\}_{n=1}^{\infty}$ and $\{v_n\}_{n=1}^{\infty}$ are two sequences of distributions on X that converge to the distributions u and v, respectively, then $\{\alpha u_n + \beta v_n\}_{n=1}^{\infty}$ converges to $\alpha u + \beta v$ on X. Here $\alpha, \beta \in \mathbb{C}$.*

Proof Indeed, let $\phi \in \mathscr{C}_0^{\infty}(X)$ be arbitrary. Then

$$u_n(\phi) \to_{n\to\infty} u(\phi), \qquad v_n(\phi) \to_{n\to\infty} v(\phi).$$

Hence,

$$\begin{aligned}
(\alpha u_n + \beta v_n)(\phi) &= (\alpha u_n)(\phi) + (\beta v_n)(\phi) \\
&= \alpha u_n(\phi) + \beta v_n(\phi) \\
&\to \alpha u(\phi) + \beta v(\phi), \quad \text{as} \quad n \to \infty.
\end{aligned}$$

This completes the proof.

Example 2.7 Let $x \in \mathbb{R}$ and

$$f_\epsilon(x) = \begin{cases} \dfrac{1}{2\epsilon} & \text{for} \quad |x| \le \epsilon, \\ 0 & \text{for} \quad |x| > \epsilon. \end{cases}$$

We will compute $\lim_{\epsilon\to 0+} f_\epsilon(x)$, $x \in \mathbb{R}$, in $\mathscr{D}'(\mathbb{R})$. Let $\phi \in \mathscr{C}_0^{\infty}(\mathbb{R})$ be arbitrary. Then

$$\begin{aligned}
\lim_{\epsilon\to 0+} f_\epsilon(\phi)(x) &= \lim_{\epsilon\to 0+} \int_{|x|\le\epsilon} \frac{1}{2\epsilon}\phi(x)dx \\
&= \frac{1}{2} \lim_{\epsilon\to 0+} \int_{|y|\le 1} \phi(\epsilon y)dy
\end{aligned}$$

$$= \phi(0)$$

$$= \delta(\phi)(x), \quad x \in \mathbb{R}.$$

Consequently $\lim_{\epsilon \to 0+} f_\epsilon(x) = \delta(x)$, $x \in \mathbb{R}$, in $\mathscr{D}'(\mathbb{R})$.

Example 2.8 We will find $\lim_{\epsilon \to 0+} \dfrac{\epsilon}{\pi(x^2 + \epsilon^2)}$, $x \in \mathbb{R}$. Let $\phi \in \mathscr{C}_0^\infty(\mathbb{R})$ be arbitrarily chosen. Then

$$\lim_{\epsilon \to 0+} \int_{-\infty}^{\infty} \frac{\epsilon}{x^2 + \epsilon^2} \phi(x) dx = \lim_{\epsilon \to 0+} \int_{-\infty}^{\infty} \frac{1}{1 + y^2} \phi(\epsilon y) dy$$

$$= \phi(0) \int_{-\infty}^{\infty} \frac{1}{1 + y^2} dy$$

$$= \phi(0) \arctan y \Big|_{y=-\infty}^{y=\infty}$$

$$= \phi(0) \left(\frac{\pi}{2} + \frac{\pi}{2} \right)$$

$$= \pi \phi(0)$$

$$= \pi \delta(\phi).$$

Exercise 2.12 Prove that

$$\lim_{\epsilon \to 0+} \frac{\epsilon}{\pi x^2} \sin^2 \frac{x}{\epsilon} = \delta(x), \quad x \in \mathbb{R}.$$

Theorem 2.7 *Let $\{f_n\}_{n=1}^\infty$ be a sequence in $\mathscr{D}'(X)$ such that $|f_n(\phi)| \le c_\phi$ for every $\phi \in \mathscr{C}_0^\infty(X)$, and $\{\phi_n\}_{n=1}^\infty \subset \mathscr{C}_0^\infty(X)$ a sequence converging to 0 in $\mathscr{C}_0^\infty(X)$ as $n \to \infty$. Then $f_n(\phi_n) \to 0$, $n \to \infty$.*

Proof We suppose the contrary. Then there exists a constant $c > 0$ such that

$$|f_n(\phi_n)| \ge c > 0,$$

for n large enough. Since $\phi_n \to 0$ in $\mathscr{C}_0^\infty(X)$ as $n \to \infty$, there exists a compact set X' such that $\mathrm{supp}\phi_n \subset X'$ for every $n \in \mathbb{N}$ and $D^\alpha \phi_n \to_{n\to\infty} 0$, for every $\alpha \in \mathbb{N}^n \cup \{0\}$. Hence,

$$|D^\alpha \phi_n(x)| \le \frac{1}{4^n}, \quad |\alpha| \le n = 0, 1, 2, \ldots,$$

for n large enough and every $x \in X'$. We set $\psi_n = 2^n \phi_n$, $n \in \mathbb{N}$. We have $\operatorname{supp} \psi_n \subset X'$ and

$$|D^\alpha \psi_n(x)| \le \frac{1}{2^n}, \quad |\alpha| \le n = 0, 1, 2, \ldots, \tag{2.9}$$

$$|f_n(\psi_n)| = 2^n |f_n(\phi_n)| \ge 2^n c \to_{n\to\infty} \infty. \tag{2.10}$$

Let us take subsequences $\{f_{k_\nu}\}_{\nu=1}^\infty$ of $\{f_n\}_{n=1}^\infty$ and $\{\psi_{k_\nu}\}_{\nu=1}^\infty$ of $\{\psi_n\}_{n=1}^\infty$ so that $|f_{k_\nu}(\psi_{k_\nu})| \ge 2^\nu$ for $\nu = 1, 2, \ldots$. As $\psi_k \to_{k\to\infty} 0$ in $\mathscr{C}_0^\infty(X)$, we have $f_{k_j}(\psi_k) \to_{k\to\infty} 0$ for $j = 1, 2, \ldots, \nu - 1$. Therefore there exists $N \in \mathbb{N}$ such that for every $k \ge N$

$$|f_{k_j}(\psi_k)| \le \frac{1}{2^{\nu-j}}, \quad j = 1, 2, \ldots, \nu - 1. \tag{2.11}$$

We note that $|f_k(\psi_{k_j})| \le c_{k_j}$, $j = 1, 2, \ldots, \nu - 1$. From (2.10), we can choose $k_\nu \ge N$ so that

$$|f_{k_\nu}(\psi_{k_\nu})| \ge \sum_{1 \le j \le \nu-1} c_{k_j} + \nu + 1. \tag{2.12}$$

From (2.11) and (2.12), we have

$$|f_{k_j}(\psi_{k_\nu})| \le \frac{1}{2^{\nu-j}}, \quad j = 1, 2, \ldots, \nu - 1, \tag{2.13}$$

$$|f_{k_\nu}(\psi_{k_\nu})| \ge \sum_{1 \le j \le \nu-1} |f_{k_\nu}(\psi_{k_j})| + \nu + 1. \tag{2.14}$$

We set $\psi = \sum_{j \ge 1} \psi_{k_j}$. From (2.9), it follows that ψ is a convergent series, $\psi \in \mathscr{C}_0^\infty(X)$ and

$$f_{k_\nu}(\psi) = f_{k_\nu}(\psi_{k_\nu}) + \sum_{j \ge 1, j \ne \nu} f_{k_\nu}(\psi_{k_j}).$$

Therefore

$$|f_{k_\nu}(\psi)| \ge |f_{k_\nu}(\psi_{k_\nu})| - \sum_{1 \le j \le \nu-1} |f_{k_\nu}(\psi_{k_j})| - \sum_{j \ge \nu+1} |f_{k_\nu}(\psi_{k_j})|$$

$$\ge \nu + 1 - \sum_{j \ge \nu+1} \frac{1}{2^{j-\nu}} = \nu,$$

and then $f_{k_\nu}(\psi) \to_{\nu \to \infty} \infty$, which contradicts with $|f_{k_\nu}(\psi)| \leq c_\psi$. This completes the proof.

Theorem 2.8 *Let* $\{f_n\}_{n=1}^\infty$ *be a sequence in* $\mathscr{D}'(X)$ *such that* $\{f_n(\phi)\}_{n=1}^\infty$ *converges for every* $\phi \in \mathscr{C}_0^\infty(X)$. *Then the functional*

$$f(\phi) = \lim_{n \to \infty} f_n(\phi), \quad \phi \in \mathscr{C}_0^\infty(X),$$

is an element of $\mathscr{D}'(X)$.

Proof Let $\alpha_1, \alpha_2 \in \mathbb{C}$ and $\phi_1, \phi_2 \in \mathscr{C}_0^\infty(X)$. Then

$$f(\alpha_1\phi_1 + \alpha_2\phi_2) = \lim_{n \to \infty} f_n(\alpha_1\phi_1 + \alpha_2\phi_2) = \lim_{n \to \infty} (\alpha_1 f_n(\phi_1) + \alpha_2 f_n(\phi_2))$$

$$= \alpha_1 \lim_{n \to \infty} f_n(\phi_1) + \alpha_2 \lim_{n \to \infty} f_n(\phi_2) = \alpha_1 f(\phi_1) + \alpha_2 f(\phi_2).$$

Therefore f is a linear map on $\mathscr{C}_0^\infty(X)$. Now, we will prove that f is a continuous functional on $\mathscr{C}_0^\infty(X)$. Let $\{\phi_n\}_{n=1}^\infty$ be a sequence in $\mathscr{C}_0^\infty(X)$ such that $\phi_n \to_{n \to \infty} 0$ in $\mathscr{C}_0^\infty(X)$. We claim $f(\phi_n) \to_{n \to \infty} 0$. Suppose the contrary. There exists a constant $a > 0$ such that $|f(\phi_\nu)| \geq a$ for every $\nu \in \mathbb{N}$. Since $f(\phi_\nu) = \lim_{k \to \infty} f_k(\phi_\nu)$, there is $k_\nu \in \mathbb{N}$ such that $|f_{k_\nu}(\phi_\nu)| \geq a$ for every $\nu \in \mathbb{N}$, which is in contradiction with Theorem 2.7. Consequently $f(\phi_n) \to_{n \to \infty} 0$ and $f \in \mathscr{D}'(X)$. This completes the proof.

Definition 2.9 Let $u_j \in \mathscr{D}'(X)$, $j \in \mathbb{N}$. An expression like

$$u_1 + u_2 + \cdots$$

or

$$\sum_{j=1}^\infty u_j \qquad (2.15)$$

is called infinite series of distributions. mth partial sum of the series (2.15) is defined by

$$S_m = \sum_{j=1}^m u_j, \quad m \in \mathbb{N}.$$

If the sequence $\{S_m\}_{m=1}^\infty$ is convergent in $\mathscr{D}'(X)$, then we say that the series (2.15) is convergent in $\mathscr{D}'(X)$. Otherwise, we say that the series (2.15) is divergent.

2.5   Support

Definition 2.10 A distribution $u \in \mathscr{D}'(X)$ is said to vanish on an open set $X_1 \subset X$ if its restriction to X_1 is the zero functional in $\mathscr{D}'(X_1)$, i.e., $u(\phi) = 0$ for all $\phi \in \mathscr{C}_0^\infty(X_1)$. This is written $u(x) = 0, x \in X_1$.

Theorem 2.9 *A distribution* $u \in \mathscr{D}'(X)$ *vanishes on* X *if and only if it vanishes on a neighbourhood of every point of* X.

Proof Suppose a distribution $u \in \mathscr{D}'(X)$ vanishes on X. Then it vanishes on a neighbourhood of every point in X. Conversely, let $u \in \mathscr{D}'(X)$ vanishes on a neighbourhood $U(x) \subset X$ of every point x in X. Consider the cover $\{U(x), x \in X\}$ of X. We will construct a locally finite cover $\{X_k\}$ such that X_k is contained in some $U(x)$. Let

$$X_1^1 \subset\subset X_2^1 \subset\subset \dots, \qquad \bigcup_{k \geq 1} X_k^1 = X.$$

By the Heine–Borel lemma, the compact set $\overline{X}_1^1$ is covered by a finite number of neighbourhoods $U(x)$, say $U(x_1), U(x_2), \dots, U(x_{N_1})$. Similarly, the compact set $\overline{X}_2^1 \backslash X_1^1$ is covered by a finite number of neighbourhoods $U(x_{N_1+1}), \dots, U(x_{N_1+N_2})$, and so on. We set

$$X_k = U(x_k) \bigcap X_1^1, \qquad k = 1, 2, \dots, N_1,$$
$$X_k = U(x_k) \bigcap (\overline{X}_2^1 \backslash X_1^1), \qquad k = N_1 + 1, \dots, N_1 + N_2,$$

and so forth. In this way, we obtain the required cover $\{X_k\}$. Let $\{e_k\}$ be the partition of unity corresponding to the cover $\{X_k\}$ of X. Then $\text{supp}(\phi e_k) = 0$ for every $\phi \in \mathscr{C}_0^\infty(X)$. This implies

$$u(\phi) = u\left(\sum_{k \geq 1} \phi e_k\right) = \sum_{k \geq 1} u(\phi e_k) = 0.$$

Consequently the distribution u vanishes on the whole X. This completes the proof.

Definition 2.11 The union of all neighbourhoods where a distribution $u \in \mathscr{D}'(X)$ vanishes forms an open set X_u, called the zero set of the distribution u. Therefore $u = 0$ on X_u and X_u is the largest open set where u vanishes.

Definition 2.12 The support of a distribution $u \in \mathscr{D}'(X)$ is the complement $\text{supp} u = X \backslash X_u$ of X_u in X.

Note that $\text{supp} u$ is a closed subset in X.

Definition 2.13 The distribution $u \in \mathscr{D}'(X)$ is said to have compact support if $\mathrm{supp}\, u \subset\subset X$.

Example 2.9 $\mathrm{supp}\, H = [0, \infty)$.

Exercise 2.13 Find $\mathrm{supp}\, 1$.

Let A be a closed set in X. With $\mathscr{D}'(X, A)$ we denote the subset of distributions on X whose supports are contained in A, endowed with the following notion of convergence: $u_k \to 0$ in $\mathscr{D}'(X, A)$, as $k \to \infty$, if $u_k \to 0$ in $\mathscr{D}'(X)$, as $k \to \infty$, and $\mathrm{supp}\, u_k \subset A$ for every $k = 1, 2, \ldots$. For simplicity, $\mathscr{D}'(A)$ will denote $\mathscr{D}'(\mathbb{R}^n, A)$.

Theorem 2.10 *Suppose that for every point $y \in X$ there is a neighbourhood $U(y) \subset\subset X$ on which a given distribution u_y is defined. Assume further that $u_{y_1}(x) = u_{y_2}(x)$ if $x \in U(y_1) \cap U(y_2) \neq \emptyset$. Then there exists a unique distribution $u \in \mathscr{D}'(X)$ so that $u = u_y$ in $U(y)$ for every $y \in X$.*

Proof To see this we construct, starting as previously with the cover $\{U(y), y \in X\}$, the locally finite cover $\{X_k\}$, $X_k \subset U(y_k)$, and the corresponding partition of unity $\{e_k\}$. We also set

$$u(\phi) = \sum_{k \geq 1} u_{y_k}(\phi e_k), \quad \phi \in \mathscr{C}_0^\infty(X). \tag{2.16}$$

The number of summands in the right-hand side of (2.16) is finite and does not depend on $\phi \in \mathscr{C}_0^\infty(X')$ for any $X' \subset\subset X$. By the definition (2.16), it follows that u is linear and continuous on $\mathscr{C}_0^\infty(X)$, i.e., $u \in \mathscr{D}'(X)$. Furthermore, if $\phi \in \mathscr{C}_0^\infty(U(y))$, then $\phi e_k \in \mathscr{C}_0^\infty(U(y_k))$. From (2.16), we get

$$u(\phi) = u_y\left(\phi \sum_{k \geq 1} e_k\right) = u_y(\phi),$$

i.e., $u = u_y$ on $U(y)$. If we suppose that there are two distributions u and $\tilde{u}$ such that $u = u_y$ and $\tilde{u} = u_y$ on $U(y)$ for every $y \in X$, then $u - \tilde{u} = 0$ on $U(y)$ for every $y \in X$. Therefore $u - \tilde{u} = 0$ in X, showing that the distribution u is unique. This completes the proof.

The set of distributions with compact support in X will be denoted by $\mathscr{E}'(X)$, and we set $\mathscr{E}'^k(X) = \mathscr{E}'(X) \cap \mathscr{D}'^k(X)$.

Theorem 2.11 *Let $\phi \in \mathscr{C}_0^\infty(X)$ and $\mathrm{supp}\, u \cap \mathrm{supp}\, \phi = \emptyset$. Then $u(\phi) = 0$.*

Proof Since $\mathrm{supp}\, u \cap \mathrm{supp}\, \phi = \emptyset$, we have $\phi \in \mathscr{C}_0^\infty(X \setminus \mathrm{supp}\, u)$. If $x \in \mathrm{supp}\, u$, then $\phi(x) = 0$, so $u(\phi) = 0$. If $x \in X \setminus \mathrm{supp}\, u$, then $u(\phi)(x) = 0$. This completes the proof.

Theorem 2.12 *Let $u \in \mathscr{D}'(X)$ and let F be a relatively closed subset of X with* $\operatorname{supp} u \subset F$. *Then there exists a unique linear map $\tilde{u}$ on*

$$\left\{ \phi : \phi \in \mathscr{C}^\infty(X), F \cap \operatorname{supp}\phi \subset\subset X \right\}$$

such that

1. $\tilde{u}(\phi) = u(\phi)$ for $\phi \in \mathscr{C}_0^\infty(X)$,
2. $\tilde{u}(\phi) = 0$ for $\phi \in \mathscr{C}^\infty(X)$, $F \cap \operatorname{supp}\phi = \emptyset$.

Proof

1. (uniqueness)   Let $\phi \in \mathscr{C}^\infty(X)$ and $F \cap \operatorname{supp}\phi = K$. As K is compact, there exists $\psi \in \mathscr{C}_0^\infty(X)$ such that $\psi \equiv 1$ on a neighbourhood of K. Let

$$\phi_0 = \psi\phi, \qquad \phi_1 = (1 - \psi)\phi.$$

Then

$$\phi = \phi_0 + \phi_1. \tag{2.17}$$

Therefore

$$\tilde{u}(\phi) = \tilde{u}(\phi_0) + \tilde{u}(\phi_1).$$

Note that $\tilde{u}(\phi_1) = 0$. So,

$$\tilde{u}(\phi) = \tilde{u}(\phi_0) = u(\phi_0).$$

Now, suppose that there are two such distributions $\tilde{u}$, $\tilde{\tilde{u}}$. Then

$$\tilde{u}(\phi) = \tilde{u}(\phi_0), \qquad \tilde{\tilde{u}}(\phi) = \tilde{\tilde{u}}(\phi_0).$$

Consequently $\tilde{u}(\phi) = \tilde{\tilde{u}}(\phi)$ for every $\phi \in \mathscr{C}^\infty(X)$ so that $F \cap \operatorname{supp}\phi = K$. Therefore $\tilde{u} = \tilde{\tilde{u}}$.

2. (existence)   Let

$$\phi = \phi_0' + \phi_1'$$

be another decomposition of kind (2.17). Define

$$\chi = \phi_0 - \phi_0'.$$

Then

$$\chi \in \mathscr{C}_0^\infty(X), \qquad F \cap \operatorname{supp}\chi = F \cap \operatorname{supp}(\phi_1 - \phi_1') = \emptyset$$

and so,

$$u(\chi) = u(\phi_0) - u(\phi_0') = 0.$$

Define $\tilde{u}(\phi)$ by

$$\tilde{u}(\phi) = u(\phi_0).$$

This completes the proof.

Theorem 2.13 *The set of distributions on X with compact support coincides with the dual space of $\mathscr{C}^\infty(X)$ with the topology*

$$\phi \mapsto \sum_{|\alpha| \le k} \sup_K \left| D^\alpha \phi \right|,$$

where K is a compact set in X.

Proof Let u be a distribution with compact support and take $\phi \in \mathscr{C}^\infty(X)$ and $\psi \in \mathscr{C}_0^\infty(X)$, $\psi \equiv 1$ on a neighbourhood of $\mathrm{supp}\,u$. Then

$$\phi = \psi\phi + (1 - \psi)\phi$$

and

$$u(\phi) = u(\psi\phi + (1 - \psi)\phi) = u(\psi\phi) + u((1 - \psi)\phi) = u(\psi\phi).$$

Define u on $\mathscr{C}^\infty(X)$ via $u(\phi) = u(\psi\phi)$ for $\phi \in \mathscr{C}^\infty(X)$. Since u is a distribution and $\psi\phi \in \mathscr{C}_0^\infty(X)$, we have

$$|u(\phi)| = |u(\psi\phi)| \le C \sum_{|\alpha| \le k} \sup_K \left| D^\alpha(\phi\psi) \right| \le C_1 \sum_{|\alpha| \le k} \left| D^\alpha \phi \right|.$$

Now, we suppose that v is a linear operator on $\mathscr{C}^\infty(X)$ for which

$$|v(\phi)| \le C \sum_{|\alpha| \le k} \sup_K \left| D^\alpha \phi \right|$$

for $\phi \in \mathscr{C}^\infty(X)$ and K a compact set. Then $v(\phi) = 0$ when $\mathrm{supp}\,\phi \cap K = \emptyset$. If $\phi \in \mathscr{C}_0^\infty(X) \subset \mathscr{C}^\infty(X)$, v is a distribution. Therefore there exists a unique distribution $u \in \mathscr{D}'(X)$ such that $u(\phi) = v(\phi)$ for every $\phi \in \mathscr{C}^\infty(X)$. This completes the proof.

Theorem 2.14 *Let u be a distribution with a compact support of order $\le k$, ϕ a $\mathscr{C}^k$ map with $D^\alpha \phi(x) = 0$ for $|\alpha| \le k$, $x \in \mathrm{supp}\,\phi$. Then $u(\phi) = 0$.*

Proof Let $\chi_\epsilon \in \mathscr{C}_0^\infty(X)$, $\chi_\epsilon \equiv 1$ on a neighbourhood U of $\mathrm{supp}\,u$, while $\chi_\epsilon \equiv 0$ on $X\setminus U$. Define the set M_ϵ, $\epsilon > 0$, by

$$M_\epsilon = \left\{ y : |x - y| \le \epsilon, \quad x \in \mathrm{supp}\,u \right\},$$

making M_ϵ an ϵ-neighbourhood of $\mathrm{supp}\,u$. Moreover,

$$\left| D^\alpha \chi_\epsilon \right| \le C\epsilon^{|\alpha|}, \quad |\alpha| \le k,$$

for some positive constant C. Since

$$\mathrm{supp}\,u \cap \mathrm{supp}(1 - \chi_\epsilon)\phi = \varnothing,$$

we have

$$u(\phi) = u(\phi\chi_\epsilon) + u((1 - \chi_\epsilon)\phi) = u(\phi\chi_\epsilon),$$

$$|u(\phi)| \le C \left| \sum_{|\alpha| \le k} \sup \left(D^\alpha(\phi\chi_\epsilon) \right) \right|$$

$$\le C_1 \sum_{|\alpha| + |\beta| \le k} \sup \left| D^\alpha \phi \right| \left| D^\beta \chi_\epsilon \right|$$

$$\le C_2 \sum_{|\alpha| + |\beta| \le k} \sup \left| D^\alpha \phi \right| \epsilon^{k - |\alpha|} \to_{\epsilon \to 0} 0, \quad |\alpha| \le k.$$

Consequently $u(\phi) = 0$. This completes the proof.

Theorem 2.15 *Let u be a distribution of order k with support $\{y\}$. Then $u(\phi) = \sum_{|\alpha| \le k} a_\alpha D^\alpha \phi(y)$, $\phi \in \mathscr{C}^k$.*

Proof Fix $x \in \mathbb{R}^n$. For $\phi \in \mathscr{C}^k(\mathbb{R}^n)$, we have

$$\phi(x) = \sum_{|\alpha| \le k} D^\alpha \phi(y) \frac{(x - y)^\alpha}{\alpha!} + \psi(x),$$

where $D^\alpha \psi(y) = 0$ for $|\alpha| \le k$. Hence, $u(\psi) = 0$. Therefore

$$u(\phi(x)) = u\left(\sum_{|\alpha| \le k} D^\alpha \phi(y) \frac{(x - y)^\alpha}{\alpha!} + \psi(x) \right)$$

$$= u\left(\sum_{|\alpha| \le k} D^\alpha \phi(y) \frac{(x - y)^\alpha}{\alpha!} \right) + u(\psi(x))$$

$$= \sum_{|\alpha| \le k} u\left(\frac{(x - y)^\alpha}{\alpha!} \right) D^\alpha \phi(y).$$

Let

$$a_\alpha = u\left(\frac{(x-y)^\alpha}{\alpha!}\right).$$

Then

$$u(\phi) = \sum_{|\alpha| \le k} a_\alpha D^\alpha \phi(y).$$

This completes the proof.

Theorem 2.16 *Write $x = (x', x'') \in \mathbb{R}^n$. Then for every distribution $u \in \mathscr{D}'(\mathbb{R}^n)$ of order k with compact support contained in the plane $x' = 0$, we have*

$$u(\phi) = \sum_{|\alpha| \le k} u_\alpha(\phi_\alpha), \tag{2.18}$$

where $\alpha = (\alpha', 0)$, u_α is a distribution in the variables x'' of order $k - |\alpha|$ with compact support and $\phi_\alpha(x'') = D^\alpha \phi(x', x'')|_{x'=0}$.

Proof For $\phi \in \mathscr{C}^\infty(\mathbb{R}^n)$, we have

$$\phi(x) = \sum_{|\alpha'| \le k, \alpha''=0} D^\alpha \phi(0, x'') \frac{x'^\alpha}{\alpha!} + \Phi(x),$$

where $D^\alpha \Phi(x)|_{x'=0} = 0$ for $|\alpha| \le k$. This implies $u(\Phi) = 0$. Since u is a distribution,

$$u(\phi) = \sum_{|\alpha'| \le k, \alpha''=0} u\left(D^\alpha \phi(0, x'') \frac{x'^\alpha}{\alpha!}\right).$$

Now, let

$$u_\alpha(\phi) = u\left(D^\alpha \phi(0, x'') \frac{x'^\alpha}{\alpha!}\right).$$

We want to show that u_α is a distribution of order $k - |\alpha|$. Set

$$\psi(x) = D^\alpha \phi(0, x'') \frac{x'^\alpha}{\alpha!} + O(|x'|^{k+1}) \quad \text{for} \quad x' \to 0.$$

Then

$$u(\psi) = u_\alpha(\phi) \quad \text{for} \quad \psi \in \mathscr{C}^\infty(\mathbb{R}^n) \tag{2.19}$$

and

$$\sum_{|\gamma|\leq k} \sup\left|D^\gamma\phi\right| \leq C \sum_{|\beta|\leq k-|\alpha|} \sup\left|D^\beta\psi\right|,$$

so

$$\sup\left|D^\alpha\phi\right| \leq C \sum_{|\beta|\leq k-|\alpha|} \sup\left|D^\beta\psi\right|.$$

Consequently

$$u_\alpha(\psi) \leq C' \sum_{|\beta|\leq k-|\alpha|} \sup\left|D^\beta\psi\right|$$

for every $\psi \in \mathscr{C}_0^\infty(\mathbb{R}^n)$, proving u_α is a distribution of order $k - |\alpha|$ in the variable x''. From (2.19), it follows that u_α has compact support. This completes the proof.

2.6   Singular Support

Definition 2.14 The set of points of X not admitting neighbourhoods where $u \in \mathscr{D}'(X)$ coincides with a $\mathscr{C}^\infty$ function is called the singular support of u, written singsupp$u$.

Hence, u coincides with a $\mathscr{C}^\infty$ function on $X\backslash$singsupp$u$.

Example 2.10 Let $f \in \mathscr{C}^\infty(X)$. We define the functional u in the following manner

$$u(\phi) = \int_X f(x)\phi(x)dx, \quad \phi \in \mathscr{C}_0^\infty(X).$$

For $\phi_1, \phi_2 \in \mathscr{C}_0^\infty(X)$ and $\alpha_1, \alpha_2 \in \mathbb{C}$, we have

$$u(\alpha_1\phi_1 + \alpha_2\phi_2) = \int_X f(x)(\alpha_1\phi_1(x) + \alpha_2\phi_2(x))dx$$

$$= \int_X (\alpha_1 f(x)\phi_1(x) + \alpha_2 f(x)\phi_2(x))dx$$

$$= \alpha_1\int_X f(x)\phi_1(x)dx + \alpha_2\int_X f(x)\phi_2(x)dx$$

$$= \alpha_1 u(\phi_1) + \alpha_2 u(\phi_2).$$

Therefore u is a linear functional on $\mathscr{C}_0^\infty(X)$. For $\phi \in \mathscr{C}_0^\infty(X)$, there exists a compact subset K of X such that supp$\phi \subset K$ and

$$|u(\phi)| = \left| \int_X f(x)\phi(x)dx \right| = \left| \int_K f(x)\phi(x)dx \right|$$

$$\leq \int_K |f(x)||\phi(x)|dx \leq \int_K |f(x)|dx \sup_{x \in K} |\phi(x)| < \infty.$$

Consequently the linear functional $u : \mathscr{C}_0^\infty(X) \mapsto \mathbb{C}$ is well defined. Let $\{\phi_n\}_{n=1}^\infty$ be a sequence in $\mathscr{C}_0^\infty(X)$ such that $\phi_n \to \phi, n \to \infty, \phi \in \mathscr{C}_0^\infty(X)$, in $\mathscr{C}_0^\infty(X)$. Then

$$u(\phi_n) = \int_X f(x)\phi_n(x)dx \to_{n\to\infty} u(\phi) = \int_X f(x)\phi(x)dx.$$

Therefore $u : \mathscr{C}_0^\infty(X) \mapsto \mathbb{C}$ is a linear continuous functional, i.e., $u \in \mathscr{D}'(X)$. Note that $u \equiv f \in \mathscr{C}^\infty(X)$ and therefore singsupp$u = \emptyset$.

Exercise 2.14 Find singsupp$P\dfrac{1}{x}$ for $x \in \mathbb{R}\setminus\{0\}$.

Exercise 2.15 Find singsupp$P\dfrac{1}{x}$ for $x \in \mathbb{R}$.

Exercise 2.16 Find singsupp$P\dfrac{1}{x^2}$ for $x \in \mathbb{R}\setminus\{0\}$.

Exercise 2.17 Find singsupp$P\dfrac{1}{x^2}$ for $x \in \mathbb{R}$.

Definition 2.15 The distribution $u \in \mathscr{D}'(X)$ is called regular if there exists $f \in L^1_{\text{loc}}(X)$ such that

$$u(\phi) = \int_X f(x)\phi(x)dx \quad \text{for} \quad \forall \phi \in \mathscr{C}_0^\infty(X).$$

In this case, we will write $u = u_f$. If no such f exists, u is called singular.

Example 2.11 Let $f = \dfrac{1}{1+x^2}, x \in \mathbb{R}$. The map $u : \mathscr{C}_0^\infty(X) \mapsto \mathbb{C}$,

$$u(\phi) = \int_\mathbb{R} f(x)\phi(x)dx, \quad \phi \in \mathscr{C}_0^\infty(\mathbb{R}),$$

is a regular distribution.

Example 2.12 Consider $\delta(x)$, $x \in \mathbb{R}$, and suppose that δ is a regular distribution. Then there exists $f \in L^1_{\text{loc}}(\mathbb{R})$ such that $u_f = \delta$. Choose $\rho \in \mathscr{C}_0^\infty(\mathbb{R})$ for which $\text{supp}(\rho) \subset \overline{U_1}$, $\rho(0) = 1$. Define the sequence $\{\rho_n\}_{n=1}^\infty$ by

$$\rho_n(x) = \rho(nx), \quad x \in \mathbb{R}, \quad n = 1, 2, \ldots.$$

Then $\text{supp}(\rho_n) \subset \overline{U_{\frac{1}{n}}}$ and $\rho_n(0) = 1$. In addition,

$$\delta(\rho_n) = \rho_n(0) = 1$$

and

$$1 = |\delta(\rho_n)| = \left| \int_{U_{\frac{1}{n}}} f(x)\rho(nx)dx \right| \le \int_{U_{\frac{1}{n}}} |f(x)||\rho(nx)|dx$$

$$\le \sup_{x \in \mathbb{R}} |\rho(x)| \int_{U_{\frac{1}{n}}} |f(x)|dx \to_{n \to \infty} 0,$$

which is a contradiction. Therefore $\delta \in \mathscr{D}'(\mathbb{R})$ is a singular distribution.

Exercise 2.18 Let $u_1, u_2 \in \mathscr{D}'(X)$ be regular distributions. Prove that $\alpha_1 u_1 + \alpha_2 u_2$ is a regular distribution for every $\alpha_1, \alpha_2 \in \mathbb{C}$.

2.7   Measures

Definition 2.16 A measure on a Borel set A is a complex-valued additive function

$$\mu(E) = \int_E \mu(dx),$$

that is finite ($|\mu(E)| < \infty$) on any bounded Borel subset E of A.

The measure μ of A can be represented in a unique way in terms of four nonnegative measures $\mu_i \ge 0$, $i = 1, 2, 3, 4$, on A in the following way

$$\mu = (\mu_1 - \mu_2) + i(\mu_3 - \mu_4)$$

and

$$\int_E \mu(dx) = \int_E \mu_1(dx) - \int_E \mu_2(dx) + i \int_E \mu_3(dx) - i \int_E \mu_4(dx)$$

for any bounded Borel subset E of A. The measure μ on the open set X determines a distribution μ on X as follows

$$\mu(\phi) = \int_X \phi(x)\mu(dx), \quad \phi \in \mathscr{C}_0^\infty(X),$$

where $\int$ is the Lebesgue–Stieltjes integral. From the integral properties, it follows that $\mu \in \mathscr{D}'(X)$. Every measure μ of X for which $\mu(dx) = f(x)dx$, $f \in L_{\text{loc}}^1(X)$, defines a regular distribution.

Theorem 2.17 *A distribution $u \in \mathscr{D}'(X)$ defines a measure μ on X if and only if $u \in \mathscr{D}'^0(X)$.*

Proof Let $u \in \mathscr{D}'(X)$ defines a measure μ of X. Then

$$|u(\phi)| = \left| \int_{X_1} \phi(x)\mu(dx) \right| \leq \int_{X_1} \mu(dx) \sup_{x \in X_1} |\phi(x)|$$

for every $X_1 \subset\subset X$ and every $\phi \in \mathscr{C}_0^\infty(X_1)$. Hence, $u \in \mathscr{D}'^0(X)$. Now, we suppose $u \in \mathscr{D}'^0(X)$, i.e., for every $X_1 \subset\subset X$

$$|u(\phi)| \leq C(X_1) \sup_{x \in X_1} |\phi(x)|,$$

where $C(X_1)$ is a constant which depends on X_1. Let $\{X_k\}_{k=1}^\infty$ be a family of sets such that $X_k \subset\subset X_{k+1}$, $\cup_k X_k = X$. Since $\mathscr{C}_0^\infty(X_k)$ is dense in $\mathscr{C}_0(\overline{X_k})$, the Riesz-Radon theorem implies that there exists a measure μ_k of $\overline{X_k}$ such that

$$u(\phi) = \int_{X_k} \phi(x)\mu_k(dx), \quad \phi \in \mathscr{C}_0(\overline{X_k}), \quad k = 1, 2, \ldots.$$

Therefore the measures μ_k and μ_{k+1} coincide on X_k. From this, we conclude that there is a measure μ on X which coincides with μ_k on X_k and with the distribution u on X. This completes the proof.

Definition 2.17 The distribution $u \in \mathscr{D}'(X)$ is called nonnegative on X if $u(\phi) \geq 0$ for every $\phi \in \mathscr{C}_0^\infty(X)$, $\phi(x) \geq 0$, $x \in X$.

Example 2.13 The distribution 1 is nonnegative.

Exercise 2.19 Prove that the distribution H is nonnegative.

Exercise 2.20 Prove that the distribution 1 is a measure.

2.8   Multiplying Distributions by $\mathscr{C}^\infty$ Functions

Definition 2.18 The product of a distribution $u \in \mathscr{D}'(X)$ by a function $b \in \mathscr{C}^\infty(X)$ is defined by

$$bu(\phi) = u(b\phi) \quad \text{for} \quad \phi \in \mathscr{C}_0^\infty(X).$$

Note that the map $\phi \mapsto b\phi$ is a linear and continuous map from $\mathscr{C}_0^\infty(X)$ to $\mathscr{C}_0^\infty(X)$. We have

$$
\begin{aligned}
bu(\alpha_1\phi_1 + \alpha_2\phi_2) &= u(b(\alpha_1\phi_1 + \alpha_2\phi_2)) \\
&= u(\alpha_1 b\phi_1 + \alpha_2 b\phi_2) \\
&= \alpha_1 u(b\phi_1) + \alpha_2 u(b\phi_2) \\
&= \alpha_1 bu(\phi_1) + \alpha_2 bu(\phi_2)
\end{aligned}
$$

for $\alpha_1, \alpha_2 \in \mathbb{C}$, $\phi_1, \phi_2 \in \mathscr{C}_0^\infty(X)$, i.e., bu is a linear map on $\mathscr{C}_0^\infty(X)$. Let $\{\phi_n\}_{n=1}^\infty$ be a sequence in $\mathscr{C}_0^\infty(X)$ such that $\phi_n \to_{n\to\infty} \phi$, $\phi \in \mathscr{C}_0^\infty(X)$, in $\mathscr{C}_0^\infty(X)$. Then $b\phi_n \to_{n\to\infty} b\phi$ in $\mathscr{C}_0^\infty(X)$. Since $u \in \mathscr{D}'(X)$, we have

$$u(b\phi_n) \to_{n\to\infty} u(b\phi),$$

so

$$bu(\phi_n) \to_{n\to\infty} bu(\phi).$$

Consequently bu is a continuous functional on $\mathscr{C}_0^\infty(X)$ and $bu \in \mathscr{D}'(X)$.

Example 2.14 Take $x^2\delta$. Then

$$x^2\delta(\phi) = \delta(x^2\phi) = 0^2\phi(0) = 0$$

for $\phi \in \mathscr{C}_0^\infty(X)$. Therefore $x^2\delta = 0$.

Exercise 2.21 Compute $(x^2 + 1)\delta$.

Answer δ.

Let $\alpha_1, \alpha_2 \in \mathbb{C}$, $b_1, b_2 \in \mathscr{C}^\infty(X)$ and $u_1, u_2 \in \mathscr{D}'(X)$. Then

1. $(\alpha_1 b_1(x) + \alpha_2 b_2(x))u_1 = \alpha_1 b_1(x)u_1 + \alpha_2 b_2(x)u_1,$
2. $b_1(x)(\alpha_1 u_1 + \alpha_2 u_2) = \alpha_1 b_1(x)u_1 + \alpha_2 b_1(x)u_2, \quad x \in X.$

Let us prove that this multiplication is neither associative nor commutative. Suppose the contrary, so

$$x\delta(\phi) = \delta(x\phi) = 0\phi(0) = 0(\phi),$$

$$xP\frac{1}{x}(\phi) = P\frac{1}{x}(x\phi) = P.V. \int_{\mathbb{R}} \phi(x)dx = 1(\phi), \quad x \in \mathbb{R},$$

for $\phi \in \mathscr{C}_0^\infty(\mathbb{R})$. Hence,

$$0 = 0P\frac{1}{x} = (x\delta(x))P\frac{1}{x} = (\delta(x)x)P\frac{1}{x} = \delta(x)(xP\frac{1}{x}) = \delta(x)1 = \delta(x), \quad x \in \mathbb{R},$$

a contradiction.

2.9 Advanced Practical Problems

Problem 2.1 Let $P\dfrac{1}{x^2}$ be defined on $\mathscr{C}_0^\infty(\mathbb{R})$ by

$$P\frac{1}{x^2}(\phi) = P.V. \int_{-\infty}^{\infty} \frac{\phi(x) - \phi(0)}{x^2} dx.$$

Prove that $P\dfrac{1}{x^2} \in \mathscr{D}'(\mathbb{R})$.

Problem 2.2 Define u by

$$u(\phi) = \int_{|x|\le 1} \phi(x)dx \quad \forall \phi \in \mathscr{C}_0^\infty(\mathbb{R}^n).$$

Prove that $u \in \mathscr{D}'(\mathbb{R}^n)$.

Problem 2.3 Define

$$u(\phi) = \int_{|x|\le 1} D^\alpha \phi(x)dx \quad \forall \phi \in \mathscr{C}_0^\infty(\mathbb{R}^n),$$

where α is a multi-index. Show that $u \in \mathscr{D}'(\mathbb{R}^n)$.

Problem 2.4 Let α be a multi-index and set $u(\phi) = D^\alpha \phi(x_0)$, $\phi \in \mathscr{C}_0^\infty(X)$, for a given $x_0 \in X$. Prove that u is a distribution of order $|\alpha|$.

Solution Let $\phi_1, \phi_2 \in \mathscr{C}_0^\infty(X)$ and $a, b \in \mathbb{C}$. Then

$$u(a\phi_1+b\phi_2) = D^\alpha(a\phi_1+b\phi_2)(x_0) = aD^\alpha\phi_1(x_0)+bD^\alpha\phi_2(x_0) = au(\phi_1)+bu(\phi_2).$$

Consequently u is a linear map on $\mathscr{C}_0^\infty(X)$. Let K be a compact subset of X and $\phi \in \mathscr{C}_0^\infty(K)$. Since supp $\phi \subset K$ we have to consider two cases: $x_0 \in K$ and $x_0 \notin K$. If $x_0 \in K$,

$$|u(\phi)| \leq C \sum_{|\beta|\leq|\alpha|} \sup_K \left|D^\beta\phi(x)\right| \qquad (2.20)$$

for $C \geq 1$. If $x_0 \notin K$, then $u(\phi) = 0$. Therefore the inequality (2.20) holds, and then $u \in \mathscr{D}'(X)$. Using the definition of u and (2.20), we conclude that u has order $|\alpha|$.

Problem 2.5 Take $f \in \mathscr{C}(\mathbb{R}^n)$ and a multi-index α. Let $D^\alpha f$ be defined on $\mathscr{C}_0^\infty(\mathbb{R}^n)$ as follows:

$$D^\alpha f(\phi) = (-1)^{|\alpha|} \int_{\mathbb{R}^n} f(x)D^\alpha\phi(x)dx.$$

Prove that $D^\alpha f$ is a distribution of order $|\alpha|$.

Problem 2.6 Prove that $H \in \mathscr{D}'^0(\mathbb{R})$.

Problem 2.7 Let $P(x, D) = \sum_{|\alpha|\leq q} a_\alpha(x)D^\alpha$, where $q \in \mathbb{N} \cup \{0\}$ is fixed, and $a \in \mathscr{C}(\mathbb{R}^n)$. Let u be defined on $\mathscr{C}_0^\infty(\mathbb{R}^n)$ by

$$u(\phi) = \int_{\mathbb{R}^n} u(x)P(x, D)\phi(x)dx.$$

Prove that $u \in \mathscr{D}'^q(\mathbb{R}^n)$.

Problem 2.8 Let $u \in \mathscr{D}'(X)$ and suppose $u(\phi) \geq 0$ for every nonnegative function $\phi \in \mathscr{C}_0^\infty(X)$. Prove that u is a measure, i.e., a distribution of order 0.

Solution Let $K \subset X$ be a compact set. Then there exists a function $\chi \in \mathscr{C}_0^\infty(X)$ such that $0 \leq \chi \leq 1$ on X and $\chi = 1$ on K. Then

$$\chi \sup_K |\phi| \pm \phi \geq 0$$

for every $\phi \in \mathscr{C}_0^\infty(K)$, and therefore

$$u(\chi \sup_K |\phi| \pm \phi) \geq 0. \qquad (2.21)$$

On the other hand,

$$u(\chi \sup_K |\phi| \pm \phi) = \sup_K |\phi| u(\chi) \pm u(\phi).$$

Consequently, using (2.21),

$$\pm u(\phi) \le u(\chi) \sup_K |\phi|.$$

Therefore $u \in \mathscr{D}'^0(X)$, i.e., u is a measure.

Problem 2.9 Take $\phi \in \mathscr{C}^\infty(X \times Y)$, where Y is an open set in $\mathbb{R}^m$, $m \ge 1$. Suppose that there is a compact set $K \subset X$ such that $\phi(x, y) = 0$ for every $x \notin K$ and for every $y \in Y$. Prove that the map

$$y \mapsto u(\phi(\cdot, y))$$

is a $\mathscr{C}^\infty$ function for every $u \in \mathscr{D}'(X)$ and

$$D_y^\alpha u(\phi(\cdot, y)) = u(D_y^\alpha \phi(\cdot, y))$$

for every multi-index α.

Solution Since $u \in \mathscr{D}'(X)$ and $\phi \in \mathscr{C}_0^\infty(X \times Y)$, we have that $u(\phi)$ is continuous with respect to the second variable. We will prove

$$\frac{\partial}{\partial y_j} u(\phi(x, y)) = u\left(\frac{\partial}{\partial y_j} \phi(x, y)\right) \quad \text{for} \quad x \in K, \quad y \in Y,$$

and $j = 1, 2, \ldots, m$. For $y \in Y$ given,

$$\phi(x, y + h) = \phi(x, y) + \sum_{j=1}^m h_j \frac{\partial \phi}{\partial y_j}(x, y) + o(|h|^2)$$

for $\phi \in \mathscr{C}_0^\infty(K \times Y)$. Let $h = (0, \ldots, 0, h_j, 0, \ldots, 0)$. Then

$$\frac{\phi(x, y + h) - \phi(x, y)}{h_j} = \frac{\partial \phi}{\partial y_j}(x, y) + \frac{1}{h_j} o(h_j^2).$$

Since u is linear, we have

$$u\left(\frac{\phi(x, y + h) - \phi(x, y)}{h_j}\right) = u\left(\frac{\partial \phi}{\partial y_j}(x, y)\right) + \frac{1}{h_j} u\left(o(h_j^2)\right).$$

From this equality, we obtain

$$u\left(\frac{\partial}{\partial y_j}\phi(x, y)\right) = \frac{\partial}{\partial y_j}u(\phi(x, y)),$$

as $h_j \to 0$. By induction, we get

$$u\left(D_y^{\alpha}\phi(x, y)\right) = D_y^{\alpha}u(\phi(x, y)), \quad x \in K, \quad y \in Y.$$

Problem 2.10 Let $u_n \in \mathscr{D}'(X)$, $u_n(\phi) \geq 0$ for every nonnegative $\phi \in \mathscr{C}_0^{\infty}(X)$ and $u_n \to_{n\to\infty} u$ in $\mathscr{D}'(X)$. Prove that $u \geq 0$ and $u_n(\phi) \to_{n\to\infty} u(\phi)$ for every $\phi \in \mathscr{C}_0^0(X)$.

Problem 2.11 Prove that the functions

1. $f = e^{\frac{1}{x^2}}$,
2. $f = e^{\frac{1}{x^m}}$, $m \in \mathbb{N}$

do not define distributions, i.e. $f \notin \mathscr{D}'(\mathbb{R}\setminus\{0\})$ in all cases.

1. **Hint** Use

$$\phi_k(x) = e^{-\frac{k^2}{4}}k\phi_0(kx).$$

2. **Hint** Use

$$\phi_k(x) = e^{-\left(\frac{k}{2}\right)^m}k\phi_0(kx).$$

Problem 2.12 Given constants $m \in \mathbb{N}$, a_i, $i = 1, 2, \ldots, m$, prove that

$$f = a_1 e^{\frac{1}{x}} + a_2 e^{\frac{1}{x^2}} + \cdots + a_m e^{\frac{1}{x^m}} \notin \mathscr{D}'(\mathbb{R}\setminus\{0\}).$$

Hint Use the previous problem.

Problem 2.13 (Sochozki Formulas) Show that

1. $\displaystyle\lim_{\epsilon\to 0}\int_{\mathbb{R}}\frac{\phi(x)}{x - i\epsilon}dx = i\pi\phi(0) + P.V.\int_{\mathbb{R}}\frac{\phi(x)}{x}dx, \phi \in \mathscr{C}_0^{\infty}(\mathbb{R}),$

2. $\displaystyle\lim_{\epsilon\to 0}\int_{\mathbb{R}}\frac{\phi(x)}{x + i\epsilon}dx = -i\pi\phi(0) + P.V.\int_{\mathbb{R}}\frac{\phi(x)}{x}dx, \phi \in \mathscr{C}_0^{\infty}(\mathbb{R}).$

1. *Solution* Take $\phi \in \mathscr{C}_0^\infty(\mathbb{R})$ with supp $\phi \subset [-R, R]$. Then

$$\int_{\mathbb{R}} \frac{\phi(x)}{x - i\epsilon} dx = \int_{-R}^{R} \frac{(x + i\epsilon)\phi(x)}{x^2 + \epsilon^2} dx$$

$$= \int_{-R}^{R} \frac{(x + i\epsilon)(\phi(x) - \phi(0))}{x^2 + \epsilon^2} dx + \int_{-R}^{R} \frac{(x + i\epsilon)\phi(0)}{x^2 + \epsilon^2} dx.$$

From this,

$$\lim_{\epsilon \to 0} \int_{-R}^{R} \frac{(x + i\epsilon)(\phi(x) - \phi(0))}{x^2 + \epsilon^2} dx = P.V. \int_{-\infty}^{\infty} \frac{\phi(x)}{x} dx.$$

What is more,

$$\lim_{\epsilon \to 0} \int_{-R}^{R} \frac{(x + i\epsilon)\phi(0)}{x^2 + \epsilon^2} dx = 2i\phi(0) \lim_{\epsilon \to 0} \arctan \frac{R}{\epsilon} = i\pi\phi(0) = i\pi\delta(\phi).$$

2. **Hint** Use the solution of part 1.

Problem 2.14 (Sochozki Formulas) Prove that

$$\frac{1}{x - i0} = i\pi\delta + P\frac{1}{x}, \qquad \frac{1}{x + i0} = -i\pi\delta + P\frac{1}{x}.$$

Hint Use the previous problem.

Problem 2.15 Prove that

1. $\lim\limits_{\epsilon \to 0+} \dfrac{1}{2\sqrt{\pi\epsilon}e^{-\frac{x^2}{4\epsilon}}} = \delta(x),\ \lim\limits_{\epsilon \to 0+} \dfrac{1}{\pi x} \sin \dfrac{x}{\epsilon} = \delta(x),$

2. $\lim\limits_{t \to \infty} \dfrac{e^{ixt}}{x - i0} = 2\pi i\delta(x),\ \lim\limits_{t \to \infty} \dfrac{e^{-ixt}}{x - i0} = 0,$

3. $\lim\limits_{t \to \infty} \dfrac{e^{ixt}}{x + i0} = 0,\ \lim\limits_{t \to \infty} \dfrac{e^{-ixt}}{x + i0} = -2\pi i\delta(x),$

4. $\lim\limits_{t \to \infty} t^m e^{ixt} = 0,\ m \geq 0,\ \lim\limits_{t \to \infty} P\left(\dfrac{\cos(tx)}{x}\right) = 0,$

5. $\lim\limits_{\epsilon \to 0+} \dfrac{1}{\epsilon}\omega\left(\dfrac{x}{\epsilon}\right) = \delta(x),\ \lim\limits_{n \to \infty} \dfrac{2n^3x^2}{\pi(1 + n^2x^2)^2} = \delta(x),$

6. $\lim\limits_{n \to \infty} \dfrac{n}{\pi(1 + n^2x^2)} = \delta(x),\ \lim\limits_{n \to \infty} \dfrac{1}{n\pi} \dfrac{\sin^2(nx)}{x^2} = \delta(x),$

7. $\lim\limits_{n\to\infty} f_n(x) = \delta(x)$, where $f_n(x) = \begin{cases} \dfrac{n}{2} & \text{for } |x| \le \dfrac{1}{n} \\ 0 & \text{otherwise}; \end{cases}$

8. $\lim\limits_{n\to\infty} \dfrac{n}{\sqrt{2\pi}} e^{-\frac{n^2 x^2}{2}} = \delta(x)$, $\lim\limits_{n\to\infty} \dfrac{\sin(nx)}{\pi x} = \delta(x)$,

9. $\lim\limits_{n\to\infty} \dfrac{1}{2} n e^{-n|x|} = \delta(x)$, $\lim\limits_{n\to\infty} \dfrac{1}{\pi} \dfrac{n}{e^{nx} + e^{-nx}} = \delta(x)$,

10. $\lim\limits_{n\to\infty} \sqrt{\dfrac{n}{\pi}} e^{-nx^2} = \delta(x)$, $\lim\limits_{n\to\infty} \dfrac{n}{n^2 x^2 + 1} = \delta(x)\pi$.

1. *Solution* Take $\phi \in \mathscr{C}_0^\infty(\mathbb{R})$. Then there exists $R > 0$ such that $\mathrm{supp}\phi \subset [-R, R]$. Now,

$$\left(\frac{1}{2\sqrt{\pi\epsilon}} e^{-\frac{x^2}{4\epsilon}}, \phi(x)\right) = \int_{-R}^{R} \frac{e^{-\frac{x^2}{4\epsilon}}}{2\sqrt{\pi\epsilon}} \phi(x)dx$$

$$= \frac{1}{\sqrt{\pi}} \int_{-R}^{R} \frac{e^{-\left(\frac{x}{2\sqrt{\epsilon}}\right)^2}}{2\sqrt{\epsilon}} \left[\phi(x) - \phi(0)\right] dx + \frac{\phi(0)}{\sqrt{\pi}} \int_{-R}^{R} \frac{e^{-\left(\frac{x}{2\sqrt{\epsilon}}\right)^2}}{2\sqrt{\epsilon}} dx$$

$$= \frac{1}{\sqrt{\pi}} \int_{-R}^{R} \frac{e^{-\left(\frac{x}{2\sqrt{\epsilon}}\right)^2}}{2\sqrt{\epsilon}} x \frac{\phi(x) - \phi(0)}{x} dx + \frac{\phi(0)}{\sqrt{\pi}} \int_{-R}^{R} e^{-\left(\frac{x}{2\sqrt{\epsilon}}\right)^2} d\left(\frac{x}{2\sqrt{\epsilon}}\right)$$

$$= \frac{1}{\sqrt{\pi}} \int_{-R}^{R} \frac{e^{-\left(\frac{x}{2\sqrt{\epsilon}}\right)^2}}{2\sqrt{\epsilon}} x \frac{\phi(x) - \phi(0)}{x} dx + \frac{\phi(0)}{\sqrt{\pi}} \int_{-\frac{R}{2\sqrt{\epsilon}}}^{\frac{R}{2\sqrt{\epsilon}}} e^{-y^2} dy.$$

Therefore

$$\lim_{\epsilon\to 0+} \left(\frac{1}{2\sqrt{\pi\epsilon}} e^{-\frac{x^2}{4\epsilon}}, \phi(x)\right)$$

$$= \lim_{\epsilon\to 0+} \frac{1}{\sqrt{\pi}} \int_{-R}^{R} \frac{e^{-\left(\frac{x}{2\sqrt{\epsilon}}\right)^2}}{2\sqrt{\epsilon}} x \frac{\phi(x) - \phi(0)}{x} dx + \frac{\phi(0)}{\sqrt{\pi}} \lim_{\epsilon\to 0+} \int_{-\frac{R}{2\sqrt{\epsilon}}}^{\frac{R}{2\sqrt{\epsilon}}} e^{-y^2} dy$$

$$= \frac{\phi(0)}{\sqrt{\pi}} \int_{-\infty}^{\infty} e^{-y^2} dy = \phi(0) = \delta(\phi).$$

Problem 2.16 Prove that $\mathrm{supp}\delta = \{0\}$.

Problem 2.17 Let K be a compact set in $\mathbb{R}^n$ which cannot be written as union of finitely many compact connected domains. Prove that there exists a distribution $u \in \mathscr{E}'(K)$ of order 1 that does not satisfy

$$u(\phi) \leq C \sum_{|\alpha| \leq k} \sup_K \left| \partial^\alpha \phi \right|, \quad \phi \in \mathscr{C}^\infty(X),$$

for any constants C and k.

Problem 2.18 Let K be a compact set in $\mathbb{R}^n$ and u_α, $|\alpha| \leq k$, be continuous functions on K. For $|\alpha| \leq k$, we set

$$U_\alpha(x, y) = \left| u_\alpha(x) - \sum_{|\beta| \leq k - |\alpha|} u_{\alpha+\beta}(y) \frac{(x-y)^\beta}{\beta!} \right| |x - y|^{|\alpha| - k},$$

for $x, y \in K$, $x \neq y$, and $U_\alpha(x, x) = 0$ for $x \in K$. Suppose that every function U_α, $|\alpha| \leq k$, is continuous on $K \times K$. Prove that there exists $v \in \mathscr{C}^k(\mathbb{R}^n)$ such that $\partial^\alpha v(x) = u_\alpha(x)$ for $x \in K$, $|\alpha| \leq k$. Then prove that v can be chosen so that

$$\sum_{|\alpha| \leq k} \sup \left| \partial^\alpha v \right| \leq C \left(\sum_{|\alpha| \leq k} \sup_{K \times K} U_\alpha + \sum_{|\alpha| \leq k} \sup_K u_\alpha \right),$$

where C is a constant depending on K only.

Problem 2.19 Prove that

$$|u(\phi)| \leq C \left(\sum_{|\alpha| \leq k} \sup_{x, y \in K, x \neq y} \left| \partial^\alpha \phi(x) - \sum_{|\beta| \leq k - |\alpha|} \partial^{\alpha+\beta} \phi(y) \frac{(x-y)^\beta}{\beta!} \right| |x - y|^{|\alpha| - k} \right.$$
$$\left. + \sum_{|\alpha| \leq k} \sup_K \left| \partial^\alpha \phi \right| \right), \quad \phi \in \mathscr{C}^\infty(\mathbb{R}^n),$$

for every distribution u of order k with compact support $K \subset \mathbb{R}^n$.

Problem 2.20 Let K be a compact set in $\mathbb{R}^n$ with finitely many connected components such that every two points x and y in the same component can be joined by a rectifiable curve in K of length $\leq C|x - y|$. Prove that for every distribution u of order k with $\operatorname{supp} u \subset K$ the estimate

$$|u(\phi)| \leq C \sum_{|\alpha| \leq k} \sup_K \left| \partial^\alpha \phi \right|, \quad \phi \in \mathscr{C}^k(\mathbb{R}^n),$$

holds.

Problem 2.21 Let f_n, $f \in L^1_{loc}(X)$ and

$$\int_K |f_n(x) - f(x)| dx \to_{n \to \infty} 0$$

for every compact subset K of X. Prove that

$$f_n \to_{n \to \infty} f$$

in $\mathscr{D}'(X)$.

Problem 2.22 Prove that

1. $\delta(x^2 - a^2) = \dfrac{1}{2a}\Big[\delta(x - a) + \delta(x + a)\Big], a \neq 0,$

2. $\delta(\sin x) = \displaystyle\sum_{k=-\infty}^{\infty} \delta(x - k\pi).$

Problem 2.23 Prove that $\delta(x)$, $x \in \mathbb{R}$, is a measure.

Problem 2.24 Prove that $H(x)$, $x \in \mathbb{R}$, is a measure.

Problem 2.25 Let $u \in \mathscr{D}'(X)$ and $b \in \mathscr{C}^\infty(X)$ be such that $b \equiv 1$ on a neighbourhood of supp$u$. Show that

$$u = bu.$$

Solution For the function $1 - b$ we have that $1 - b \equiv 0$ on supp$u$. Then for $\phi \in \mathscr{C}_0^\infty(X)$, we have

$$0 = u((1 - b)\phi) = u(\phi - b\phi) = u(\phi) - u(b\phi) = u(\phi) - bu(\phi),$$

so

$$u(\phi) = bu(\phi)$$

for every $\phi \in \mathscr{C}_0^\infty(X)$. Therefore $u = bu$.

Problem 2.26 Compute

$$(x^4 + x^2 + 3)\delta(x) + xP\frac{1}{x}, \quad x \in \mathbb{R}.$$

Answer $3\delta + 1$.

Problem 2.27 Let $b \in \mathscr{C}^\infty(\mathbb{R})$. Compute

$$b(x)\delta(x), \quad x \in \mathbb{R}.$$

Answer $b(0)\delta$.

Problem 2.28 Let $a \in \mathscr{C}^{\infty}(X)$, $u \in \mathscr{D}'(X)$. Prove that $\mathrm{supp}(au) \subset \mathrm{supp}\,a \cap \mathrm{supp}\,u$.

Problem 2.29 Let $f, u \in \mathscr{D}'(X)$ and $\mathrm{singsupp}\,u \cap \mathrm{singsupp}\,f = \varnothing$. Prove that $f \circ u \in \mathscr{D}'(X)$.

Problem 2.30 Let $f \in \mathscr{C}^{\infty}(X)$, $u \in \mathscr{D}'(X)$ and $\mathrm{supp}\,u \cap \mathrm{supp}\,f \subset\subset X$. Prove that $u(f)$ can be defined by $u(f) = (fu)(1)$.

Problem 2.31 Let $f \in \mathscr{C}^{k}(X)$, $u \in \mathscr{D}'^{k}(X)$. Prove that $fu \in \mathscr{D}'^{k}(X)$.

Problem 2.32 Solve the equation

$$(x - 3)u = 0$$

in $\mathscr{D}'(X)$.

Solution Let $\phi \in \mathscr{C}_0^{\infty}(\mathbb{R})$. Then we have

$$(x - 3)u(\phi) = 0 \quad \text{or} \quad u((x - 3)\phi) = 0. \tag{2.22}$$

Let now $\psi \in \mathscr{C}_0^{\infty}(\mathbb{R})$. We choose $\eta \in \mathscr{C}_0^{\infty}(\mathbb{R})$ so that $\eta \equiv 1$ on $[3 - \epsilon, 3 + \epsilon]$ and $\eta \equiv 0$ on $\mathbb{R} \backslash [3 - \epsilon, 3 + \epsilon]$, for a small enough $\epsilon > 0$. Let $g(x) = \dfrac{\psi(x) - \eta(x)\psi(3)}{x - 3}$, $x \in \mathbb{R}$. Then $g \in \mathscr{C}_0^{\infty}(\mathbb{R})$. From this and (2.22), we have that

$$u\left((x - 3)\frac{\psi(x) - \eta(x)\psi(3)}{x - 3}\right) = 0.$$

Hence,

$$\begin{aligned}
u(\psi) &= u\left((x - 3)\frac{\psi(x) - \eta(x)\psi(3)}{x - 3} + \eta(x)\psi(3)\right) \\
&= u\left((x - 3)\frac{\psi(x) - \eta(x)\psi(3)}{x - 3}\right) + u(\eta(x)\psi(3)) \\
&= \psi(3)u(\eta) = C\psi(3) = C\delta(x - 3)(\psi).
\end{aligned}$$

Here $C = u(\eta) = \mathrm{const}$. Since $\psi \in \mathscr{C}_0^{\infty}(\mathbb{R})$ was arbitrarily chosen, we conclude that $u = C\delta(x - 3)$.

Problem 2.33 Solve the equation

$$(x - 3)u = P\frac{1}{x - 3}$$

in $\mathscr{D}'(\mathbb{R})$.

Solution By using the previous problem, the corresponding homogeneous equation $(x - 3)u = 0$ is solved by $u = C\delta(x - 3)$, $C = $ const, and a particular solution is $P \dfrac{1}{(x - 3)^2}$. Therefore

$$u = C\delta(x - 3) + P\frac{1}{(x - 3)^2}.$$

Problem 2.34 Solve the equations

1. $(x - 1)(x - 2)u = 0$,
2. $(\sin x)u = 0$.

Answer

1. $u = C_1\delta(x - 1) + c_2\delta(x - 2)$, $C_1, C_2 = $ const,
2. $\displaystyle\sum_{k=-\infty}^{\infty} C_k\delta(x - k\pi)$, $C_k = $ const.

2.10   Notes and References

In this chapter are collected the definitions for distributions, their order, sequences and series, support and singular support, and multiplication by $\mathscr{C}^\infty$ functions. They are deducted some of their basic properties. Additional materials can be found in [7, 16, 17, 20, 21, 24, 25] and references therein.

Chapter 3
Differentiation

3.1 Derivatives

Let X be an open set in $\mathbb{R}^n$.

Definition 3.1 For $u \in \mathscr{D}'(X)$ and $\alpha \in \mathbb{N}^n \cup \{0\}$, we define $D^\alpha u$ as follows

$$D^\alpha u(\phi) = (-1)^{|\alpha|} u(D^\alpha \phi) \tag{3.1}$$

for every $\phi \in \mathscr{C}_0^\infty(X)$.

Since the operation $\phi \mapsto D^\alpha \phi$ is linear and continuous on $\mathscr{C}_0^\infty(X)$, the functional $D^\alpha u$, determined by (3.1), is linear and continuous, i.e., $D^\alpha u \in \mathscr{D}'(X)$. Below, we will list some of the properties of the derivatives of the distributions.

Theorem 3.1 *The operation* $u \mapsto D^\alpha u : \mathscr{D}'(X) \mapsto \mathscr{D}'(X)$ *is linear and continuous.*

Proof We start by showing linearity. Let $\alpha_1, \alpha_2 \in \mathbb{C}$, $u_1, u_2 \in \mathscr{D}'(X)$ and $\phi \in \mathscr{C}_0^\infty(X)$. Then

$$
\begin{aligned}
(\alpha_1 u_1 + \alpha_2 u_2)(\phi) &\mapsto D^\alpha(\alpha_1 u_1 + \alpha_2 u_2)(\phi) \\
&= (-1)^{|\alpha|}(\alpha_1 u_1 + \alpha_2 u_2)(D^\alpha \phi) \\
&= (-1)^{|\alpha|}\alpha_1 u_1(D^\alpha \phi) + (-1)^{|\alpha|}\alpha_2 u_2(D^\alpha \phi) \\
&= \alpha_1 D^\alpha u_1(\phi) + \alpha_2 D^\alpha u_2(\phi).
\end{aligned}
$$

Now, we will prove the continuity. Let $\{u_n\}_{n=1}^\infty$ be a sequence in $\mathscr{D}'(X)$ such that $u_n \to_{n\to\infty} 0$ in $\mathscr{D}'(X)$. Then

$$D^\alpha u_n(\phi) = (-1)^{|\alpha|} u_n(D^\alpha \phi) \to_{n\to\infty} 0$$

© The Author(s), under exclusive license to Springer Nature Switzerland AG 2021
S. G. Georgiev, *Theory of Distributions*,
https://doi.org/10.1007/978-3-030-81265-2_3

for $\phi \in \mathscr{C}_0^\infty(X)$. Consequently $D^\alpha u_n \to_{n\to\infty} 0$ in $\mathscr{D}'(X)$. This completes the proof.

Example 3.1 Let us consider the Heaviside function $H(x)$, $x \in \mathbb{R}$. For its derivatives, we have

$$D^\alpha H(\phi) = (-)^\alpha H(D^\alpha \phi)$$
$$= (-1)^\alpha \int_{-\infty}^{\infty} H(x) D^\alpha \phi(x) dx$$
$$= (-1)^\alpha \int_{0}^{\infty} D^\alpha \phi(x) dx$$
$$= (-1)^\alpha D^{\alpha-1} \phi(x)|_{x=0}^{x=\infty}$$
$$= (-1)^{\alpha+1} D^{\alpha-1} \phi(0)$$

for $\phi \in \mathscr{C}_0^\infty(\mathbb{R})$ and $\alpha \in \mathbb{N}$.

Example 3.2 Let us compute

$$\lim_{\epsilon \to 0} D^\alpha \omega_\epsilon \quad \text{in} \quad \mathscr{D}'(\mathbb{R}), \quad \text{for} \quad \alpha \in \mathbb{N}.$$

Let $\phi \in \mathscr{C}_0^\infty(\mathbb{R})$. Then

$$\lim_{\epsilon \to 0} D^\alpha \omega_\epsilon(\phi) = (-1)^\alpha \lim_{\epsilon \to 0} \omega_\epsilon(D^\alpha \phi)$$
$$= (-)^\alpha \lim_{\epsilon \to 0} \int_{-\infty}^{\infty} \omega_\epsilon(x) D^\alpha \phi(x) dx$$
$$= (-1)^\alpha \lim_{\epsilon \to 0} \int_{|x| \le \epsilon} e^{-\frac{\epsilon^2}{\epsilon^2-|x|^2}} D^\alpha \phi(x) dx$$
$$= (-1)^\alpha D^\alpha \phi(0)$$
$$= (-1)^\alpha \delta(D^\alpha \phi)$$
$$= D^\alpha \delta(\phi).$$

Therefore $\lim_{\epsilon \to 0} D^\alpha \omega_\epsilon = D^\alpha \delta$ in $\mathscr{D}'(X)$.

Exercise 3.1 Find $\delta^{(k)}(x)$, $x \in \mathbb{R}$, $k \in \mathbb{N}$.

Answer $(-1)^k \phi^{(k)}(0)$, $\phi \in \mathscr{C}_0^\infty(\mathbb{R})$.

By Theorem 3.1, it follows that if the series

$$\sum_{k\geq 1} u_k(x) = S(x), \quad u_k \in L^1_{\mathrm{loc}}(X),$$

is uniformly convergent on every compact subset K of X, it may be differentiated term by term any number of times, and the resulting series will converge in $\mathscr{D}'(X)$. We have

$$\sum_{k\geq 1} D^\alpha u_k(x) = D^\alpha S(x).$$

Theorem 3.2 *Every distribution $u \in \mathscr{D}'(X)$ is differentiable infinitely many times.*

Proof Since $u(D^\alpha \phi)$ exists for every $\alpha \in \mathbb{N}^n \cup \{0\}$ and every $\phi \in \mathscr{C}_0^\infty(X)$, we conclude that u is differentiable infinitely many times. This completes the proof.

Theorem 3.3 *We have*

$$D^{\alpha+\beta} u = D^\alpha (D^\beta u)$$

for every $\alpha, \beta \in \mathbb{N}^n \cup \{0\}$ and every $u \in \mathscr{D}'(X)$.

Proof Let in fact $\phi \in \mathscr{C}_0^\infty(X)$ be arbitrary. Then

$$\begin{aligned} D^{\alpha+\beta} u(\phi) &= (-1)^{|\alpha+\beta|} u(D^{\alpha+\beta}\phi) \\ &= (-1)^{|\alpha|}(-1)^{|\beta|} u(D^\beta(D^\alpha\phi)) \\ &= (-1)^{|\alpha|} D^\beta u(D^\alpha\phi) \\ &= D^\alpha(D^\beta u)(\phi). \end{aligned}$$

This completes the proof.

Theorem 3.4 *We have*

$$D^\alpha(\alpha_1 u_1 + \alpha_2 u_2) = \alpha_1 D^\alpha u_1 + \alpha_2 D^\alpha u_2$$

for $u_1, u_2 \in \mathscr{D}'(X)$, $\alpha_1, \alpha_2 \in \mathbb{C}$, $\alpha \in \mathbb{N}^n \cup \{0\}$.

Proof Let $\phi \in \mathscr{C}_0^\infty(X)$ be arbitrarily chosen. Then

$$\begin{aligned} D^\alpha(\alpha_1 u_1 + \alpha_2 u_2)(\phi) &= (-1)^{|\alpha|}(\alpha_1 u_1 + \alpha_2 u_2)(D^\alpha\phi) \\ &= (-1)^{|\alpha|}\alpha_1 u_1(D^\alpha\phi) + (-1)^{|\alpha|}\alpha_2 u_2(D^\alpha\phi) \\ &= \alpha_1 D^\alpha u_1(\phi) + \alpha_2 D^\alpha u_2(\phi) \\ &= \left(\alpha_1 D^\alpha u_1 + \alpha_2 D^\alpha u_2\right)(\phi). \end{aligned}$$

This completes the proof.

Theorem 3.5 *If $u \in \mathcal{D}'(X)$, $a \in \mathcal{C}^\infty(X)$, we have*

$$D^\alpha(au) = \sum_{\beta \le \alpha} \binom{\alpha}{\beta} D^\beta a D^{\alpha-\beta} u,$$

for every $\alpha \in \mathbb{N}^n \cup \{0\}$.

Proof We will prove the simple case

$$\frac{\partial}{\partial x_i}(au) = a\frac{\partial u}{\partial x_i} + \frac{\partial a}{\partial x_i}u$$

for some $i \in \{1, 2, \ldots, n\}$. Using induction, the reader can deduce the general case. If $\phi \in \mathcal{C}_0^\infty(X)$, then

$$\begin{aligned}
\frac{\partial}{\partial x_i}(au)(\phi) &= -au\left(\frac{\partial}{\partial x_i}\phi\right) \\
&= -u\left(a\frac{\partial}{\partial x_i}\phi\right) \\
&= -u\left(\frac{\partial}{\partial x_i}(a\phi) - \frac{\partial a}{\partial x_i}\phi\right) \\
&= -u\left(\frac{\partial}{\partial x_i}(a\phi)\right) + u\left(\frac{\partial a}{\partial x_i}\phi\right) \\
&= \frac{\partial u}{\partial x_i}(a\phi) + \frac{\partial a}{\partial x_i}u(\phi) \\
&= a\frac{\partial u}{\partial x_i}(\phi) + \frac{\partial a}{\partial x_i}u(\phi) \\
&= \left(a\frac{\partial u}{\partial x_i} + \frac{\partial a}{\partial x_i}u\right)(\phi),
\end{aligned}$$

so $\dfrac{\partial}{\partial x_i}(au) = \dfrac{\partial a}{\partial x_i}u + a\dfrac{\partial u}{\partial x_i}$ in $\mathcal{D}'(X)$. This completes the proof.

Theorem 3.6 *For every $u \in \mathcal{D}'(X)$ and every $\alpha \in \mathbb{N}^n \cup \{0\}$,*

$$\mathrm{supp}D^\alpha u \subset \mathrm{supp}u.$$

Proof To see this, let $u \in \mathcal{D}'(X)$, $\phi \in \mathcal{C}_0^\infty(X)$ and $\mathrm{supp}\phi \cap \mathrm{supp}u = \emptyset$. Then $D^\alpha \phi \in \mathcal{C}_0^\infty(X)$ and $\mathrm{supp}D^\alpha \phi \cap \mathrm{supp}u = \emptyset$. Hence,

$$D^\alpha u(\phi) = (-1)^{|\alpha|}u(D^\alpha \phi) = 0,$$

and the assertion follows.

Theorem 3.7 *Let f be a function defined on (a, b) that is piecewise-differentiable with continuity. Call $\{x_k\}$ the points in (a, b) where f or its derivative have jump discontinuities. Write*

$$[f]_{x_k} = f(x_k + 0) - f(x_k - 0),$$

and denote by $\{f'\}$ the classical derivative of f at $x \in (a, b)$. Then

$$f' = \{f'\} + \sum_k [f]_{x_k} \delta(x - x_k).$$

Proof For $\phi \in \mathscr{C}_0^\infty(a, b)$, we have

$$f'(\phi) = -f(\phi')$$

$$= -\sum_k \int_{x_k}^{x_{k+1}} f(x)\phi'(x)dx$$

$$= \sum_k \int_{x_k}^{x_{k+1}} \{f'\}(x)\phi(x)dx - \sum_k \Big[f(x_{k+1} - 0)\phi(x_{k+1}) - f(x_k + 0)\phi(x_k) \Big]$$

$$= \int_a^b \{f'\}(x)\phi(x)dx + \sum_k \Big[f(x_k + 0) - f(x_k - 0) \Big]\phi(x_k)$$

$$= \{f'\}(\phi) + \sum_k [f]_{x_k} \delta(x - x_k)(\phi).$$

This completes the proof.

Theorem 3.8 *The trigonometric series*

$$\sum_{k=-\infty}^{k=\infty} a_k e^{ikx}, \quad x \in \mathbb{R},$$

where $a_k \in \mathbb{C}, k \in \mathbb{Z}$, satisfy

$$|a_k| \le A(1 + |k|)^m, \quad k \in \mathbb{Z},$$

for some $m \in \mathbb{N} \cup \{0\}$, is convergent in $\mathscr{D}'(\mathbb{R})$.

Proof Since the series

$$\frac{a_0 x^{2m+2}}{(m+2)!} + \sum_{k \neq 0} \frac{a_k}{(ik)^{m+2}} e^{ikx}, \quad x \in \mathbb{R},$$

is uniformly convergent in $\mathbb{R}$, the series which is a derivative of it of order $m + 2$ is convergent in $\mathscr{D}'(\mathbb{R})$. This completes the proof.

Theorem 3.9 *We have*

$$\frac{1}{2\pi} \sum_k e^{ikx} = \sum_k \delta(x - 2k\pi).$$

Proof The function

$$f_0(x) = \frac{x}{2} - \frac{x^2}{4\pi}, \quad 0 \le x < 2\pi,$$

has Fourier series

$$f_0(x) = \frac{\pi}{6} - \frac{1}{2\pi} \sum_{k \ne 0} \frac{e^{ikx}}{k^2}, \quad x \in \mathbb{R}.$$

Using Theorem 3.7, we have

$$f_0'(x) = -\frac{i}{2\pi} \sum_{k \ne 0} \frac{e^{ikx}}{k} = \frac{1}{2} - \frac{x}{2\pi}$$

and

$$f_0''(x) = \frac{1}{2\pi} \sum_{k \ne 0} e^{ikx} = -\frac{1}{2\pi} + \sum_k \delta(x - 2k\pi).$$

From the last equation, we obtain

$$\frac{1}{2\pi} \sum_k e^{ikx} = \sum_k \delta(x - 2k\pi).$$

This completes the proof.

3.2   The Local Structure of Distributions

Theorem 3.10 *Let $u \in \mathscr{D}'(X)$ and $X_1 \subset X$ be a bounded set. Then there exist a function $f \in L^\infty(X_1)$ and an integer $m \ge 0$ such that*

$$u(x) = D_1^m \dots D_n^m f(x), \quad x \in X_1.$$

Proof Since $u \in \mathscr{D}'(X)$ and $X_1 \subset X$, there exist constants C and k so that

$$|u(\phi)| \leq C \sum_{|\alpha| \leq k} \sup_{x \in X_1} |D^\alpha \phi(x)|, \quad \phi \in \mathscr{C}_0^\infty(X_1).$$

With d we will denote the diameter of X_1. Applying the inequality

$$|\phi(x)| = \left| \int_{-\infty}^{x_j} \frac{\partial \phi}{\partial x_j}(x_1, \ldots, x_{j-1}, t, x_{j+1}, \ldots, x_n) dt \right|$$

$$\leq \int_{-\infty}^{x_j} \left| \frac{\partial \phi}{\partial x_j}(x_1, \ldots, x_{j-1}, t, .x_{j+1}, \ldots, x_n) \right| dt$$

$$\leq d \max_{x \in \overline{X_1}} \left| \frac{\partial \phi}{\partial x_j}(x) \right|, \quad x \in X_1, \quad j \in \{1, \ldots, n\},$$

a sufficiently number of times, we obtain

$$|u(\phi)| \leq c_1 \max_{x \in \overline{X_1}} |D_1^k \ldots D_n^k \phi(x)|, \quad \phi \in \mathscr{C}_0^\infty(X_1), \tag{3.2}$$

for some constant $c_1 > 0$. Note that, for any $\psi \in \mathscr{C}_0^\infty(X_1)$, we have

$$\psi(x) = \int_{-\infty}^{x_1} \ldots \int_{-\infty}^{x_n} \frac{\partial^n \psi}{\partial y_1 \ldots \partial y_n}(y) dy_n \ldots dy_1, \quad x \in X_1,$$

and then

$$|\psi(x)| \leq \int_{X_1} |D_1 \ldots D_n \psi(y)| dy, \quad x \in X_1.$$

From here and from (3.2), for $m = k + 1$, we find

$$|u(\phi)| \leq c_2 \int_{X_1} |D_1^m \ldots D_n^m \phi(x)| dx, \quad \phi \in \mathscr{C}_0^\infty(X_1),$$

for some constant $c_2 > 0$. For $\phi \in \mathscr{C}_0^\infty(X_1)$, denote

$$\chi(x) = (-1)^{mn} D_1^m \ldots D_n^m \phi(x), \quad x \in X_1,$$

and define the functional

$$u^*(\chi) = u(\phi).$$

By the Hahn–Banach theorem, it follows that there exist an extension $\tilde{u}$ of u^* on $L^1(X_1)$ such that

$$|\tilde{u}(\phi)| = |u(\phi)| \le c_3 \|\chi\|_1$$

for some constant $c_3 > 0$. By a F. Riesz theorem, we find that there is a function $f \in L^\infty(X_1)$ so that $\|f\|_\infty \le c_4$ and

$$\tilde{u}(\chi) = \int_{X_1} f(x)\chi(x)dx.$$

Hence and the definitions of χ, u^* and $\tilde{u}$, we arrive at

$$u(\phi) = (-1)^{mn} \int_{X_1} f(x) D_1^m \dots D_n^m \phi(x) dx$$

$$= \int_{X_1} D_1^m \dots D_n^m f(x)\phi(x)dx, \quad \phi \in \mathscr{C}_0^\infty(X_1).$$

This completes the proof.

Definition 3.2 We say that the sequence $\{\phi_k\}_{k=1}^\infty \subset \mathscr{C}^\infty(X)$ converges to 0 in $\mathscr{C}^\infty(X)$ if $D^\alpha \phi_k \to_{k\to\infty} 0$ uniformly on any $X_1 \subset X$ for any $\alpha \in \mathbb{N}^n \cup \{0\}$.

Note that the convergence in $\mathscr{C}_0^\infty(X)$ implies the convergence in $\mathscr{C}^\infty(X)$, but not vice versa.

Theorem 3.11 *For a distribution $u \in \mathscr{D}'(X)$ to have a compact support in X, it is necessary and sufficient that it admits of a linear and continuous extension onto $\mathscr{C}^\infty(X)$.*

Proof

1. Suppose that $u \in \mathscr{D}'(X)$ has a compact support in X, i.e., $\mathrm{supp}\, u = K \subset\subset X$. Take $\eta \in \mathscr{C}_0^\infty(X)$ such that $\eta \equiv 1$ in a neighbourhood of K. Define the functional

$$\tilde{u}(\phi) = u(\eta\phi), \quad \phi \in \mathscr{C}^\infty(X). \tag{3.3}$$

Let $\phi_1, \phi_2 \in \mathscr{C}^\infty(X)$ and $\alpha_1, \alpha_2 \in \mathbb{C}$. Then

$$\tilde{u}(\alpha_1\phi_1 + \alpha_2\phi_2) = u(\eta(\alpha_1\phi_1 + \alpha_2\phi_2))$$

$$= u(\alpha_1\eta\phi_1 + \alpha_2\eta\phi_2)$$

$$= \alpha_1 u(\eta \phi_1) + \alpha_2 u(\eta \phi_2)$$

$$= \alpha_1 \widetilde{u}(\phi_1) + \alpha_2 \widetilde{u}(\phi_2),$$

i.e., $\widetilde{u}$ is a linear functional on $\mathscr{C}^\infty(X)$. Since the map $\phi \mapsto \eta \phi$ is a continuous map from $\mathscr{C}^\infty(X)$ into $\mathscr{C}_0^\infty(X)$, we conclude that $\widetilde{u}$ is a continuous functional on $\mathscr{C}^\infty(X)$. Next, for $\phi \in \mathscr{C}_0^\infty(X)$, we have

$$\widetilde{u}(\phi) = u(\eta \phi) = (\eta u)(\phi) = u(\phi).$$

Thus, $\widetilde{u}$ is an extension of u from $\mathscr{C}_0^\infty(X)$ onto $\mathscr{C}^\infty(X)$. Now, we will prove that this extension is unique. Suppose that $\widetilde{\widetilde{u}}$ is another extension of u onto $\mathscr{C}^\infty(X)$. Let X_k, $k \in \mathbb{N}$, be a sequence of bounded subsets of X such that $X = \cup_{k=1}^\infty X_k$ and $X_1 \subset X_2 \subset \ldots$. Let $\{\eta_k\}_{k=1}^\infty$ be a sequence of elements of $\mathscr{C}_0^\infty(X)$ such that $\eta_k \equiv 1$ on X_k, $k \in \mathbb{N}$, and $\eta_k \to_{k \to \infty} 1$ in $\mathscr{C}^\infty(X)$. Therefore, for any $\phi \in \mathscr{C}^\infty(X)$, we have $\eta_k \phi \to_{k \to \infty} \phi$ in $\mathscr{C}^\infty(X)$. From here, we get

$$\widetilde{u}(\phi) = \widetilde{u}\left(\lim_{k \to \infty} (\eta_k \phi) \right)$$

$$= \lim_{k \to \infty} \widetilde{u}(\eta_k \phi)$$

$$= \lim_{k \to \infty} u(\eta_k \phi)$$

$$= \lim_{k \to \infty} \widetilde{\widetilde{u}}(\eta_k \phi)$$

$$= \widetilde{\widetilde{u}}\left(\lim_{k \to \infty} (\eta_k \phi) \right)$$

$$= \widetilde{\widetilde{u}}(\phi), \quad \phi \in \mathscr{C}^\infty(X).$$

So, $\widetilde{u} = \widetilde{\widetilde{u}}$.

2. Let $u \in \mathscr{D}'(X)$ admits of a linear and continuous extension $\widetilde{u}$ onto $\mathscr{C}^\infty(X)$. Suppose that $\operatorname{supp} u$ is not a compact set. Then there exists a sequence $\{\phi_k\}_{k=1}^\infty$ of elements of $\mathscr{C}_0^\infty(X)$ such that $\operatorname{supp}\phi_k \subset X \backslash \overline{X_k}$, $X_1 \subset X_2 \subset \ldots$, $\cup_{k=1}^\infty X_k = X$, and $u(\phi_k) = 1$, $k \in \mathbb{N}$. Note that $\phi_k \to_{k \to \infty} 0$ in $\mathscr{C}^\infty(X)$ and then $\widetilde{u}(\phi_k) \to_{k \to \infty} 0$, which is a contradiction because

$$\widetilde{u}(\phi_k) = u(\phi_k) = 1, \quad k \in \mathbb{N}.$$

This completes the proof.

Theorem 3.12 *Any distribution $u \in \mathscr{D}'(X)$ with compact support has a finite order in X.*

Proof Since $u \in \mathscr{D}'(X)$ has a compact support, it admits of a linear and continuous extension $\tilde{u}$ onto $\mathscr{C}^{\infty}(X)$ and (3.3) holds. Because $\eta \in \mathscr{C}_0^{\infty}(X)$ and $\operatorname{supp}\eta \subset X$, there exists an $X_1 \subset X$ so that $\eta \in \mathscr{C}_0^{\infty}(X_1)$. Therefore $\eta\phi \in \mathscr{C}_0^{\infty}(X_1)$ for any $\phi \in \mathscr{C}^{\infty}(X)$. From the last conclusion, it follows that there exist constants $K = K(X_1)$ and $m = m(X_1)$ so that

$$|\tilde{u}(\phi)| = |u(\eta\phi)| \leq K_1 \sum_{|\alpha|\leq m} \sup_{x\in X_1} |D^{\alpha}(\eta\phi)(x)|,$$

$\phi \in \mathscr{C}_0^{\infty}(X)$. Hence,

$$|\tilde{u}(\phi)| \leq C_1 \sum_{|\alpha|\leq m} \sup_{x\in X_1} |D^{\alpha}\phi(x)|, \quad \phi \in \mathscr{C}_0^{\infty}(X),$$

for some constant $C_1 > 0$. This completes the proof.

Theorem 3.13 *Let $0 \in X$, $u \in \mathscr{D}'(X)$ and $\operatorname{supp}u = \{0\}$. Then u is uniquely representable in the following form*

$$u(x) = \sum_{|\alpha|\leq k} c_{\alpha} D^{\alpha}\delta(x), \quad x \in X, \tag{3.4}$$

where k is the order of u and $c_{\alpha} \in \mathbb{C}$, $|\alpha| \leq k$.

Proof Let $\eta \in \mathscr{C}_0^{\infty}(U_1)$ be such that $\eta \equiv 1$ in $U_{\frac{1}{2}}$. For $\phi \in \mathscr{C}_0^{\infty}(X)$, define

$$\phi_N(x) = \sum_{|\alpha|\leq N} \frac{D^{\alpha}\phi}{\alpha!}(0)x^{\alpha}, \quad x \in X.$$

Then, for any $\epsilon > 0$, we have $u = \eta\left(\dfrac{x}{\epsilon}\right)u$. Hence,

$$\begin{aligned}
u(\phi) &= \left(\eta\left(\frac{x}{\epsilon}\right)u\right)(\phi)\\
&= u\left(\eta\left(\frac{x}{\epsilon}\right)(\phi - \phi_N)\right) + u\left(\eta\left(\frac{x}{\epsilon}\right)\phi_N\right).
\end{aligned} \tag{3.5}$$

Note that $\eta\left(\dfrac{x}{\epsilon}\right)(\phi - \phi_N) \in \mathcal{C}_0^\infty(U_\epsilon)$. Then

$$\left| u\left(\eta\left(\frac{x}{\epsilon}\right)(\phi - \phi_N)\right)\right| \leq C \max_{\substack{|x| \leq \epsilon \\ |\alpha| \leq N}} \left| D^\alpha\left(\eta\left(\frac{x}{\epsilon}\right)(\phi(x) - \phi_N(x))\right)\right|$$

$$\leq C \max_{\substack{|x| \leq \epsilon \\ |\alpha| \leq N}} \sum_{\beta \leq \alpha} \binom{\alpha}{\beta}\left| D^\beta \eta\left(\frac{x}{\epsilon}\right) D^{\alpha-\beta}(\phi(x) - \phi_N(x))\right|$$

$$\leq C_1 \max_{|\alpha \leq N} \sum_{\beta \leq \alpha} \epsilon^{|\beta|}\epsilon^{N-|\alpha-\beta|}\epsilon$$

$$\leq C_1 \max_{|\alpha| \leq N} \epsilon^{N-|\alpha|+1}$$

$$\leq C_2\epsilon,$$

for some positive constants C, C_1, C_2. Let $\tilde{u}$ is an extension of u onto $\mathcal{C}^\infty(X)$. Now, employing (3.5), we arrive at

$$u(\phi) = \tilde{u}(\phi) = \sum_{|\alpha| \leq k} \frac{D^\alpha \phi}{\alpha!}(0)\tilde{u}\left(x^\alpha\right).$$

Set

$$c_\alpha = \frac{(-1)^{|\alpha|}}{\alpha!}\tilde{u}\left(x^\alpha\right).$$

Then

$$u(\phi) = \sum_{|\alpha| \leq k} (-1)^{|\alpha|} c_\alpha D^\alpha \phi(0)$$

$$= \sum_{|\alpha| \leq k} (-1)^{|\alpha|} c_\alpha \delta\left(D^\alpha \phi\right)$$

$$= \sum_{|\alpha| \leq k} c_\alpha D^\alpha \delta(\phi).$$

Now, we will prove the uniqueness of the representation (3.4). Assume that there is another such representation

$$u(x) = \sum_{|\alpha| \leq k} c'_\alpha D^\alpha \delta(x), \quad x \in X.$$

Hence,

$$0 = \sum_{|\alpha| \leq k} \left(c'_\alpha - c_\alpha \right) D^\alpha \delta(x), \quad x \in X,$$

and then

$$0 = \sum_{|\alpha| \leq k} \left(c'_\alpha - c_\alpha \right) D^\alpha \delta(x^m)$$

$$= \sum_{|\alpha| \leq k} \left(c'_\alpha - c_\alpha \right) (-1)^{|\alpha|} D^\alpha x^m \Big|_{x=0}$$

$$= (-1)^m m! \left(c'_m - c_m \right), \quad x \in X,$$

i.e., $c'_m = c_m$. This completes the proof.

3.3   The Primitive of a Distribution

Let $(a, b) \subset \mathbb{R}$, $u \in \mathscr{D}'(a, b)$, $\phi \in \mathscr{C}_0^\infty(a, b)$ and $x_0 \in (a, b)$ be arbitrary but fixed. We also fix $\epsilon > 0$ such that $\epsilon < \min\{x_0 - a, b - x_0\}$. The function ϕ can be represented as

$$\phi(x) = \psi'(x) + \omega_\epsilon(x - x_0) \int_{-\infty}^{\infty} \phi(\xi)d\xi, \quad x \in (a, b), \tag{3.6}$$

where ψ is determined by the equality

$$\psi(x) = \int_{-\infty}^{x} \left(\phi(s) - \omega_\epsilon(s - x_0) \int_{-\infty}^{\infty} \phi(\xi)d\xi \right) ds, \quad x \in (a, b). \tag{3.7}$$

Suppose $\text{supp}\phi \subset [a', b'] \subset (a, b)$. Since $\phi \in \mathscr{C}_0^\infty(a, b)$ and $\omega_\epsilon \in \mathscr{C}^\infty(a, b)$, we have that $\psi \in \mathscr{C}^\infty(a, b)$. Moreover, $\psi(x) \equiv 0$ if $x < a'' = \min\{a', x_0 - \epsilon\}$. As $\epsilon < \{x_0 - a, b - x_0\}$, it follows $\epsilon < x_0 - a$ and $a < x_0 - \epsilon$, and since $[a', b'] \subset (a, b)$, we get $a < a'$ and $a'' > a$. For $x > b'' = \max\{b', x_0 + \epsilon\} < b$, using (3.7), we have

$$\psi(x) = \int_{-\infty}^{\infty} \left(\phi(s) - \omega_\epsilon(s - x_0) \int_{-\infty}^{\infty} \phi(\xi)d\xi \right) ds$$

$$= \int_{-\infty}^{\infty} \phi(s)ds - \int_{-\infty}^{\infty} \omega_\epsilon(s - x_0) \int_{-\infty}^{\infty} \phi(\xi)d\xi ds$$

$$= \int\limits_{-\infty}^{\infty} \phi(s)ds - \int\limits_{-\infty}^{\infty} \omega_\epsilon(s - x_0)ds \int\limits_{-\infty}^{\infty} \phi(\xi)d\xi$$

$$= \int\limits_{-\infty}^{\infty} \phi(s)ds - \int\limits_{-\infty}^{\infty} \phi(\xi)d\xi = 0.$$

Therefore $\mathrm{supp}\,\psi \subset [a'', b'']$ and $\psi \in \mathscr{C}_0^\infty(a, b)$.

Definition 3.3 A distribution $u^{(-1)} \in \mathscr{D}'(a, b)$ is said to be a primitive of the distribution u if

$$\left(u^{(-1)}\right)' = u \quad \text{in} \quad \mathscr{D}'(a, b).$$

We assume that the primitive $u^{(-1)}$ of the distribution $u \in \mathscr{D}'(a, b)$ exists in $\mathscr{D}'(a, b)$. Then we have the representation

$$u^{(-1)}(\phi) = u^{(-1)}\left(\psi' + \omega_\epsilon(x - x_0) \int\limits_{-\infty}^{\infty} \phi(\xi)d\xi\right)$$

$$= u^{(-1)}(\psi') + u^{(-1)}\left(\omega_\epsilon(x - x_0) \int\limits_{-\infty}^{\infty} \phi(\xi)d\xi\right)$$

$$= -\left(u^{(-1)}\right)'(\psi) + u^{(-1)}(\omega_\epsilon(x - x_0)) \int\limits_{-\infty}^{\infty} \phi(\xi)d\xi$$

where $\phi \in \mathscr{C}_0^\infty(a, b)$ and $\psi \in \mathscr{C}_0^\infty(a, b)$ satisfy (3.7). Setting

$$C = u^{(-1)}(\omega_\epsilon(x - x_0)) = \text{const},$$

we obtain

$$u^{(-1)}(\phi) = -u(\psi) + C \int\limits_{-\infty}^{\infty} \phi(\xi)d\xi. \tag{3.8}$$

Now, we will show that if the functional $u^{(-1)}$ satisfies (3.8) for an arbitrary constant C, then it is first of all a distribution in $\mathscr{D}'(a, b)$, and also a primitive for $u \in \mathscr{D}'(a, b)$.

Let $\alpha_1, \alpha_2 \in \mathbb{C}, \phi_1, \phi_2 \in \mathscr{C}_0^\infty(a, b)$. Take $\psi_1, \psi_2 \in \mathscr{C}_0^\infty(a, b)$ such that

$$u^{(-1)}(\phi_1) = -u(\psi_1) + C \int_{-\infty}^{\infty} \phi_1(\xi)d\xi,$$

$$u^{(-1)}(\phi_2) = -u(\psi_2) + C \int_{-\infty}^{\infty} \phi_2(\xi)d\xi,$$

$$\psi_1(x) = \int_{-\infty}^{x} \left(\phi_1(s) - \omega_\epsilon(s - x_0) \int_{-\infty}^{\infty} \phi_1(\xi)d\xi \right)ds,$$

$$\psi_2(x) = \int_{-\infty}^{x} \left(\phi_2(s) - \omega_\epsilon(s - x_0) \int_{-\infty}^{\infty} \phi_2(\xi)d\xi \right)ds, \quad x \in (a, b).$$

Then we get

$$u^{(-1)}(\alpha_1\phi_1 + \alpha_2\phi_2) = -u(\alpha_1\psi_1 + \alpha_2\psi_2) + C \int_{-\infty}^{\infty} (\alpha_1\phi_1(\xi) + \alpha_2\phi_2(\xi))d\xi$$

$$= -\alpha_1 u(\psi_1) - \alpha_2 u(\psi_2) + C\alpha_1 \int_{-\infty}^{\infty} \phi_1(\xi)d\xi + C\alpha_2 \int_{-\infty}^{\infty} \phi_2(\xi)d\xi$$

$$= \alpha_1 u^{(-1)}(\phi_1) + \alpha_2 u^{(-1)}(\phi_2).$$

Consequently $u^{(-1)}$ is a linear functional on $\mathscr{C}_0^\infty(a, b)$.
Let now $\{\phi_n\}_{n=1}^\infty$ be a sequence in $\mathscr{C}_0^\infty(a, b)$ with $\phi_n \to 0, n \to \infty$, in $\mathscr{C}_0^\infty(a, b)$.
Choose $\psi_n \in \mathscr{C}_0^\infty(a, b)$ so that

$$u^{(-1)}(\phi_n) = -u(\psi_n) + C \int_{-\infty}^{\infty} \phi_n(\xi)d\xi,$$

$$\psi_n(x) = \int_{-\infty}^{x} \left(\phi_n(s) - \omega_\epsilon(s - x_0) \int_{-\infty}^{\infty} \phi_n(\xi)d\xi \right)ds.$$

Then $\psi_n \to 0, n \to \infty$, in $\mathscr{C}_0^\infty(a, b)$. From here,

$$u^{(-1)}(\phi_n) \to_{n \to \infty} 0.$$

Therefore $u^{(-1)}$ is a linear continuous functional on $\mathscr{C}_0^\infty(a, b)$, i.e., $u^{(-1)} \in \mathscr{D}'(a, b)$.

Now, we will show that $u^{(-1)}$ is a primitive of u. We replace ϕ by ϕ' in (3.8) and we get

$$u^{(-1)}(\phi') = -u(\psi) + C \int_{-\infty}^{\infty} \phi'(\xi)d\xi = -u(\psi), \qquad (3.9)$$

where

$$\psi(x) = \int_{-\infty}^{x} \Big(\phi'(s) - \omega_\epsilon(s - x_0) \int_{-\infty}^{\infty} \phi'(\xi)d\xi\Big)ds = \int_{-\infty}^{x} \phi'(s)ds = \phi(x), \quad x \in (a, b).$$

The last relation and (3.9) imply $u^{(-1)}(\phi') = -u(\phi)$, so $-\Big(u^{(-1)}\Big)'(\phi) = -u(\phi)$, i.e., $\Big(u^{(-1)}\Big)'(\phi) = u(\phi)$. Since $\phi \in \mathscr{C}_0^\infty(a, b)$ was arbitrary, we conclude $\Big(u^{(-1)}\Big)' = u$. The solution to the equation

$$u' = f, \qquad (3.10)$$

for $u, f \in \mathscr{D}'(a, b)$, can be represented in the form

$$u = f^{(-1)} + C, \qquad (3.11)$$

where C is an arbitrary constant. If $f \in \mathscr{C}(a, b)$, (3.11) is a classical solution of the Eq. (3.10).

Proceeding as above, we can define successive primitives $u^{(-n)}$ using the relationship $u^{(-n)'} = u^{(-n-1)}$.

3.4 Simple and Double Layers on Surfaces

Definition 3.4 Let S be a piecewise smooth surface in $\mathbb{R}^n$ and let f be a continuous function on S. We introduce the generalized function $f\delta_S$ in the following manner

$$f\delta_S(\phi) = \int_S f(x)\phi(x)dS, \qquad \phi \in \mathscr{C}_0^\infty(S).$$

The generalized function $f\delta_S$ is called a simple layer on the surface S.

By the definition, it follows that $f\delta_S \in \mathscr{D}'(S)$, $f\delta_S(x) = 0$ for $x \notin S$ and so, $\text{supp} f\delta_S \subset S$.

Theorem 3.14 *Let $\Omega \subset \mathbb{R}^n$ be a bounded domain with boundary $\partial\Omega = S$ and write $\Omega = \mathbb{R}^n \setminus \Omega_1$. Consider*

$$f \in \mathscr{C}^1(\overline{\Omega}) \cap \mathscr{C}^1(\overline{\Omega_1}), \, [f]_S(x) = \lim_{\substack{x' \to x \\ x' \in \Omega_1}} f(x') - \lim_{\substack{x'' \to x \\ x'' \in \Omega}} f(x''), \quad x \in S.$$

With $\left\{\dfrac{\partial f}{\partial x_i}\right\}(x)$, $i = 1, 2, \ldots, n$, we will denote the classical derivatives of f at $x \in \mathbb{R}^n$, $x \notin S$, while $\dfrac{\partial}{\partial x_i}$, $i = 1, 2, \ldots, n$, will denote derivatives in $\mathscr{D}'(\mathbb{R}^n)$. Then for every $i = 1, 2, \ldots, n$,

$$\frac{\partial f}{\partial x_i} = \left\{\frac{\partial f}{\partial x_i}\right\} + [f]_S \cos(n, x_i)\delta_S, \quad f \in \mathscr{D}'(\mathbb{R}^n),$$

where n is the outer normal to S.

Proof Let $\phi \in \mathscr{C}_0^\infty(\mathbb{R}^n)$. The Gauss theorem tells us that

$$\frac{\partial f}{\partial x_i}(\phi) = -f\left(\frac{\partial \phi}{\partial x_i}\right)$$

$$= -\int_{\mathbb{R}^n} f(x)\frac{\partial \phi}{\partial x_i}(x)dx$$

$$= -\int_{\Omega} f(x)\frac{\partial \phi}{\partial x_i}(x)dx - \int_{\Omega_1} f(x)\frac{\partial \phi}{\partial x_i}(x)dx$$

$$= -\int_{\Omega} f(x)\frac{\partial \phi}{\partial x_i}(x)dx - \int_{\Omega}\left\{\frac{\partial f}{\partial x_i}\right\}(x)\phi(x)dx + \int_{\Omega}\left\{\frac{\partial f}{\partial x_i}\right\}(x)\phi(x)dx$$

$$- \int_{\Omega_1} f(x)\frac{\partial \phi}{\partial x_i}(x)dx - \int_{\Omega_1}\left\{\frac{\partial f}{\partial x_i}\right\}(x)\phi(x)dx + \int_{\Omega_1}\left\{\frac{\partial f}{\partial x_i}\right\}(x)\phi(x)dx$$

$$= \int_{S} [f]_S \cos(n, x_i)\phi(x)dS + \int_{\mathbb{R}^n}\left\{\frac{\partial f}{\partial x_i}\right\}(x)\phi(x)dx$$

$$= [f]_S \cos(n, x_i)\delta_S(\phi) + \left\{\frac{\partial f}{\partial x_i}\right\}(\phi).$$

As $\phi \in \mathscr{C}_0^\infty(\mathbb{R}^n)$ was arbitrary, we conclude

$$\frac{\partial f}{\partial x_i} = \left\{\frac{\partial f}{\partial x_i}\right\} + [f]_S \cos(n, x_i)\delta_S.$$

This completes the proof.

Theorem 3.15 *Consider the plane* $\mathbb{R}^2$ *with complex coordinate* $z = x + iy$, *and the differential form* $dz = dx + i\,dy$ *annihilating the Cauchy–Riemann operator*

$$\frac{\partial}{\partial \bar{z}} = \frac{1}{2}\left(\frac{\partial}{\partial x} + i\frac{\partial}{\partial y}\right).$$

Let Ω *be a bounded domain in* $\mathbb{R}^2$ *with piecewise-smooth boundary* S. *We take* $f \in \mathscr{C}^1(\overline{\Omega})$ *with* $f = 0$ *on* $\mathbb{R}^2 \backslash \Omega$. *If* $\dfrac{\partial f}{\partial \bar{z}}$ *is the derivative of* f *in the sense of distributions and* $\left\{\dfrac{\partial f}{\partial \bar{z}}\right\}$ *is the classical derivative of* f *at* z, $z \notin S$, *then*

$$\frac{\partial f}{\partial \bar{z}} = \left\{\frac{\partial f}{\partial \bar{z}}\right\} + \frac{1}{2}f\left(\cos(nx) + i\cos(ny)\right)\delta_S.$$

Proof As

$$\frac{\partial f}{\partial \bar{z}} = \frac{1}{2}\frac{\partial f}{\partial x} + \frac{i}{2}\frac{\partial f}{\partial y}, \tag{3.12}$$

applying Theorem 3.14 to $\dfrac{\partial f}{\partial x}$ and $\dfrac{\partial f}{\partial y}$ gives

$$\frac{\partial f}{\partial x} = \left\{\frac{\partial f}{\partial x}\right\} + f\cos(n,x)\delta_S,$$
$$\frac{\partial f}{\partial y} = \left\{\frac{\partial f}{\partial y}\right\} + f\cos(n,y)\delta_S.$$

From here and (3.12), it follows

$$\frac{\partial f}{\partial \bar{z}} = \frac{1}{2}\left\{\frac{\partial f}{\partial x}\right\} + \frac{i}{2}\left\{\frac{\partial f}{\partial y}\right\} + \frac{1}{2}f\cos(n,x)\delta_S + \frac{i}{2}f\cos(n,y)\delta_S$$
$$= \left\{\frac{\partial f}{\partial \bar{z}}\right\} + \frac{1}{2}f\left(\cos(n,x) + i\cos(n,y)\right)\delta_S.$$

This completes the proof.

Definition 3.5 Let S be a piecewise smooth two-sided surface with normal n, and ν a continuous function on S. Define the functional $-\dfrac{\partial}{\partial n}(f\delta_S)$ on $\mathscr{C}_0^\infty(X)$ in the following way

$$-\frac{\partial}{\partial n}(f\delta_S)(\phi) = \int\limits_S f(x)\frac{\partial \phi(x)}{\partial n}dS, \quad \phi \in \mathscr{C}_0^\infty(X).$$

Exercise 3.2 Prove that $-\dfrac{\partial}{\partial n}(f\delta_S) \in \mathcal{D}'(X)$.

Exercise 3.3 Prove that $\text{supp}\left(-\dfrac{\partial}{\partial n}(f\delta_S)\right) \subset S$.

Definition 3.6 Let S be a piecewise smooth two-sided surface with normal n, f is a continuous function on S. The distribution $-\dfrac{\partial}{\partial n}(f\delta_S)$ is called a double layer on the surface S.

Theorem 3.16 *Let* $\Omega \subset \mathbb{R}^n$ *be a bounded domain with boundary* $\partial\Omega = S$, $\Omega = \mathbb{R}^n\backslash\Omega_1$,

$$f \in \mathscr{C}^1(\overline{\Omega})\cap\mathscr{C}^1(\overline{\Omega_1}), \quad [f]_S(x) = \lim_{x'\to x, x'\in\Omega_1} f(x') - \lim_{x''\to x, x''\in\Omega} f(x''), \quad x \in S.$$

By $\left\{\dfrac{\partial f}{\partial x_i}\right\}(x)$, $i = 1, 2, \ldots, n$, *we denote the classical derivatives of* f *at the point* $x \in \mathbb{R}^n$, $x \notin S$, *and by* $\dfrac{\partial}{\partial x_i}$, $i = 1, 2, \ldots, n$, *the derivatives in* $\mathscr{D}'(\mathbb{R}^n)$. *If* n *is the outer normal to* S, $\dfrac{\partial}{\partial n}$ *is the normal derivative,* $\left[\dfrac{\partial f}{\partial n}\right]_S(x) = \lim_{x'\to x, x'\in\Omega_1} \dfrac{\partial f}{\partial n}(x') - \lim_{x''\to x, x''\in\Omega} \dfrac{\partial f}{\partial n}(x'')$ *and* $\left[\left\{\dfrac{\partial f}{\partial x_i}\right\}\right]_S(x) = \lim_{x'\to x, x'\in\Omega_1} \left\{\dfrac{\partial f}{\partial x_i}\right\}(x') - \lim_{x''\to x, x''\in\Omega} \left\{\dfrac{\partial f}{\partial x_i}\right\}(x'')$, $x \in S$, $i = 1, 2, \ldots, n$. *Then*

$$\sum_{i=1}^{n} \frac{\partial}{\partial x_i}\left([f]_S \cos(n, x_i)\delta_S\right) = \frac{\partial}{\partial n}\left([f]_S\,\delta_S\right).$$

Proof Let $\phi \in \mathscr{C}_0^\infty(\mathbb{R}^n)$ be arbitrarily chosen. Then

$$\sum_{i=1}^{n}\left(\frac{\partial}{\partial x_i}\left([f]_S\cos(n, x_i)\delta_S\right), \phi\right) = -\sum_{i=1}^{n}\left([f]_S\cos(n, x_i)\delta_S, \frac{\partial\phi}{\partial x_i}\right)$$

$$= -\sum_{i=1}^{n}\int_S [f]_S\cos(n, x_i)\frac{\partial\phi}{\partial x_i}dS$$

$$= -\sum_{i=1}^{n}\int_S [f]_S\frac{\partial\phi}{\partial n_i}dS$$

$$= -\sum_{i=1}^{n}\left([f]_S\,\delta_S, \frac{\partial\phi}{\partial n_i}\right)$$

$$= \sum_{i=1}^{n} \left(\frac{\partial}{\partial n_i} \left([f]_S \, \delta_S \right), \phi \right)$$

$$= \left(\sum_{i=1}^{n} \frac{\partial}{\partial n_i} \left([f]_S \, \delta_S \right), \phi \right)$$

$$= \left(\frac{\partial}{\partial n} \left([f]_S \, \delta_S \right), \phi \right).$$

This completes the proof.

3.5   Advanced Practical Problems

Problem 3.1 Compute

$$\frac{d^3}{dx^3} |x|, \quad x \in \mathbb{R}.$$

Answer $2\delta'(x)$.

Problem 3.2 Compute

$$x^m \delta^{(k)}(x).$$

Answer

$$x^m \delta^{(k)}(x) = \begin{cases} 0 & \text{for} \quad k < m; \\ (-1)^m m! \delta & \text{for} \quad k = m; \\ (-1)^m \binom{k}{m} m! \delta^{(k-m)} & \text{for} \quad k > m. \end{cases}$$

Problem 3.3 Prove that

$$\delta(x) + \delta'(x-1) + \delta''(x-2) + \cdots$$

converges in $\mathscr{D}'(X)$ and that it has a finite order.

Solution Let $\phi \in \mathscr{C}_0^\infty(X)$. Then

$$\sum_{j=0}^{\infty} \delta_j^{(j)}(\phi) = \sum_{j=0}^{\infty} (-1)^j \phi^{(j)}(j),$$

$$\left| \sum_{j=0}^{\infty} \delta_j^{(j)}(\phi) \right| \leq \sum_{j=0}^{\infty} \left| \phi^{(j)}(j) \right| < \infty.$$

Since ϕ has compact support, only a finite number of terms in $\displaystyle\sum_{j=0}^{\infty} \left| \phi^{(j)}(j) \right|$ are nonzero. Therefore $\delta(x) + \delta'(x - 1) + \cdots$ has a finite order.

Problem 3.4 Prove

1. $\displaystyle\frac{2}{\pi} \sum_{k=0}^{\infty} \cos(2k + 1)x = \sum_{k=-\infty}^{\infty} (-1)^k \delta(x - k\pi),$

2. $|\sin x|'' + |\sin x| = 2 \displaystyle\sum_{k=-\infty}^{\infty} \delta(x - k\pi).$

Problem 3.5 Prove

1. $\displaystyle\frac{d}{dx} \log |x| = P\frac{1}{x},$

2. $\displaystyle\frac{d}{dx} P\frac{1}{x} = -P\frac{1}{x^2},$

3. $\displaystyle\frac{d}{dx} \frac{1}{x - i0} = i\pi \delta'(x) - P\frac{1}{x^2},$

4. $\displaystyle\frac{d}{dx} \frac{1}{x + i0} = -i\pi \delta'(x) - P\frac{1}{x^2}.$

Hint
3. Use

$$\frac{1}{x - i.0} = P\frac{1}{x} + i\pi \delta.$$

4. Use

$$\frac{1}{x + i.0} = P\frac{1}{x} - i\pi \delta.$$

Problem 3.6 Compute the first and the second derivative of the following functions in $\mathscr{D}'(\mathbb{R})$

1. $u(x) = \begin{cases} \sin x & x \geq 0, \\ \cos x - 1 & x \leq 0, \end{cases}$

2. $u(x) = \begin{cases} x - 1 & x \geq 0, \\ -1 & -1 \leq x \leq 0, \\ -x^2 & x \leq -1, \end{cases}$

3. $u(x) = \begin{cases} x^4 & -1 \leq x \leq 1, \\ 0 & |x| > 1, \end{cases}$

4. $u(x) = \begin{cases} x^2 + x + 1 & -1 \leq x \leq 1, \\ 0 & |x| > 1. \end{cases}$

1. *Solution* Let $\phi \in \mathscr{C}_0^\infty(\mathbb{R})$. Then

$$u'(\phi) = -u(\phi')$$

$$= -\int_{-\infty}^{\infty} u(x)\phi'(x)dx$$

$$= -\int_0^{\infty} \sin x \phi'(x)dx - \int_{-\infty}^0 (\cos x - 1)\phi'(x)dx$$

$$= -\sin x \phi(x)\Big|_{x=0}^{x=\infty} + \int_0^{\infty} \cos x \phi(x)dx - (\cos x - 1)\phi(x)\Big|_{x=-\infty}^{x=0} - \int_{-\infty}^0 \sin x \phi(x)dx$$

$$= H(x)(\cos x\phi) - H(-x)(\sin x\phi) = \cos x H(x)(\phi) - \sin x H(-x)(\phi).$$

Since $\phi \in \mathscr{C}_0^\infty(\mathbb{R})$ is arbitrary, we conclude that

$$u' = \cos x H(x) - H(-x) \sin x.$$

As $H'(x) = \delta(x)$, $H'(-x) = \delta(-x)$, $\cos x\delta(x) = \delta(x)$, $\sin x\delta(x) = 0$, the second derivative reads

$$u'' = -\sin x H(x) + \cos x H'(x) - \cos x H(-x) - \sin x H'(-x)$$

$$= -\sin x H(x) + \cos x\delta(x) - \cos x H(-x) + \sin x\delta(x)$$

$$= -\sin x H(x) - \cos x H(-x) + \delta(x).$$

2. **Answer**

$$u' = -2x H(-x - 1) + H(x),$$

$$u'' = -2H(-x - 1) + 2\delta(-x - 1) + \delta(x),$$

3. **Answer**

$$u' = -\delta(x - 1) + \delta(x + 1) + 4x^3 H(x + 1) - 4x^3 H(x - 1),$$

$$u'' = -\delta'(x - 1) + \delta'(x + 1) + 12x^2 H(x + 1)$$

$$-12x^2 H(x - 1) - 4\delta(x + 1) - 4\delta(x - 1),$$

4. **Answer**

$$u' = -3\delta(x - 1) + \delta(x + 1) + (2x + 1)H(x + 1) - (2x + 1)H(x - 1),$$

$$u'' = -3\delta'(x - 1) + \delta'(x + 1) + 2H(x + 1) - 2H(x - 1) - \delta(x + 1) - 3\delta(x - 1).$$

Problem 3.7 Compute

1. $\dfrac{d^3}{dx^3} H(x + 1),$

2. $\dfrac{d}{dx}(x\,\mathrm{sign}x),$

3. $\dfrac{d}{dx}\Big((\sin x + \cos x)H(x + 2)\Big)$

in the space $\mathscr{D}'(\mathbb{R})$.

Answer
1. $\delta''(x + 1),$
2. $H(x) - H(-x),$
3. $(\cos 2 - \sin 2)\delta(x + 2) + (\cos x - \sin x)H(x + 2).$

Problem 3.8 Compute the first, second and third derivatives of the function

$$u(x) = |x| \sin(2x)$$

in the space $\mathscr{D}'(\mathbb{R})$.

Answer
1. $u' = -(\sin(2x) + 2x \cos(2x))H(-x) + (\sin(2x) + 2x \cos(2x))H(x),$
2. $u'' = -4(\cos(2x) - x \sin(2x))H(-x) + 4(\cos(2x) - x \sin(2x))H(x),$
3. $u''' = 4(3 \sin(2x) + 2x \cos(2x))H(-x) - 4(3 \sin(2x) + 2x \cos(2x))H(x) + 8\delta(x).$

Problem 3.9 Let $\Omega \subset \mathbb{R}^n$ be a bounded domain with boundary $\partial\Omega = S$, $\Omega = \mathbb{R}^n \backslash \Omega_1$,

$$f \in \mathscr{C}^2(\overline{\Omega}) \cap \mathscr{C}^2(\overline{\Omega_1}), \quad [f]_S(x) = \lim_{x' \to x, x' \in \Omega_1} f(x') - \lim_{x'' \to x, x'' \in \Omega} f(x''), \quad x \in S.$$

By $\left\{\dfrac{\partial f}{\partial x_i}\right\}(x)$, $\left\{\dfrac{\partial^2 f}{\partial x_i \partial x_j}\right\}(x)$, $i, j = 1, 2, \ldots, n$, we denote the classical derivatives

of f at the point $x \in \mathbb{R}^n$, $x \notin S$, $\left[\left\{\dfrac{\partial f}{\partial x_i}\right\}\right]_S (x) = \lim_{x' \to x, x' \in \Omega_1} \left\{\dfrac{\partial f}{\partial x_i}\right\}(x') - \lim_{x'' \to x, x'' \in \Omega}$

$\left\{\dfrac{\partial f}{\partial x_i}\right\}(x'')$, $x \in S$, $i = 1, 2, \ldots, n$, and by $\dfrac{\partial^2}{\partial x_i \partial x_j}$, $\dfrac{\partial}{\partial x_i}$, $i, j = 1, 2, \ldots, n$, the corresponding derivatives in $\mathscr{D}'(\mathbb{R}^n)$. Prove that for every $i, j \in \{1, 2, \ldots, n\}$,

1. $\dfrac{\partial^2 f}{\partial x_i \partial x_j} = \left\{\dfrac{\partial^2 f}{\partial x_i \partial x_j}\right\} + \dfrac{\partial}{\partial x_i}\left([f]_S \cos(n, x_j)\delta_S\right) + \left[\left\{\dfrac{\partial f}{\partial x_j}\right\}\right]_S \cos(n, x_i)\delta_S,$

2. $\Delta f = \left\{\Delta f\right\} + \displaystyle\sum_{i=1}^{n} \dfrac{\partial}{\partial x_i}\left([f]_S \cos(n, x_i)\delta_S\right) + \displaystyle\sum_{i=1}^{n}\left[\left\{\dfrac{\partial f}{\partial x_i}\right\}\right]_S \cos(n, x_i)\delta_S.$

Here n is the outer normal to S.

Problem 3.10 Let $\Omega \subset \mathbb{R}^n$ be a bounded domain with boundary $\partial\Omega = S$, $\Omega = \mathbb{R}^n \setminus \Omega_1$,

$$f \in \mathscr{C}^1(\overline{\Omega}) \cap \mathscr{C}^1(\overline{\Omega_1}), \quad [f]_S(x) = \lim_{x' \to x, x' \in \Omega_1} f(x') - \lim_{x'' \to x, x'' \in \Omega} f(x''), \quad x \in S.$$

By $\left\{\dfrac{\partial f}{\partial x_i}\right\}(x)$, $i = 1, 2, \ldots, n$, we denote the classical derivatives of f at the point $x \in \mathbb{R}^n$, $x \notin S$, and by $\dfrac{\partial}{\partial x_i}$, $i = 1, 2, \ldots, n$, the derivatives in $\mathscr{D}'(\mathbb{R}^n)$. If n is the outer normal to S, $\dfrac{\partial}{\partial n}$ is the normal derivative, $\left[\dfrac{\partial f}{\partial n}\right]_S(x) = \lim_{x' \to x, x' \in \Omega_1} \dfrac{\partial f}{\partial n}(x') - \lim_{x'' \to x, x'' \in \Omega} \dfrac{\partial f}{\partial n}(x'')$ and $\left[\left\{\dfrac{\partial f}{\partial x_i}\right\}\right]_S(x) = \lim_{x' \to x, x' \in \Omega_1} \left\{\dfrac{\partial f}{\partial x_i}\right\}(x') - \lim_{x'' \to x, x'' \in \Omega} \left\{\dfrac{\partial f}{\partial x_i}\right\}(x'')$, $x \in S$, $i = 1, 2, \ldots, n$. Prove

$$\sum_{i=1}^{n}\left[\left\{\dfrac{\partial f}{\partial x_i}\right\}\right]_S \cos(n, x_i)\delta_S = \left[\dfrac{\partial f}{\partial n}\right]_S \delta_S.$$

Problem 3.11 Let $\Omega \subset \mathbb{R}^n$ be a bounded domain with boundary $\partial\Omega = S$, $\Omega = \mathbb{R}^n \setminus \Omega_1$,

$$f \in \mathscr{C}^2(\overline{\Omega}) \cap \mathscr{C}^2(\overline{\Omega_1}), \quad [f]_S(x) = \lim_{x' \to x, x' \in \Omega_1} f(x') - \lim_{x'' \to x, x'' \in \Omega} f(x''), \quad x \in S.$$

By $\left\{\dfrac{\partial f}{\partial x_i}\right\}(x)$, $\left\{\dfrac{\partial^2 f}{\partial x_i \partial x_j}\right\}(x)$, $i, j = 1, 2, \ldots, n$, we will denote the classical derivatives of f at the point $x \in \mathbb{R}^n$, $x \notin S$, and by $\dfrac{\partial^2}{\partial x_i \partial x_j}$, $\dfrac{\partial}{\partial x_i}$, $i, j = 1, 2, \ldots, n$, the corresponding derivatives in $\mathscr{D}'(\mathbb{R}^n)$. If n is the outer normal to S, $\dfrac{\partial}{\partial n}$ is the normal derivative, $\left[\dfrac{\partial f}{\partial n}\right]_S(x) = \lim_{x' \to x, x' \in \Omega_1} \dfrac{\partial f}{\partial n}(x') - \lim_{x'' \to x, x'' \in \Omega} \dfrac{\partial f}{\partial n}(x'')$ for $x \in S$.

Prove

$$\Delta f = \left\{ \Delta f \right\} + \left[\frac{\partial f}{\partial n} \right]_s \delta s + \frac{\partial}{\partial n} \big([f]_s \delta s \big), \quad f \in \mathscr{D}'(\mathbb{R}^n).$$

Problem 3.12 Let $\Omega \subset \mathbb{R}^n$ be a bounded domain with boundary $\partial \Omega = S$, let also, $\Omega = \mathbb{R}^n \setminus \Omega_1$,

$$f \in \mathscr{C}^2(\overline{\Omega}) \cap \mathscr{C}^2(\overline{\Omega_1}), \quad [f]_S(x) = \lim_{x' \to x, x' \in \Omega_1} f(x') - \lim_{x'' \to x, x'' \in \Omega} f(x''), \quad x \in S.$$

By $\left\{ \dfrac{\partial f}{\partial x_i} \right\}(x)$, $\left\{ \dfrac{\partial^2 f}{\partial x_i \partial x_j} \right\}(x)$, $i, j = 1, 2, \ldots, n$, we will denote the classical

derivatives of f at the point $x \in \mathbb{R}^n$, $x \notin S$, and by $\dfrac{\partial^2}{\partial x_i \partial x_j}$, $\dfrac{\partial}{\partial x_i}$, $i, j = 1, 2, \ldots, n$,

the corresponding derivatives in $\mathscr{D}'(\mathbb{R}^n)$. If n is the outer normal to S, $\dfrac{\partial}{\partial n}$ indicates

the normal derivative, $\left[\dfrac{\partial f}{\partial n} \right]_S (x) = \lim\limits_{x' \to x, x' \in \Omega_1} \dfrac{\partial f}{\partial n}(x') - \lim\limits_{x'' \to x, x'' \in \Omega} \dfrac{\partial f}{\partial n}(x'')$ for

$x \in S$. Assume $f = 0$ on Ω_1. Prove

$$\Delta f = \left\{ \Delta f \right\} - \frac{\partial f}{\partial n} \delta s - \frac{\partial}{\partial n} (f \delta s).$$

Problem 3.13 Let $u \in \mathscr{D}'(\mathbb{R})$ and $u' = 0$ in the sense of the distributions. Prove that u is a constant.

Solution Let $\phi \in \mathscr{C}_0^\infty(\mathbb{R})$. Then $u'(\phi) = 0$, whereupon $u(\phi') = 0$ for every $\phi \in \mathscr{C}_0^\infty(\mathbb{R})$. Take $\psi \in \mathscr{C}_0^\infty(\mathbb{R})$. There exists a $\psi_1 \in \mathscr{C}_0^\infty(\mathbb{R})$ such that

$$\psi(x) = \psi_0(x) \int_{-\infty}^{\infty} \psi(s) ds + \psi_1'(x), \quad x \in \mathbb{R},$$

for $\psi_0 \in \mathscr{C}_0^\infty(\mathbb{R})$ such that $\int_{-\infty}^{\infty} \psi_0(x) dx = 1$. Then $u(\psi_1') = 0$ and

$$u(\psi) = u\left(\psi_0 \int_{-\infty}^{\infty} \psi(x) dx + \psi_1' \right)$$

$$= \int_{-\infty}^{\infty} \psi(x) dx\, u(\psi_0) + u(\psi_1')$$

$$= \int_{-\infty}^{\infty} \psi(x)dx u(\psi_0)$$

$$= C \int_{-\infty}^{\infty} \psi(x)dx$$

$$= C1(\psi)$$

$$= C(\psi),$$

where $C = u(\psi_0)$. Consequently $u = C$. If $C \in \mathbb{C}$ is arbitrary, then $u = C$ solves $u' = 0$.

Problem 3.14 Take $u \in \mathscr{D}'(X)$, where $X \subset \mathbb{R}$ is an open set. Consider

$$u' + au = f$$

for given $f \in \mathscr{C}(X)$, $a \in \mathscr{C}^{\infty}(X)$. Prove that $u \in \mathscr{C}^1(X)$.

Solution (1st case)   Let $a \equiv 0$, so

$$u' = f.$$

Since $f \in \mathscr{C}(X)$, f has a primitive $v \in \mathscr{C}^1(X)$, so $v' = f$ and

$$(u - v)' = u' - v' = f - f = 0.$$

Using the previous problem, we conclude that $u - v = C = \text{const}$, and

$$u = v + C \in \mathscr{C}^1(X).$$

(2nd case)   Suppose a is not identically zero and define

$$E(x) = e^{\int a(x)dx}.$$

Since $a \in \mathscr{C}^{\infty}(X)$, then $E \in \mathscr{C}^{\infty}(X)$. The product Eu is well defined and the chain rule says

$$(Eu)' = E'u + Eu' = E(-au + f) + Eau = Ef \in \mathscr{C}(X),$$

because $f \in \mathscr{C}(X)$. Therefore $Eu \in \mathscr{C}^1(X)$. Since $E \in \mathscr{C}^{\infty}(X)$, we obtain that $u \in \mathscr{C}^1(X)$.

Problem 3.15 Let $X \subset \mathbb{R}$ be an open set, and suppose $u = (u_1, u_2, \ldots, u_n) \in \mathscr{D}'(X) \times \mathscr{D}'(X) \times \ldots \times \mathscr{D}'(X)$, $f = (f_1, f_2, \ldots, f_n) \in \mathscr{C}(X) \times \mathscr{C}(X) \times \ldots \times \mathscr{C}(X)$, $a = \{a_{ij}\}_{i,j=1}^n$, $a_{ij} \in \mathscr{C}^\infty(X)$, satisfy

$$u' + au = f.$$

Prove that $u \in \mathscr{C}^1(X) \times \mathscr{C}^1(X) \times \ldots \times \mathscr{C}^1(X)$.

Hint Use the previous problem.

Problem 3.16 Let $X \subset \mathbb{R}$ be an open set where $u \in \mathscr{D}'(X)$, $a_i \in \mathscr{C}^\infty(X)$, $i = 0, 1, \ldots, m-1$, $f \in \mathscr{C}(X)$ satisfy

$$u^{(m)} + a_{m-1}u^{(m-1)} + \cdots + a_1 u' + a_0 u = f.$$

Prove that $u \in \mathscr{C}^m(X)$.

Solution Setting

$$u_j = u^{(j-1)}, \quad j = 1, 2, \ldots, m,$$

we have

$$u_j' = u^{(j)} = u_{j+1} \quad \text{for} \quad j = 1, 2, \ldots, m-1,$$

and

$$u^{(m)} + a_{m-1}u_m + \cdots + a_1 u_2 + a_0 u_1 = f.$$

The previous problem tells us that $u_j \in \mathscr{C}^1(X)$ for $j = 1, 2, \ldots, m$. Using

$$u^{(m)} = -a_{m-1}u_m - \cdots - a_1 u_2 - a_0 u_1 + f,$$

we conclude that $u^{(m)} \in \mathscr{C}(X)$, so $u \in \mathscr{C}^{(m)}(X)$.

Problem 3.17 Solve the equation

$$u'' = 0$$

in $\mathscr{D}'(X)$.

Solution Set $u' = v$, so that $v' = 0$ and therefore $v = C_0$, $C_0 = \text{const}$. Hence, $u' = C_0$. The solution of the homogeneous equation $u' = 0$ is $u = C_1$, $C_1 = \text{const}$. A particular solution of $u' = C_0$ is $u = C_0 x$. Therefore the general solution reads

$$u = C_0 + C_1 x.$$

Problem 3.18 Solve

$$u^{(m)} = 0, \quad m \geq 3,$$

in $\mathscr{D}'(\mathbb{R})$.

Answer $u = C_{m-1}x^{m-1} + \cdots + C_1 x + C_0$, where $C_i = \text{const}, i = 0, 1, \ldots, m-1$.

Problem 3.19 Solve the following equations

1. $xu' = 1$,
2. $xu' = P\dfrac{1}{x}$,
3. $x^2 u' = 0$,
4. $xu = \text{sign}x$,
5. $(\sin x)u = 0$,
6. $(\cos x)u = 0$,
7. $x^n u^{(m)} = 0, n > m$,
8. $u'' = \delta(x)$,
9. $x^2 u = 1$.

Answer

1. $c_1 + c_2 H(x) + \log|x|, \quad c_i = \text{const}, i = 1, 2$,
2. $c_1 + c_2 H(x) - P\dfrac{1}{x}, \quad c_i = \text{const}, i = 1, 2$,
3. $c_1 + c_2 H(x) + c_3 \delta(x), c_i = \text{const}, i = 1, 2, 3$,
4. $c\delta(x) + P\dfrac{1}{|x|}, c = \text{const}$,
5. $\displaystyle\sum_k c_k \delta(x - k\pi), c_k = \text{const}$,
6. $\displaystyle\sum_k c_k \delta\left(x - \dfrac{(2k+1)\pi}{2}\right), c_k = \text{const}$,
7. $\displaystyle\sum_{k=0}^{m-1} c_k H(x)x^{m-1-k} + \sum_{k=m}^{n-1} c_k \delta^{(k-m)}(x) + \sum_{k=0}^{m-1} d_k x^k, c_k, d_k = \text{const}$,
8. $xH(x) + c_1 x + c_2, c_i = \text{const}, i = 1, 2$,
9. $P\dfrac{1}{x^2} + c_1 \delta(x), c_1 = \text{const}$.

Problem 3.20 Let $u \in \mathscr{D}'(\mathbb{R})$, $u(x) = 0$ when $x < x_0$ for some given x_0 in $\mathbb{R}$. Prove that there exists a unique primitive U^{-1} of u for which $U^{-1} = 0$ when $x < x_0$.

Problem 3.21 Let $\{f_n\}_{n=1}^{\infty} \subset \mathscr{D}'(\mathbb{R})$ converges to $f \in \mathscr{D}'(\mathbb{R})$. Prove

$$\int_a^b f_n(x+t)dt \to_{n\to\infty} \int_a^b f(x+t)dt$$

in $\mathscr{D}'$, where $a < b$ are arbitrary fixed constants.

Problem 3.22 Let $\sum_{n=1}^{\infty} g_n(x)$ be convergent. Prove

$$\int_a^b \sum_{n=1}^{\infty} g_n(x+t)dt = \sum_{n=1}^{\infty} \int_a^b g_n(x+t)dt,$$

where $a < b$ are arbitrary fixed numbers.

Problem 3.23 Prove that the functions $D^{\alpha}\delta(x)$, $|\alpha| = m$, $m = 0, 1, \ldots$, are linearly independent.

Problem 3.24 Let Y be an open set in $\mathbb{R}^{n-1}$, I an open interval of $\mathbb{R}$, and take $u \in \mathscr{D}'(Y \times I)$ with $\dfrac{\partial}{\partial x_n}u = 0$. Prove that

$$u(\phi) = \int_I u_0(\phi(x_1, x_2, \ldots, x_n))dx_n, \quad \phi \in \mathscr{C}_0^{\infty}(Y \times I), u_0 \in \mathscr{D}'(Y).$$

Solution Choose $\psi_0 \in \mathscr{C}_0^{\infty}(I)$ so that $\int_I \psi_0(x)dx = 1$. For a given $g \in \mathscr{C}_0^{\infty}(Y)$ we write

$$g_0(x) = g(x_1, , \ldots, x_{n-1})\psi_0(x_n), \quad x = (x_1, x_2, \ldots, x_n),$$
$$u_0(g) = u(g_0).$$

We have that $u_0 \in \mathscr{D}'(Y)$. Let $\phi \in \mathscr{C}_0^{\infty}(Y \times I)$ and

$$I\phi(x_1, \ldots, x_{n-1}) = \int_I \phi(x_1, \ldots, x_{n-1}, x_n)dx_n.$$

The function $\phi - (I\phi)\psi_0$ has a primitive Φ with respect to the variable x_n, i.e.,

$$\phi(x) - I\phi(x_1, \ldots, x_{n-1})\psi_0(x_n) = \frac{\partial}{\partial x_n}\Phi(x_1, \ldots, x_n).$$

Therefore

$$\phi(x) - I\phi_0(x) = \frac{\partial}{\partial x_n}\Phi(x),$$

so $u(\phi - I\phi_0) = u\left(\frac{\partial}{\partial x_n}\Phi\right)$, whence $u(\phi) - u(I\phi_0) = -\frac{\partial}{\partial x_n}u(\Phi)$ and then $u(\phi) - u_0(\phi) = 0$, i.e.,

$$u(\phi) = u_0\left(\int_I \phi(x_1, x_2, \ldots, x_{n-1}, x_n)dx_n\right) = \int_I u_0(\phi(x_1, x_2, \ldots, x_{n-1}, x_n))dx_n.$$

Problem 3.25 Let $X \subset \mathbb{R}^n$ be an open set, $u, f \in \mathscr{C}(X)$ with $\frac{\partial}{\partial x_j}u = f$, for some $j = 1, 2, \ldots, n$, in $\mathscr{D}'(X)$. Prove that $\frac{\partial}{\partial x_j}u$ exists at every point $x \in X$ and $\frac{\partial u}{\partial x_j} = f$, $j = 1, 2, \ldots, n$.

Solution We will prove the assertion for $j = n$. Since $f \in \mathscr{C}(X)$, it has a primitive v with respect to the variable x_n and

$$\frac{\partial}{\partial x_n}v = f.$$

We consider $u - v$. Then

$$\frac{\partial}{\partial x_n}(u - v) = \frac{\partial}{\partial x_n}u - \frac{\partial}{\partial x_n}v = f - f = 0.$$

Let $X = Y \times I$, where Y is open in $\mathbb{R}^{n-1}$ and I is a real open interval. Then

$$(u - v)(\phi) = \int_I u_0(\phi(x_1, x_2, \ldots, x_{n-1}, x_n))dx_n, \quad u_0 \in \mathscr{D}'(Y).$$

As v and u_0 are piecewise-differentiable in x_n and $\frac{\partial}{\partial x_n}v = f$, we conclude that u is piecewise-differentiable in x_n and $\frac{\partial}{\partial x_n}u = f$.

3.6   Notes and References

In this chapter we introduce derivatives of distributions and we deduct some of their properties. It is investigated the local structure of distributions. In the chapter is defined antiderivative of a distribution. They are defined simple and double layers on surfaces and they are given some of their applications. Additional materials can be found in [7, 16, 17, 20, 21, 24, 25] and references therein.

Chapter 4
Homogeneous Distributions

4.1 Definition

Definition 4.1 A distribution $u \in \mathscr{D}'(\mathbb{R}^n \backslash \{0\})$ is said homogeneous of degree a if

$$u(\phi(x)) = t^a u\left(t^n \phi(tx)\right), \quad x \in \mathbb{R}^n \backslash \{0\}, \quad t > 0,$$

for every $\phi \in \mathscr{C}_0^\infty(\mathbb{R}^n \backslash \{0\})$. We introduce the notation $\phi_t(x) = t^n \phi(tx)$ for $x \in \mathbb{R}^n \backslash \{0\}$, $\phi \in \mathscr{C}_0^\infty(\mathbb{R}^n \backslash \{0\})$. Take $\phi \in \mathscr{C}_0^\infty(\mathbb{R} \backslash \{0\})$ and $a \in \mathbb{C}$, $\mathrm{Re}\, a > -1$. Define the function

$$x_+^a = \begin{cases} x^a & x > 0, \\ 0 & x \le 0, \end{cases}$$

and the functional

$$I_a(\phi) = x_+^a(\phi) = \int_0^\infty x^a \phi(x) dx.$$

We have $x_+^a \in \mathscr{D}'(\mathbb{R} \backslash \{0\})$ for $a \in \mathbb{C}$, $\mathrm{Re}\, a > -1$, and

$$x_+^a(\phi) = \int_0^\infty x^a \phi(x) dx = \int_0^\infty t^a y^a t \phi(ty) dy$$

$$= t^a \int_0^\infty y^a t \phi(ty) dy = t^a x_+^a(t\phi(tx)) = t^a x_+^a(\phi_t)$$

for $t > 0$. Consequently x_+^a is a homogeneous distribution of degree a.

© The Author(s), under exclusive license to Springer Nature Switzerland AG 2021
S. G. Georgiev, *Theory of Distributions*,
https://doi.org/10.1007/978-3-030-81265-2_4

Exercise 4.1 Let $a \in \mathbb{C}$ with $\mathrm{Re}\, a > -1$. Prove that the function x_+^a is a locally integrable function.

Exercise 4.2 Take $a \in \mathbb{C}$ with $\mathrm{Re}(a) > -1$, and distinct points $x_1, x_2 \in \mathbb{R}\backslash\{0\}$. Prove that $x_{1+}^a \neq x_{2+}^a$.

4.2 Properties

In this section, we will deduct some of the properties of the homogeneous distributions.

Theorem 4.1 *Let* $\mathrm{Re}(a) > 0$. *Then for every* $\phi \in \mathscr{C}_0^\infty(\mathbb{R}\backslash\{0\})$

$$I_a(\phi') = -a I_{a-1}(\phi).$$

Proof Let $\phi \in \mathscr{C}_0^\infty(\mathbb{R}\backslash\{0\})$. Then

$$I_a(\phi') = \int_0^\infty x^a d\phi(x) = -a \int_0^\infty x^{a-1}\phi(x)dx = -a I_{a-1}(\phi).$$

This completes the proof.

Theorem 4.2 *Let* $\mathrm{Re}(a) > -1$ *and* $k \in \mathbb{N}$. *Then for every* $\phi \in \mathscr{C}_0^\infty(\mathbb{R}\backslash\{0\})$

$$I_a(\phi) = (-1)^k I_{a+k}\left(\phi^{(k)}\right) \frac{1}{(a+1)\dots(a+k)}. \tag{4.1}$$

Proof Let $\phi \in \mathscr{C}_0^\infty(\mathbb{R}\backslash\{0\})$. From the previous theorem, we get

$$I_a(\phi) = -\frac{I_{a+1}(\phi')}{a+1}$$

and

$$I_{a+1}(\phi') = -\frac{I_{a+2}(\phi'')}{a+2}.$$

Therefore

$$I_a(\phi) = (-1)^2 \frac{I_{a+2}(\phi'')}{(a+1)(a+2)}.$$

Using induction, we obtain

$$I_a(\phi) = (-1)^k \frac{I_{a+k}(\phi^{(k)})}{(a+1)(a+2)\ldots(a+k)}.$$

This completes the proof.

By (4.1), it follows that if $a \in \mathbb{C}$, $\mathrm{Re}(a) > -1$, then $I_a \in \mathcal{D}'^k(\mathbb{R}\backslash\{0\})$.

Theorem 4.3 *Given $k \in \mathbb{N}$, we have*

$$\lim_{a \to -k}(a+k)x_+^a = (-1)^{k-1}\frac{\delta^{(k-1)}(x)}{(k-1)!}, \quad x \in \mathbb{R}\backslash\{0\}. \tag{4.2}$$

Proof Let $\phi \in \mathscr{C}_0^\infty(\mathbb{R}\backslash\{0\})$. Then, using (4.1), we have

$$\lim_{a\to -k}(a+k)I_a(\phi) = \lim_{a\to -k}(-1)^k I_{a+k}\Big(\phi^{(k)}\Big)\frac{1}{(a+1)\ldots(a+k-1)}$$

$$= (-1)^k \frac{I_0\Big(\phi^{(k)}\Big)}{(1-k)(1-k+1)\ldots(-1)} = -\frac{1}{(k-1)!}I_0\Big(\phi^{(k)}\Big)$$

$$= -\frac{1}{(k-1)!}\int_0^\infty \phi^{(k)}(x)dx = -\frac{1}{(k-1)!}\phi^{(k-1)}(x)\Big|_{x=0}^{x=\infty}$$

$$= \frac{\phi^{(k-1)}(0)}{(k-1)!} = \frac{1}{(k-1)!}\delta\Big(\phi^{(k-1)}\Big) = \frac{(-1)^{k-1}}{(k-1)!}\delta^{(k-1)}(\phi).$$

Since $\phi \in \mathscr{C}_0^\infty(\mathbb{R}\backslash\{0\})$ was chosen arbitrarily, we conclude that

$$\lim_{a\to -k}(a+k)x_+^a = \frac{(-1)^{k-1}}{(k-1)!}\delta^{(k-1)}(x), \quad x \in \mathbb{R}\backslash\{0\}.$$

This completes the proof.

Theorem 4.4 *Let $k \in \mathbb{N}$. Then*

$$\lim_{\epsilon \to 0}\left(I_a(\phi) - \frac{\phi^{(k-1)}(0)}{(k-1)!\epsilon}\right) = -\int_0^\infty \frac{(\log x)\phi^{(k)}(x)}{(k-1)!}dx + \frac{\phi^{(k-1)}(0)}{(k-1)!}\sum_{j=1}^{k-1}\frac{1}{j}, \tag{4.3}$$

$x \in \mathbb{R}\backslash\{0\}$, *for $\phi \in \mathscr{C}_0^\infty(\mathbb{R}\backslash\{0\})$. Here $a + k = \epsilon$.*

Proof Let $\phi \in \mathscr{C}_0^\infty(\mathbb{R}\backslash\{0\})$. Using (4.1), we can represent $I_a(\phi)$ in the following form

$$I_a(\phi) = (-1)^k \frac{I_\epsilon\Big(\phi^{(k)}\Big)}{(\epsilon+1-k)\ldots\epsilon}.$$

Then

$$
\lim_{\epsilon \to 0} \left(I_a(\phi) - \frac{\phi^{(k-1)}(0)}{(k-1)!\epsilon} \right) = \lim_{\epsilon \to 0} \left((-1)^k \frac{I_\epsilon\left(\phi^{(k)}\right)}{(\epsilon+1-k)\dots\epsilon} - \frac{\phi^{(k-1)}(0)}{(k-1)!\epsilon} \right)
$$

$$
= \lim_{\epsilon \to 0} \left((-1)^k \int_0^\infty \frac{x^\epsilon \phi^{(k)}(x)}{(\epsilon+1-k)\dots\epsilon} dx - (-1)^k \int_0^\infty \frac{\phi^{(k)}(x)}{(\epsilon+1-k)\dots\epsilon} dx \right.
$$

$$
\left. + \phi^{(k-1)}(0)\left(\frac{1}{(k-1-\epsilon)\dots(1-\epsilon)} - \frac{1}{(k-1)!} \right)\frac{1}{\epsilon} \right)
$$

$$
= \lim_{\epsilon \to 0} \left((-1)^k \int_0^\infty \frac{\left(x^\epsilon-1\right)\phi^{(k)}(x)}{(\epsilon+1-k)\dots\epsilon} dx + \phi^{(k-1)}(0)\left(\frac{1}{(k-1-\epsilon)\dots(1-\epsilon)} - \frac{1}{(k-1)!} \right)\frac{1}{\epsilon} \right)
$$

$$
= -\frac{1}{(k-1)!} \int_0^\infty \log x \phi^{(k)}(x) dx + \phi^{(k-1)}(0) \sum_{j=1}^{k-1} \frac{1}{j} \frac{1}{(k-1)!}.
$$

This completes the proof.

Definition 4.2 Define

$$
x_+^{-k}(\phi) = -\frac{1}{(k-1)!} \int_0^\infty \log x \phi^{(k)}(x) dx + \phi^{(k-1)}(0) \sum_{j=1}^{k-1} \frac{1}{j} \frac{1}{(k-1)!}, \qquad (4.4)
$$

for $k \in \mathbb{N}$ and $\phi \in \mathscr{C}_0^\infty(\mathbb{R}\backslash\{0\})$.

Note that, for $k \in \mathbb{N}$, x_+^{-k}, defined by (4.4), is an element of $\mathscr{D}'^k(\mathbb{R}\backslash\{0\})$.

Theorem 4.5 *Let $k \in \mathbb{N}$. Then*

$$
\lim_{a \to -k} \left(\frac{d}{dx} x_+^a + kx_+^{a-1} \right) = (-1)^k \frac{\delta^{(k)}(x)}{k!}.
$$

Proof We have, using (4.2),

$$
\lim_{a \to -k} \left(\frac{d}{dx} x_+^a + kx_+^{a-1} \right) = \lim_{a \to -k} \left(ax_+^{a-1} + kx_+^{a-1} \right)
$$

$$
= \lim_{a \to -k} (a+k)x_+^{a-1} = \lim_{a \to -k-1} (a+k+1)x_+^a = (-1)^k \frac{\delta^{(k)}(x)}{k!}.
$$

This completes the proof.

Theorem 4.6 *Let $k \in \mathbb{N}$. Then*

$$
x_+^{-k}(\phi) = t^{-k} x_+^{-k}(\phi_t) + \log t \frac{\phi^{(k-1)}(0)}{(k-1)!} \qquad \text{for} \quad \phi \in \mathscr{C}_0^\infty(\mathbb{R}\backslash\{0\}), \quad t > 0.
$$

Proof Let $\phi \in \mathscr{C}_0^\infty(\mathbb{R}\setminus\{0\})$. Using (4.4), for $t > 0$, we have

$$t^{-k}x_+^{-k}(\phi_t) = -\frac{t^{-k}}{(k-1)!}\int_0^\infty \log x \Big(\phi_t(x)\Big)_x^{(k)} dx + \phi_t^{(k-1)}(0)\frac{t^{-k}}{(k-1)!}\sum_{j=1}^{k-1}\frac{1}{j}$$

$$= -\frac{1}{(k-1)!}\int_0^\infty (\log x)t\phi^{(k)}(tx)dx + \phi^{(k-1)}(0)\frac{1}{(k-1)!}\sum_{j=1}^{k-1}\frac{1}{j}$$

$$= -\frac{1}{(k-1)!}\int_0^\infty (\log(tx) - \log t)\phi^{(k)}(tx)d(tx) + \phi^{(k-1)}(0)\frac{1}{(k-1)!}\sum_{j=1}^{k}\frac{1}{j}$$

$$= -\frac{1}{(k-1)!}\int_0^\infty \log(tx)\phi^{(k)}(tx)d(tx) + \phi^{(k-1)}(0)\frac{1}{(k-1)!}\sum_{j=1}^{k}\frac{1}{j}$$

$$+ \log t\frac{1}{(k-1)!}\int_0^\infty \phi^{(k)}(tx)d(tx)$$

$$= -\frac{1}{(k-1)!}\int_0^\infty \log y\phi^{(k)}(y)dy + \phi^{(k-1)}(0)\frac{1}{(k-1)!}\sum_{j=1}^{k-1}\frac{1}{j} + \frac{\log t}{(k-1)!}\int_0^\infty \phi^{(k)}(y)dy$$

$$= x_+^{-k}(\phi) + \frac{\log t}{(k-1)!}\int_0^\infty \phi^{(k)}(y)dy$$

$$= x_+^{-k}(\phi) - \frac{\log t}{(k-1)!}\phi^{(k-1)}(0).$$

This completes the proof.

Theorem 4.7 *Fix $a \notin \mathbb{Z}^-$ and take the smallest $k \in \mathbb{N}$ so that $k + \mathrm{Re}(a) > -1$. We define, for $\epsilon > 0$ and $\phi \in \mathscr{C}_0^\infty(\mathbb{R}\setminus\{0\})$,*

$$H_{a,\epsilon}(\phi) = \int_\epsilon^\infty x^a\phi(x)dx.$$

Then there exist unique constants C_0, B_j, $j = 0, 1, \ldots, k-1$, such that

$$H_{a,\epsilon}(\phi) = C_0 + \sum_{j=0}^{k-1} B_j\epsilon^{-\lambda_j} + o(1) \quad \text{when} \quad \epsilon \to 0,$$

where $\lambda_j = -(a + j + 1)$.

Proof Let $\phi \in \mathscr{C}_0^\infty(\mathbb{R} \setminus \{0\})$. We integrate $H_{a,\epsilon}(\phi)$ by parts k times, to the effect that

$$H_{a,\epsilon}(\phi) = \int_\epsilon^\infty x^a \phi(x)\, dx = \frac{1}{a+1} \int_\epsilon^\infty \phi(x)\, dx^{a+1}$$

$$= \frac{1}{a+1} x^{a+1} \phi(x) \Big|_{x=\epsilon}^{x=\infty} - \frac{1}{a+1} \int_\epsilon^\infty x^{a+1} \phi'(x)\, dx$$

$$= -\frac{1}{a+1} \epsilon^{a+1} \phi(\epsilon) - \frac{1}{(a+1)(a+2)} \int_\epsilon^\infty \phi'(x)\, dx^{a+2}$$

$$= -\frac{1}{a+1} \epsilon^{a+1} \phi(\epsilon) - \frac{1}{(a+1)(a+2)} x^{a+2} \phi'(x) \Big|_{x=\epsilon}^{x=\infty} + \frac{1}{(a+1)(a+2)} \int_\epsilon^\infty x^{a+2} \phi''(x)\, dx$$

$$= -\frac{1}{a+1} \epsilon^{a+1} \phi(\epsilon) + \frac{1}{(a+1)(a+2)} \epsilon^{a+2} \phi'(\epsilon) + \frac{1}{(a+1)(a+2)(a+3)} \int_\epsilon^\infty \phi''(x)\, dx^{a+3}$$

$$\vdots$$

$$= \frac{(-1)^k}{(a+1)(a+2)\ldots(a+k)} \int_\epsilon^\infty x^{a+k} \phi^{(k)}(x)\, dx$$

$$+ \sum_{j=0}^{k-1} \frac{(-1)^{j+1} \phi^{(j)}(\epsilon)}{(a+1)(a+2)\ldots(a+j+1)} \epsilon^{a+j+1}$$

$$= \frac{(-1)^k}{(a+1)(a+2)\ldots(a+k)} \int_0^\infty x^{a+k} \phi^{(k)}(x)\, dx$$

$$+ \sum_{j=0}^{k-1} \frac{(-1)^{j+1} \phi^{(j)}(0)}{(a+1)(a+2)\ldots(a+j+1)} \epsilon^{a+j+1} + o(1).$$

Now, let

$$C_0 = \frac{(-1)^k}{(a+1)(a+2)\ldots(a+k)} \int_0^\infty x^{a+k} \phi^{(k)}(x)\, dx,$$

$$B_j = \frac{(-1)^{j+1} \phi^{(j)}(0)}{(a+1)(a+2)\ldots(a+j+1)}, \quad j = 0, 1, \ldots, k-1.$$

Using the above expression of $H_{a,\epsilon}(\phi)$, we obtain

$$H_{a,\epsilon}(\phi) = C_0 + \sum_{j=0}^{k-1} B_j \epsilon^{-\lambda_j} + o(1).$$

Suppose

$$H_{a,\epsilon}(\phi) = D_0 + \sum_{j=0}^{k-1} Q_j \epsilon^{-\lambda_j} + o(1),$$

where $D_0, Q_j, j = 0, 1, 2, \ldots, k - 1$, are constants. Then

$$C_0 - D_0 + \sum_{j=0}^{k-1} (B_j - Q_j)\epsilon^{-\lambda_j} \to_{\epsilon \to 0} 0.$$

Since $\lambda_i \neq \lambda_j$ for $i \neq j$, $\mathrm{Re}\lambda_j \geq 0, i, j = 0, 1, \ldots, k - 1$, the above limit exists if and only if $C_0 - D_0 = 0$, $B_j - Q_j = 0$, $j = 0, 1, 2, \ldots, k - 1$. This completes the proof.

Theorem 4.8 *We have*

$$e^{\mp i\pi a} x_+^a + \frac{\delta^{(k-1)}}{(k+a)(k-1)!} \mp i\pi \frac{\delta^{(k-1)}(x)}{(k-1)!} \to_{a \to -k} (-1)^k x_+^{-k}.$$

Proof By (4.2) and the definition of x_+^{-k}, we have

$$\lim_{a \to -k} \left(x_+^a - \frac{(-1)^{k-1}\delta^{(k-1)}(x)}{(k-1)!(a+k)} \right) = x_+^{-k}. \tag{4.5}$$

Using (4.5) and

$$e^{\mp \pi a} = (-1)^k(1 \mp \pi i(a+k) + O(a+k)^2) \quad \text{when} \quad a \to -k, \tag{4.6}$$

we have

$$\lim_{a \to -k} e^{\mp i\pi a} \left(x_+^a - \frac{(-1)^{k-1}\delta^{(k-1)}(x)}{(k-1)!(a+k)} \right) = (-1)^k x_+^{-k}.$$

Now, using (4.6), we get

$$e^{\mp i\pi a} x_+^a - e^{\mp i\pi a} \frac{(-1)^{k-1}\delta^{(k-1)}(x)}{(k-1)!(a+k)}$$
$$= e^{\mp i\pi a} x_+^a - (-1)^k \frac{(-1)^{k-1}\delta^{(k-1)}(x)}{(k-1)!(a+k)}$$
$$\pm (-1)^k \frac{(-1)^{k-1}\delta^{(k-1)}(x)}{(k-1)!(a+k)} i\pi(a+k) - \frac{(-1)^{k-1}\delta^{(k-1)}(x)}{(k-1)!(a+k)} O(a+k)^2.$$

Therefore

$$\lim_{a \to -k} \left(e^{\mp i\pi a} x_+^a - e^{\mp i\pi a} \frac{(-1)^{k-1} \delta^{(k-1)}(x)}{(k-1)!(a+k)} \right)$$
$$= \lim_{a \to -k} \left(e^{\mp i\pi a} x_+^a - (-1)^k \frac{(-1)^{k-1} \delta^{(k-1)}(x)}{(k-1)!(a+k)} \right.$$
$$\pm (-1)^k \frac{(-1)^{k-1} \delta^{(k-1)}(x)}{(k-1)!(a+k)} i\pi(a+k) - \frac{(-1)^{k-1} \delta^{(k-1)}(x)}{(k-1)!(a+k)} O(a+k)^2 \right)$$
$$= \lim_{a \to -k} \left(e^{\mp i\pi a} x_+^a - (-1)^k \frac{(-1)^{k-1} \delta^{(k-1)}(x)}{(k-1)!(a+k)} \pm (-1)^k \frac{(-1)^{k-1} \delta^{(k-1)}(x)}{(k-1)!} i\pi \right)$$
$$= (-1)^k x_+^{-k}.$$

This completes the proof.

Theorem 4.9 *We have*

$$(x \pm i.0)^{-k} = x_+^{-k} + (-1)^k x_-^{-k} \pm i\pi (-1)^k \frac{\delta^{(k-1)}(x)}{(k-1)!}.$$

Proof Since

$$(x \pm i.0)^a = x_+^a + e^{\pm i\pi a} x_-^a,$$

we have

$$e^{\mp i\pi a} (x \pm i.0)^a = e^{\mp i\pi a} x_+^a + x_-^a. \tag{4.7}$$

Moreover,

$$e^{\mp i\pi a} (x \pm i.0)^a \to_{a \to -k} (-1)^k (x \pm i.0)^{-k} \tag{4.8}$$

and

$$e^{\mp i\pi a} x_+^a + x_-^a \to_{a \to -k} (-1)^k x_+^{-k} \pm i\pi \frac{\delta^{(k-1)}(x)}{(k-1)!} + x_-^{-k}. \tag{4.9}$$

From (4.7), (4.8) and (4.9), we then get

$$(-1)^k (x \pm i.0)^{-k} = x_-^{-k} + (-1)^k x_+^{-k} \pm i\pi \frac{\delta^{(k-1)}(x)}{(k-1)!}.$$

This completes the proof.

Theorem 4.10 *We have*

$$(x + i.0)^{-k} - (x - i.0)^{-k} = 2i\pi (-1)^k \frac{\delta^{(k-1)}(x)}{(k-1)!}.$$

Proof From the previous theorem, we have

$$(x + i.0)^{-k} = x_+^{-k} + (-1)^k x_-^{-k} + (-1)^k i\pi \frac{\delta^{(k-1)}(x)}{(k-1)!},$$
$$(x - i.0)^{-k} = x_+^{-k} + (-1)^k x_-^{-k} - (-1)^k i\pi \frac{\delta^{(k-1)}(x)}{(k-1)!},$$

and immediately

$$(x + i.0)^{-k} - (x - i.0)^{-k} = 2i\pi(-1)^k \frac{\delta^{(k-1)}(x)}{(k-1)!}.$$

This completes the proof.

4.3 Advanced Practical Problems

Problem 4.1 Prove that the function $a \to I_a(\phi)$, $\phi \in \mathscr{C}_0^\infty(\mathbb{R}\backslash\{0\})$, is analytic when $\mathrm{Re}(a) > -1$.

Problem 4.2 Let $a \in \mathbb{N} \cup \{0\}$. Prove that I_a can alternatively be defined as the analytic continuation with respect to a.

Hint Use (4.1).

Problem 4.3 Let $k \in \mathbb{N}$. Prove

$$\frac{d}{dx} x_+^{-k} = -k x_+^{-k-1} + (-1)^k \frac{\delta^{(k)}(x)}{k!} \quad \text{on} \quad \mathbb{R}\backslash\{0\}.$$

Problem 4.4 Let $k \in \mathbb{N}$, $k \geq 2$. Prove that there exist unique constants A_j, $j = 0, 1, \ldots, k-2$, such that

$$H_{-k,\epsilon}(\phi) = -\frac{1}{(k-1)!} \int_0^\infty \log x \phi^{(k)}(x) dx + \phi^{(k-1)}(0) \frac{1}{(k-1)!} \sum_{j=1}^{k-1} \frac{1}{j}$$
$$+ \sum_{j=0}^{k-2} A_j \phi^{(j)}(0) \epsilon^{j+1-k} - \frac{\log \epsilon}{(k-1)!} \phi^{(k-1)}(0) + o(1) \quad \text{when} \quad \epsilon \to 0,$$

for every $\phi \in \mathscr{C}_0^\infty(\mathbb{R}\backslash\{0\})$.

Problem 4.5 Let $a \in \mathbb{C}$, $\mathrm{Re}(a) > -1$ and define

$$x_-^a = \begin{cases} 0 & x > 0, \\ |x|^a & x < 0. \end{cases}$$

Prove

$$x_-^a(\phi) = x_+^a(\check{\phi}), \quad \check{\phi}(x) = \phi(-x),$$

for every $\phi \in \mathscr{C}_0^\infty(\mathbb{R}\backslash\{0\})$.

Problem 4.6 Let $a \in \mathbb{C}$ and $a \notin \mathbb{Z}^- \cup \{0\}$. Prove that

$$(x \pm i.0)^a = \lim_{\epsilon \to 0}(x \pm i.\epsilon)^a = x_+^a + e^{\pm i\pi a}x_-^a, \quad x \in \mathbb{R}.$$

Problem 4.7 Prove

$$\frac{d}{dx}(x \pm i.0)^a = a(x \pm i.0)^{a-1}.$$

Problem 4.8 Define

$$\underline{x}^{-k} = \frac{(x + i.0)^{-k} + (x - i.0)^{-k}}{2}.$$

Prove that

1. $\underline{x}^{-k} = x_+^{-k} + (-1)^k x_-^{-k}$,
2. $\dfrac{d}{dx}\left(\underline{x}^{-k}\right) = -k\underline{x}^{-k-1}$,
3. $x\underline{x}^{-k} = \underline{x}^{1-k}$.

Problem 4.9 Show that

$$\underline{x}^{-1} = \frac{d}{dx}\log|x|.$$

Problem 4.10 Define the function χ_+^a as follows

$$\chi_+^a = \frac{x_+^a}{\Gamma(a+1)}$$

for $a \in \mathbb{C}, \operatorname{Re}(a) > -1$. Prove

1. $\chi_+^a(\phi') = -\chi_+^{a-1}(\phi), \quad \phi \in \mathscr{C}_0^\infty(\mathbb{R})$,
2. $\chi_+^{-k} = \delta^{(k-1)}(x)$.

Problem 4.11 Let u be a homogeneous distribution of degree a on $\mathbb{R}^n\backslash\{0\}$ and

$$\lambda = \sum_j x_j \partial_j.$$

Prove

$$au - \lambda u = 0.$$

Hint Differentiate with respect to t the equality

$$u(\phi(x)) = t^a u(t^n \phi(tx))$$

for $\phi \in \mathscr{C}_0^\infty(\mathbb{R}^n \backslash \{0\})$.

Problem 4.12 Let $\psi \in \mathscr{C}^\infty(\mathbb{R}^n \backslash \{0\})$ be a homogeneous function of degree b and $u \in \mathscr{D}'(\mathbb{R}^n \backslash \{0\})$ a homogeneous distribution of degree a. Prove that ψu is a homogeneous distribution of degree $a + b$ in $\mathbb{R}^n \backslash \{0\}$.

Problem 4.13 Let u be a homogeneous distribution of degree a on $\mathbb{R}^n \backslash \{0\}$, $\psi \in \mathscr{C}_0^\infty(\mathbb{R}^n \backslash \{0\})$ and

$$\int_0^\infty r^{a+n-1} \psi(rx) dr = 0, \quad x = rw \in \mathbb{R}^n \backslash \{0\}.$$

Prove

$$u(\psi) = 0.$$

Hint Use $au = \lambda u$. Deduce $u((a + n)\phi(x) + \lambda\phi(x)) = 0$ for every $\phi \in \mathscr{C}_0^\infty(\mathbb{R}^n \backslash \{0\})$. Then rewrite the last equality in polar coordinates and multiply by r^{a+n-1}.

Problem 4.14 Let $u \in \mathscr{D}'(\mathbb{R}^n \backslash \{0\})$ be homogeneous of degree a. Prove that $\dfrac{\partial u}{\partial x_j}$, $j = 1, 2, \ldots, n$, are homogeneous distributions of degree $a - 1$ on $\mathbb{R}^n \backslash \{0\}$.

Problem 4.15 Let $u \in \mathscr{D}'(\mathbb{R}^n \backslash \{0\})$ be homogeneous of degree a and $\alpha \in \mathbb{N}^n$. Prove that $D^\alpha u$ is homogeneous of degree $a - |\alpha|$ on $\mathbb{R}^n \backslash \{0\}$.

Problem 4.16 Let $u_j \in \mathscr{D}'(\mathbb{R}^n \backslash \{0\})$, $j = 1, 2$, be homogeneous of degree a_j. Find conditions on a_1, a_2 so that the combination $\alpha_1 u_1 + \alpha_2 u_2$ becomes homogeneous on $\mathbb{R}^n \backslash \{0\}$, for any $\alpha_1, \alpha_2 \in \mathbb{C}$.

Answer $a_1 = a_2$.

Problem 4.17 Let $u \in \mathscr{D}'(\mathbb{R}^n \backslash \{0\})$ be homogeneous of degree a. Prove that $x_j u$ is a homogeneous distribution of degree $a + 1$ on $\mathbb{R}^n \backslash \{0\}$), for any $j = 1, 2, \ldots, n$.

4.4 Notes and References

In this chapter we define homogeneous distributions and we deduct some of their properties. In the chapter are given some applications of homogeneous distributions. Additional materials can be found in [7, 16, 17, 20, 21, 24, 25] and references therein.

Chapter 5
The Direct Product of Distributions

5.1 Definition

Definition 5.1 Let $X_1 \subset \mathbb{R}^n$, $X_2 \subset \mathbb{R}^m$ be open sets. The direct product of the distributions $u_1 \in \mathscr{D}'(X_1)$, $u_2 \in \mathscr{D}'(X_2)$ is defined through

$$u_1(x) \times u_2(y)(\phi) = u_1(x)(u_2(y)(\phi(x, y))),$$
$$u_2(y) \times u_1(x)(\phi) = u_2(y)(u_1(x)(\phi(x, y))), \quad \phi \in \mathscr{C}_0^\infty(X_1 \times X_2).$$

Take $X' \subset\subset X_1 \times X_2$ and $\phi \in \mathscr{C}_0^\infty(X')$. Since $\operatorname{supp}\phi \subset X' \subset\subset X_1 \times X_2$ is compact, there exist open sets $X_1' \subset\subset X_1$, $X_2' \subset\subset X_2$ such that $X' \subset\subset X_1' \times X_2'$. Let $x \in X_1 \backslash X_1'$. Then $\phi(x, y) = 0$ for every $y \in X_2$. Hence, $\psi(x) = u_2(y)(\phi(x, y)) = 0$, i.e., $\psi \equiv 0$ on $X_1 \backslash X_1'$. We may choose an open set $\tilde{X}_1$ so that $X_1' \subset\subset \tilde{X}_1 \subset\subset X_1$, and consequently $\operatorname{supp}\psi \subset \tilde{X}_1$.

Take $x \in X_1$ and let $\{x_k\}_{k=1}^\infty$ be a sequence in X_1 tending to x, as $k \to \infty$. Then $\phi(x_k, y) \to_{k\to\infty} \phi(x, y)$ in $\mathscr{C}_0^\infty(X_2)$ for every $\phi \in \mathscr{C}_0^\infty(X_1 \times X_2)$. In fact, $\operatorname{supp}\phi(x_k, y) \subset X_2' \subset\subset X_2$ and $D_y^\alpha \phi(x_k, y) \overset{y\in X_2}{\to} D_y^\alpha \phi(x, y)$, $k \to \infty$, for every multi-index $\alpha \in \mathbb{N}^m$. Because $u_2 \in \mathscr{D}'(X_2)$, we have

$$\psi(x_k) = u_2(y)(\phi(x_k, y)) \to_{k\to\infty} u_2(y)(\phi(x, y)) = \psi(x),$$

i.e., ψ is continuous at x. Since $x \in X_1$ was completely arbitrary, we conclude that $\psi \in \mathscr{C}(X_1)$.

Let now $e_1 = (1, 0, \ldots, 0)$ and consider the function

$$\chi_h(y) = \frac{1}{h}(\phi(x + he_1, y) - \phi(x, y))$$

© The Author(s), under exclusive license to Springer Nature Switzerland AG 2021
S. G. Georgiev, *Theory of Distributions*,
https://doi.org/10.1007/978-3-030-81265-2_5

for $x \in X_1$. For it we have $\text{supp}\chi_h \subset X_2' \subset\subset X_2$ and

$$D^\alpha \chi_h(y) \stackrel{y \in X_2}{\to} D_y^\alpha \frac{\partial \phi(x, y)}{\partial x_1}, \quad h \to 0,$$

for every $\alpha \in \mathbb{N}^m$. Because $u_2 \in \mathscr{D}'(X_2)$, we have

$$\frac{\psi(x + he_1) - \psi(x)}{h} = \frac{1}{h}\Big(u_2(y)(\phi(x + he_1, y)) - u_2(y)(\phi(x, y))\Big)$$

$$= u_2(y)\Big(\frac{\phi(x + he_1, y) - \phi(x, y)}{h}\Big)$$

$$= u_2(y)(\chi_h(y)) \to_{h \to 0} u_2(y)\Big(\frac{\partial \phi}{\partial x_1}(x, y)\Big).$$

By induction, we conclude that

$$D^\alpha \psi(x) = u_2(y)\Big(D_x^\alpha \phi(x, y)\Big)$$

for every $\alpha \in \mathbb{N}^n \cup \{0\}$ and $\phi \in \mathscr{C}_0^\infty(X_1 \times X_2)$. Therefore $\psi \in \mathscr{C}_0^\infty(\tilde{X}_1)$ for $\phi \in \mathscr{C}_0^\infty(X')$.

Let $\phi \in \mathscr{C}_0^\infty(X')$ and $x \in X_1$. Then $D_x^\alpha \phi(x, y) \in \mathscr{C}_0^\infty(X_2')$. Since $u_2 \in \mathscr{D}'(X_2)$, there exist constants $C \geq 0$ and $m \in \mathbb{N} \cup \{0\}$, $C = C(u_2)$, $m = m(u_2)$, such that

$$|D^\alpha \psi(x)||u_2(y)(D_x^\alpha \phi(x, y))| \leq C \max_{y \in X_2', |\beta| \leq m} |D_y^\beta D_x^\alpha \phi(x, y)|$$

for $x \in X_1$. Now, we consider the operation

$$\phi(x, y) \mapsto \psi(x) = u_2(y)(\phi(x, y)) \tag{5.1}$$

from $\mathscr{C}_0^\infty(X_1 \times X_2)$ to $\mathscr{C}_0^\infty(X_1)$. If $\phi_1, \phi_2 \in \mathscr{C}_0^\infty(X_1 \times X_2)$ and $\alpha_1, \alpha_2 \in \mathbb{C}$, then

$$\alpha_1 \phi_1 + \alpha_2 \phi_2 \mapsto u_2(y)(\alpha_1 \phi_1(x, y) + \alpha_2 \phi_2(x, y))$$
$$= \alpha_1 u_2(y)(\phi_1(x, y)) + \alpha_2 u_2(y)(\phi_2(x, y))$$
$$= \alpha_1 \psi_1(x) + \alpha_2 \psi_2(x),$$

i.e., the operation $\phi \mapsto u_2(y)(\phi)$ from $\mathscr{C}_0^\infty(X_1 \times X_2)$ to $\mathscr{C}_0^\infty(X_1)$ is linear.

Let now $\{\phi_n\}_{n=1}^\infty$ be a sequence in $\mathscr{C}_0^\infty(X_1 \times X_2)$ such that $\phi_n \to_{n \to \infty} 0$ in $\mathscr{C}_0^\infty(X_1 \times X_2)$. Then there exists a compact set $X_3' \subset X_1 \times X_2$ such that $\text{supp}\phi_n \subset X_3'$ for every $n \in \mathbb{N}$ and

$$D_x^\alpha D_y^\beta \phi_n(x, y) \stackrel{(x, y)}{\to}_{n \to \infty} 0$$

for every $\alpha \in \mathbb{N}^n$, $\beta \in \mathbb{N}^m$. From here, we conclude that there exists a compact set $X_4' \subset X_1$ such that $\operatorname{supp}\psi_n \subset X_4'$ for every $n \in \mathbb{N}$ and $D^\alpha \psi_n(x) \xrightarrow{x} 0$, $n \to \infty$, and $\psi_n \to_{n \to \infty} 0$ in $\mathscr{C}_0^\infty(X_1)$. Therefore (5.1) is a linear and continuous operation from $\mathscr{C}_0^\infty(X_1 \times X_2)$ to $\mathscr{C}_0^\infty(X_1)$. It follows that $u_1(x)(u_2(y)(\cdot))$ is a linear and continuous functional on $\mathscr{C}_0^\infty(X_1 \times X_2)$, so $u_1 \times u_2 \in \mathscr{D}'(X_1 \times X_2)$.

In a similar way it can be proved that $u_2 \times u_1 \in \mathscr{D}'(X_2 \times X_1)$.

Example 5.1 Let us consider $\delta(x) \times \delta(y)(\phi(x, y))$ for $\phi \in \mathscr{C}_0^\infty(\mathbb{R} \times \mathbb{R})$. We have

$$\delta(x) \times \delta(y)(\phi(x, y)) = \delta(x)(\delta(y)(\phi(x, y))) = \delta(x)(\phi(x, 0)) = \phi(0, 0).$$

Exercise 5.1 Compute

$$H'(x) \times \delta(y)(\phi(x, y)), \quad \phi \in \mathscr{C}_0^\infty(\mathbb{R} \times \mathbb{R}).$$

Answer $\phi(0, 0)$.

Exercise 5.2 Compute

$$\delta(x - 2) \times H'(y)(\phi(x, y)), \quad \phi \in \mathscr{C}_0^\infty(\mathbb{R} \times \mathbb{R}).$$

Answer $\phi(2, 0)$.

5.2 Properties

Theorem 5.1 *We have*

$$u_1(x) \times u_2(y) = u_2(y) \times u_1(x), \quad u_1 \in \mathscr{D}'(X_1), u_2 \in \mathscr{D}'(X_2).$$

Proof To prove the property we take $\phi \in \mathscr{C}_0^\infty(X_1 \times X_2)$. So, there exist sequences $\{\psi_k\}_{k=1}^\infty$ in $\mathscr{C}_0^\infty(X_1 \times X_2)$ and $\{N_k\}_{k=1}^\infty$ in $\mathbb{N} \cup \{0\}$ such that

$$\phi_k(x, y) = \sum_{i=1}^{N_k} \phi_{ik}(x)\psi_{ik}(y)$$

and $\phi_k \to_{k \to \infty} \phi$ in $\mathscr{C}_0^\infty(X_1 \times X_2)$. From here, for $(x, y) \in X_1 \times X_2$, we get

$$u_1(x) \times u_2(y)(\phi(x, y)) = u_1(x)(u_2(y)(\phi(x, y)))$$

$$= \lim_{k \to \infty} u_1(x)(u_2(y)(\phi_k(x, y))) = \lim_{k \to \infty} u_1(x)(u_2(y)(\sum_{i=1}^{N_k} \phi_{ik}(x)\psi_{ik}(y)))$$

$$= \lim_{k\to\infty} u_1(x)(\sum_{i=1}^{N_k}\phi_{ik}(x)u_2(y)(\psi_{ik}(y))) = \lim_{k\to\infty}\sum_{i=1}^{N_k}u_1(x)(\phi_{ik}(x))u_2(y)(\psi_{ik}(y))$$

$$= \lim_{k\to\infty}\sum_{i=1}^{N_k}u_2(y)(\psi_{ik}(y))u_1(x)(\phi_{ik}(x)) = \lim_{k\to\infty}\sum_{i=1}^{N_k}u_2(y)(\psi_{ik}(y)u_1(x)(\phi_{ik}(x)))$$

$$= \lim_{k\to\infty}\sum_{i=1}^{N_k}u_2(y)(u_1(x)(\psi_{ik}(y)\phi_{ik}(x))) = \lim_{k\to\infty}\sum_{i=1}^{N_k}u_2(y)(u_1(x)(\phi_{ik}(x)\psi_{ik}(y)))$$

$$= \lim_{k\to\infty} u_2(y)(u_1(x)(\sum_{i=1}^{N_k}\phi_{ik}(x)\psi_{ik}(y))) = \lim_{k\to\infty} u_2(y)(u_1(x)(\phi_k(x,y)))$$

$$= u_2(y)(u_1(x)(\phi(x,y))) = u_2(y)\times u_1(x)(\phi(x,y)), \quad (x,y)\in X_1\times X_2.$$

Since $\phi \in \mathscr{C}_0^\infty(X_1\times X_2)$ was arbitrary, $u_1(x)\times u_2(y) = u_2(y)\times u_1(x)$. This completes the proof.

Theorem 5.2 *We have*

$$(u_1(x)\times u_2(y))\times u_3(z) = u_1(x)\times (u_2(y)\times u_3(z))$$
$$\text{for}\quad u_1\in\mathscr{D}'(X_1), u_2\in\mathscr{D}'(X_2), u_3\in\mathscr{D}'(X_3), \quad (x,y,z)\in X_1\times X_2\times X_3,$$

where $X_3\subset\mathbb{R}^k$ is an open set.

Proof Let $\phi\in\mathscr{C}_0^\infty(X_1\times X_2\times X_3)$. Then

$$(u_1(x)\times u_2(y))\times u_3(z)(\phi(x,y,z)) = (u_1(x)\times u_2(y))(u_3(z)(\phi(x,y,z)))$$
$$= u_1(x)(u_2(y)(u_3(z)(\phi(x,y,z)))) = u_1(x)((u_2(y)\times u_3(z))(\phi(x,y,z)))$$
$$= u_1(x)\times (u_2(y)\times u_3(z))(\phi(x,y,z)), \quad (x,y,z)\in X_1\times X_2\times X_3.$$

Since $\phi\in\mathscr{C}_0^\infty(X_1\times X_2\times X_3)$ was arbitrary, $(u_1(x)\times u_2(y))\times u_3(z) = u_1(x)\times (u_2(y)\times u_3(z))$ for $(x,y,z)\in X_1\times X_2\times X_3$. This completes the proof.

Exercise 5.3 Let $u_1\in\mathscr{D}'(X_1), u_2\in\mathscr{D}'(X_2)$. Prove that the operator

$$u_1(x)\mapsto u_1(x)\times u_2(y), \quad (x,y)\in X_1\times X_2,$$

defined from $\mathscr{D}'(X_1)$ to $\mathscr{D}'(X_1\times X_2)$ is linear and continuous.

Definition 5.2 We will say that the distribution $u(x,y)\in\mathscr{D}'(X_1\times X_2)$ does not depend on the variable y if there exists a distribution $u_1(x)\in\mathscr{D}'(X_1)$ such that

$$u(x,y) = u_1(x)\times 1(y).$$

If this is the case, $u \in \mathscr{D}'(X_1 \times \mathbb{R}^m)$ and for $\phi \in \mathscr{C}_0^\infty(X_1 \times \mathbb{R}^m)$

$$u(x, y)(\phi(x, y)) = u_1(x) \times 1(y)(\phi(x, y))$$
$$= u_1(x)(1(y)(\phi(x, y))) = u_1(x)\left(\int_{\mathbb{R}^m} \phi(x, y)dy\right)$$
$$= 1(y) \times u_1(x)(\phi(x, y)) = 1(y)(u_1(x)(\phi(x, y))) = \int_{\mathbb{R}^m} u_1(x)(\phi(x, y))dy,$$

i.e.,

$$u_1(x)\left(\int_{\mathbb{R}^m} \phi(x, y)dy\right) = \int_{\mathbb{R}^m} u_1(x)(\phi(x, y))dy.$$

Exercise 5.4 Let $(a, b) \subset \mathbb{R}$, $a < b$, and take $u(x, y) \in \mathscr{D}'(X_1 \times (a, b))$ not depending on y. Prove that

$$u(x, y + h) = u(x, y), \quad \forall x \in X_1, \forall y, y + h \in (a, b).$$

Solution There exists a distribution $u_1(x) \in \mathscr{D}'(X_1)$ such that

$$u(x, y) = u_1(x) \times 1(y).$$

Since $1(y) = 1(y + h)$ for every y, $y + h \in (a, b)$, we have

$$u(x, y) = u_1(x) \times 1(y + h) = u(x, y + h).$$

For $u \in \mathscr{D}'(X_1 \times X_2)$ and $\phi \in \mathscr{C}_0^\infty(X_1)$, we define the distribution u_ϕ on $\mathscr{C}_0^\infty(X_2)$ by

$$u_\phi(\psi) = u(\phi(x)\psi(y)) \quad \text{for} \quad \psi \in \mathscr{C}_0^\infty(X_2).$$

Definition 5.3 The distribution $u \in \mathscr{D}'(X_1 \times X_2)$ is said to be an element of $\mathscr{C}^p(X_2)$, $p = 0, 1, 2, \ldots$, if for every $\phi \in \mathscr{C}_0^\infty(X_1)$ we have $u_\phi \in \mathscr{C}^p(X_2)$.

Exercise 5.5 Prove $D^\alpha u_\phi = \left(D_y^\alpha u\right)_\phi$.

Solution Choose $\psi \in \mathscr{C}_0^\infty(X_2)$ arbitrarily. Then

$$D^\alpha u_\phi(\psi) = (-1)^{|\alpha|} u_\phi(D^\alpha \psi(y)) = (-1)^{|\alpha|} u(D_y^\alpha(\phi(x)\psi(y)))$$
$$= D_y^\alpha u(\phi(x)\psi(y)) = \left(D_y^\alpha u\right)_\phi(\psi).$$

5.3   Advanced Practical Problems

Problem 5.1 Let $X_1 \subset \mathbb{R}^n$, $X_2 \subset \mathbb{R}^m$ be open sets and take $u_1 \in \mathscr{D}'(X_1)$, $u_2 \in \mathscr{D}'(X_2)$. Prove that

$$\mathrm{supp}(u_1 \times u_2) = \mathrm{supp} u_1 \times \mathrm{supp} u_2.$$

Solution Let $(x_0, y_0) \in \mathrm{supp} u_1 \times \mathrm{supp} u_2$ be arbitrary, and suppose U is a neighbourhood of the point (x_0, y_0) contained in $X_1 \times X_2$. Let $U_1 \subset X_1$ be a neighbourhood of x_0, $U_2 \subset X_2$ a neighbourhood of y_0. As $(x_0, y_0) \in \mathrm{supp} u_1 \times \mathrm{supp} u_2$, there exist $\phi_1 \in \mathscr{C}_0^\infty(U_1)$, $\phi_2 \in \mathscr{C}_0^\infty(U_2)$ such that $u_1(\phi_1) \neq 0$, $u_2(\phi_2) \neq 0$. Therefore, $u_1 \times u_2(\phi_1 \phi_2)(x_0, y_0) = u_1(\phi_1)(x_0)u_2(\phi_2)(y_0) \neq 0$ by the definition of the direct product. Consequently $(x_0, y_0) \in \mathrm{supp}(u_1 \times u_2)$, so we conclude

$$\mathrm{supp} u_1 \times \mathrm{supp} u_2 \subset \mathrm{supp}(u_1 \times u_2). \tag{5.2}$$

Let now $\phi \in \mathscr{C}_0^\infty(X_1 \times X_2)$ be chosen so that $\mathrm{supp}\phi \subset (X_1 \times X_2)\backslash(\mathrm{supp} u_1 \times \mathrm{supp} u_2)$. Then there exists a neighbourhood U_3 of $\mathrm{supp} u_1$ such that $\mathrm{supp}\phi(x, y) \subset X_2\backslash\mathrm{supp} u_2$ for every $x \in U_3$. Consequently, $\psi(x) = u_2(y)(\phi(x, y)) = 0$ for $x \in U_3$. As $\mathrm{supp}\psi \cap \mathrm{supp} u_1 = \emptyset$,

$$(X_1 \times X_2)\backslash(\mathrm{supp} u_1 \times \mathrm{supp} u_2) \subset (X_1 \times X_2)\backslash(\mathrm{supp}(u_1 \times u_2)),$$

from which

$$\mathrm{supp}(u_1 \times u_2) \subset \mathrm{supp} u_1 \times \mathrm{supp} u_2.$$

From here and (5.2), we get

$$\mathrm{supp}(u_1 \times u_2) = \mathrm{supp} u_1 \times \mathrm{supp} u_2.$$

Problem 5.2 Let $X_1 \subset \mathbb{R}^n$, $X_2 \subset \mathbb{R}^m$ be open sets, $u_1 \in \mathscr{D}'(X_1)$, $u_2 \in \mathscr{D}'(X_2)$. Prove

1.

$$D_{x_1}^\alpha u_1(x_1) \times D_{x_2}^\beta u_2(x_2) = D_{x_1}^\alpha D_{x_2}^\beta (u_1(x_1) \times u_2(x_2))$$

for any $\alpha \in \mathbb{N}^n \cup \{0\}$, $\beta \in \mathbb{N}^m \cup \{0\}$.

2.

$$a(x_1)b(x_2)(u_1(x_1) \times u_2(x_2)) = (a_1(x_1)u_1(x_1)) \times (b(x_2)u_2(x_2)),$$

where $a \in \mathscr{C}^\infty(X_1)$, $b \in \mathscr{C}^\infty(X_2)$.

Problem 5.3 Let $X_1 \subset \mathbb{R}^n$ be an open set and $(a, b) \subset \mathbb{R}$, $a < b$. Take $u \in \mathscr{D}'(X_1 \times (a, b))$ satisfying $u(x, y) = u(x, y+h)$ for every $x \in X_1$, y, $y+h \in (a, b)$. Prove

$$\frac{\partial u}{\partial y}(x, y) = 0, \quad (x, y) \in X_1 \times (a, b).$$

Solution Since for every (x, y), $(x, y + h) \in X_1 \times (a, b)$, $h \neq 0$, we have

$$\lim_{h \to 0} \frac{u(x, y + h) - u(x, y)}{h} = 0.$$

Then

$$\frac{\partial u}{\partial y} = 0 \quad \text{on} \quad X_1 \times (a, b).$$

Problem 5.4 Let $X_1 \subset \mathbb{R}^n$ be an open set, $(a, b) \subset \mathbb{R}$, $u \in \mathscr{D}'(X_1 \times (a, b))$ and $\frac{\partial u}{\partial y} = 0$ on $X_1 \times (a, b)$. Prove that u does not depend on y.

Solution Let $\phi \in \mathscr{C}_0^\infty(X_1 \times (a, b))$. From here,

$$\frac{\partial u}{\partial y}(\phi) = 0$$

i.e.

$$u\left(\frac{\partial \phi}{\partial y}\right) = 0 \tag{5.3}$$

for every $\phi \in \mathscr{C}_0^\infty(X_1 \times (a, b))$. Let $\psi \in \mathscr{C}_0^\infty(X_1 \times (a, b))$. Then there exists $\psi_1 \in \mathscr{C}_0^\infty(X_1 \times (a, b))$ such that

$$\psi(x, y) = \frac{\partial \psi_1(x, y)}{\partial y} + \omega_\epsilon(y - y_0) \int_a^b \psi(x, \xi) d\xi.$$

We define the distribution $u_1 \in \mathscr{D}'(X_1)$ by

$$u_1(\psi_2) = u(\omega_\epsilon(y - y_0)\psi_2(x)) \quad \text{for} \quad \psi_2 \in \mathscr{C}_0^\infty(X_1).$$

Using (5.3),

$$u(\psi) = u\left(\frac{\partial \psi_1(x, y)}{\partial y} + \omega_\epsilon(y - y_0)\int_a^b \psi(x, \xi)d\xi\right)$$

$$= u\left(\frac{\partial \psi_1(x, y)}{\partial y}\right) + u\left(\omega_\epsilon(y - y_0)\int_a^b \psi(x, \xi)d\xi\right)$$

$$= u_1\left(\int_a^b \psi(x, \xi)d\xi\right),$$

i.e.,

$$u(x, y) = u_1(x) \times 1(y).$$

Problem 5.5 Let $X_1 \subset \mathbb{R}^n$ be an open set and $F \in \mathscr{D}'(X_1 \times \mathbb{R})$. Prove that the distribution $u \in \mathscr{D}'(X_1 \times \mathbb{R})$, defined by

$$u(\phi) = F(\psi) + f(x) \times \delta(y)(\phi), \quad \phi \in \mathscr{C}_0^\infty(X_1 \times \mathbb{R}),$$

satisfies the equation

$$yu(x, y) = F(x, y).$$

Here $f \in \mathscr{D}'(X_1)$,

$$\psi(x, y) = \frac{1}{y}(\phi(x, y) - \eta(y)\phi(x, 0)),$$

and $\eta \in \mathscr{C}_0^\infty(\mathbb{R})$ equals 1 on a neighbourhood of $y = 0$.

Problem 5.6 Let $X_1 \subset \mathbb{R}^n$, $X_2 \subset \mathbb{R}^m$ be open sets, $u \in \mathscr{C}(X_2)$ in y. Prove that for every $y \in X_2$ there exists $u_y(x) \in \mathscr{D}'(X_1)$ such that

$$u_\phi(y) = u_y(\phi), \quad \phi \in \mathscr{C}_0^\infty(X_1).$$

Problem 5.7 Let $X_1 \subset \mathbb{R}^n$, $X_2 \subset \mathbb{R}^m$ be open sets, $u \in \mathscr{D}'(X_1 \times X_2)$, $u \in \mathscr{C}(X_2)$ in y. Prove that for every $\phi \in \mathscr{C}_0^\infty(X_1)$, every $y \in X_2$ and every $\alpha \in \mathbb{N}^m$

$$D^\alpha(u_y(\phi)) = \left(D_y^\alpha u\right)_y(\phi).$$

Problem 5.8 Let $X_1 \subset \mathbb{R}^n$, $X_2 \subset \mathbb{R}^m$ be open sets, $u \in \mathscr{D}'(X_1 \times X_2)$, $u \in \mathscr{C}(X_2)$ in y. Prove that the operation

$$\psi \mapsto u_y(\psi(x, y))$$

is linear and continuous from $\mathscr{C}_0^\infty(X_1 \times X_2)$ to $\mathscr{C}_0(X_2)$.

Problem 5.9 Let $X_1 \subset \mathbb{R}^n$, $X_2 \subset \mathbb{R}^m$ be open sets, $u \in \mathscr{D}'(X_1 \times X_2)$, $u \in \mathscr{C}(X_2)$. Prove that

$$u(\psi) = \int_{X_2} u_y(\psi(x, y)) dy$$

for every $\psi \in \mathscr{C}_0^\infty(X_1 \times X_2)$.

Problem 5.10 Compute

1. $\dfrac{\partial^n H(x)}{\partial x_1 \ldots \partial x_n}$, where $H(x) = H(x_1) \cdots H(x_n)$,
2. $\delta(x_1) \times \cdots \times \delta(x_n)$,
3. $\dfrac{\partial^2}{\partial t^2} H(x, t)$, where $H(x, t) = H(x)H(t)$,
4. $v(x) \times \delta(t)$, where $v \in \mathscr{C}(\mathbb{R}_x^n)$,
5. $-v(x) \times \delta'(t)$, where $v \in \mathscr{C}(\mathbb{R}_x^n)$.

Answer

1. $(-1)^n \delta_{x_1} \times \delta_{x_2} \times \cdots \times \delta_{x_n}$,
2. $\delta(x)$,
3. $H(x) \times \delta'(t)$,
4. $v(x)\delta_t$,
5. $-v(x)\dfrac{d}{dt}\delta_t$.

Problem 5.11 Let $X_1 \subset \mathbb{R}^n$, $X_2 \subset \mathbb{R}^m$, $X_3 \subset \mathbb{R}^k$ be open sets, $u_1 \in \mathscr{D}'(X_1)$, $u_2 \in \mathscr{D}'(X_2)$, $u_3 \in \mathscr{D}'(X_3)$. Prove

$$u_1 \times (u_2 + u_3) = u_1 \times u_2 + u_1 \times u_3.$$

Problem 5.12 Let $u \in \mathscr{D}'(\mathbb{R}^n)$ and $\alpha \in \mathbb{N}^n$. Determine

$$D_y^\alpha(u(x) \times 1(y)).$$

Answer 0.

5.4   Notes and References

In this chapter we define the direct product of two distributions and we prove that it is well defined. In the chapter are proved the commutativity and associativity of the direct product of distributions. They are considered some applications of the direct product of distributions. Additional materials can be found in [7, 16, 17, 20, 21, 24, 25] and references therein.

Chapter 6
Convolutions

6.1 Definition

Consider $u_1, u_2 \in \mathscr{D}'(\mathbb{R}^n)$ and a sequence $\{\eta_k(x, y)\}_{k=1}^{\infty}$ in $\mathscr{C}_0^{\infty}(\mathbb{R}^{2n})$ converging to 1 in $\mathbb{R}^{2n}$. Suppose that

$$\lim_{k \to \infty} u_1(x) \times u_2(y)(\eta_k(x, y)\phi(x + y)) = u_1(x) \times u_2(y)(\phi(x + y))$$

exists for every $\phi \in \mathscr{C}_0^{\infty}(\mathbb{R}^n)$ and does not depend on the choice of the sequence $\{n_k(x, y)\}_{k=1}^{\infty}$.

Definition 6.1 The convolution of the distributions u_1 and u_2 is defined by

$$
\begin{aligned}
u_1 * u_2(\phi) &= u_1(x) \times u_2(y)(\phi(x + y)) \\
&= \lim_{k \to \infty} u_1(x) \times u_2(y)(\eta_k(x, y)\phi(x + y))
\end{aligned}
$$

for any $\phi \in \mathscr{C}_0^{\infty}(\mathbb{R}^n)$.

For $\phi_1, \phi_2 \in \mathscr{C}_0^{\infty}(\mathbb{R}^n)$ and $\alpha_1, \alpha_2 \in \mathbb{C}$, we have

$$u_1 * u_2(\alpha_1\phi_1 + \alpha_2\phi_2) = \lim_{k \to \infty} u_1(x) \times u_2(y)(\eta_k(x, y)(\alpha_1\phi_1(x + y) + \alpha_2\phi_2(x + y)))$$

$$= \lim_{k \to \infty} \left(\alpha_1 u_1(x) \times u_2(y)(\eta_k(x, y)\phi_1(x + y)) + \alpha_2 u_1(x) \times u_2(y)(\eta_k(x, y)\phi_2(x + y)) \right)$$

$$= \alpha_1 \lim_{k \to \infty} u_1(x) \times u_2(y)(\eta_k(x, y)\phi_1(x + y)) + \alpha_2 \lim_{k \to \infty} u_1(x) \times u_2(y)(\eta_k(x, y)\phi_2(x + y))$$

$$= \alpha_1 u_1 * u_2(\phi_1) + \alpha_2 u_1 * u_2(\phi_2),$$

proving the convolution $*$ is a linear functional on $\mathscr{C}_0^{\infty}(\mathbb{R}^n)$.

© The Author(s), under exclusive license to Springer Nature Switzerland AG 2021
S. G. Georgiev, *Theory of Distributions*,
https://doi.org/10.1007/978-3-030-81265-2_6

Let now $\{\phi_n\}_{n=1}^{\infty}$ be a sequence in $\mathscr{C}_0^{\infty}(\mathbb{R}^n)$ which tends to 0 in $\mathscr{C}_0^{\infty}(\mathbb{R}^n)$. Then

$$
\begin{aligned}
\lim_{n \to 0} u_1 * u_2(\phi_n) &= \lim_{n \to \infty} \lim_{k \to \infty} u_1(x) \times u_2(y)(\eta_k(x, y)\phi_n(x + y)) \\
&= \lim_{k \to \infty} \lim_{n \to \infty} u_1(x) \times u_2(y)(\eta_k(x, y)\phi_n(x + y)) \\
&= \lim_{k \to \infty} u_1(x) \times u_2(y)(0) = 0.
\end{aligned}
$$

Consequently $u_1 * u_2 \in \mathscr{D}'(\mathbb{R}^n)$.

Exercise 6.1 Let $u_1, u_2 \in \mathscr{D}'(\mathbb{R}^n)$ and assume $u_1 * u_2$ exists. Prove that

$$u_1 \mapsto u_1 * u_2 \tag{6.1}$$

is a linear map from $\mathscr{D}'(\mathbb{R}^n)$ to itself.

Example 6.1 The operation defined in (6.1) is not continuous on $\mathscr{D}'(\mathbb{R}^n)$, because

$$\delta(x - k) \to_{k \to \infty} 0$$

but

$$1 * \delta(x - k) = 1.$$

Exercise 6.2 Let $u_1, u_2 \in \mathscr{D}'(\mathbb{R}^n)$ and assume $u_1 * u_2$ exists. Prove

$$\text{supp}(u_1 * u_2) \subset \overline{\text{supp}u_1 + \text{supp}u_2}.$$

Solution Pick $\phi \in \mathscr{C}_0^{\infty}(\mathbb{R}^n)$ so that

$$\text{supp}\phi \cap \overline{\text{supp}u_1 + \text{supp}u_2} = \emptyset.$$

Then, since $\text{supp}(u_1 \times u_2) = \text{supp}u_1 \times \text{supp}u_2$,

$$
\begin{aligned}
&\text{supp}(u_1 \times u_2) \cap \text{supp}(\eta_k(x, y)\phi(x + y)) \\
&\subset (\text{supp}u_1 \times \text{supp}u_2) \cap \{(x, y) \in \mathbb{R}^{2n} : x + y \in \text{supp}\phi\} = \emptyset.
\end{aligned}
$$

Consequently,

$$\text{supp}(u_1 * u_2) \subset \overline{\text{supp}u_1 + \text{supp}u_2}.$$

6.2 Properties

Let $u_1, u_2 \in \mathscr{D}'(\mathbb{R}^n)$ and suppose $u_1 * u_2 \in \mathscr{D}'(\mathbb{R}^n)$ exists.

1. Commutativity.

$$u_1 * u_2(\phi) = \lim_{k \to \infty} u_1(x) \times u_2(y)(\eta_k(x, y)\phi(x + y))$$
$$= \lim_{k \to \infty} u_2(y) \times u_1(x)(\eta_k(x, y)\phi(x + y))$$
$$= u_2(y) \times u_1(x)(\phi(x + y)) = u_2 * u_1(\phi), \quad \phi \in \mathscr{C}_0^\infty(\mathbb{R}^n).$$

2. Convolution with the δ function.

$$u_1 * \delta(\phi) = u_1(x) \times \delta(y)(\phi(x + y))$$
$$= \lim_{k \to \infty} u_1(x) \times \delta(y)(\eta_k(x, y)\phi(x + y))$$
$$= \lim_{k \to \infty} u_1(x)(\delta(y)(\eta_k(x, y)\phi(x + y)))$$
$$= \lim_{k \to \infty} u_1(x)(\eta_k(x, 0)\phi(x))$$
$$= u_1(\phi), \quad \phi \in \mathscr{C}_0^\infty(\mathbb{R}^n),$$

   and analogously, $\delta * u_1 = u_1$.
3. Translation.

   Let $h \in \mathbb{R}^n$. Then for $\phi \in \mathscr{C}_0^\infty(\mathbb{R}^n)$, we have

$$(u_1 * u_2)(x + h)(\phi) = u_1 * u_2(\phi(x - h))$$
$$= \lim_{k \to \infty} u_1(x) \times u_2(y)(\eta_k(x - h, y)\phi(x - h + y))$$
$$= \lim_{k \to \infty} u_1(x + h) \times u_2(y)(\eta_k(x, y)\phi(x + y))$$
$$= u_1(x + h) * u_2(\phi).$$

Moreover,

$$u_1(-x) * u_2(-x)(\phi) = \lim_{k \to \infty} u_1(-x) \times u_2(-y)(\eta_k(x, y)\phi(x + y))$$
$$= \lim_{k \to \infty} u_1(x) \times u_2(y)(\eta_k(-x, -y)\phi(-x - y))$$
$$= u_1 * u_2(\phi(-x)) = (u_1 * u_2)(-x)(\phi), \quad \phi \in \mathscr{C}_0^\infty(\mathbb{R}^n).$$

4. Let $\alpha \in \mathbb{N}^n \cup \{0\}$. Then $D^\alpha u_1 * u_2, u_1 * D^\alpha u_2$ exist and satisfy

$$D^\alpha(u_1 * u_2) = D^\alpha u_1 * u_2 = u_1 * D^\alpha u_2.$$

In fact, for $\phi \in \mathscr{C}_0^\infty(\mathbb{R}^n)$, we have

$$
\begin{aligned}
D^\alpha(u_1 * u_2)(\phi) &= (-1)^{|\alpha|} u_1 * u_2(D^\alpha \phi) \\
&= (-1)^{|\alpha|} \lim_{k \to \infty} u_1(x) \times u_2(y)(\eta_k(x, y) D_x^\alpha \phi(x + y)) \\
&= (-1)^{|\alpha|} \lim_{k \to \infty} u_1(x) \times u_2(y)\Big(D_x^\alpha(\eta_k(x, y)\phi(x + y)) \\
&\quad - \sum_{\beta < \alpha} \binom{\alpha}{\beta} D_x^{\alpha - \beta} \eta_k(x, y) D_x^\beta \phi(x + y)\Big) \\
&= (-1)^{|\alpha|} \lim_{k \to \infty} u_1(x) \times u_2(y)(D_x^\alpha(\eta_k(x, y)\phi(x + y))) \\
&\quad - (-1)^{|\alpha|} \lim_{k \to \infty} u_1(x) \times u_2(y)\Big(\sum_{\beta < \alpha} \binom{\alpha}{\beta} D_x^{\alpha - \beta} \eta_k(x, y) D_x^\beta \phi(x + y)\Big) \\
&= (-1)^{|\alpha|} \lim_{k \to \infty} u_1(x) \times u_2(y)(D_x^\alpha(\eta_k(x, y)\phi(x + y)) \\
&= \lim_{k \to \infty} D^\alpha u_1(x) \times u_2(y)(\eta_k(x, y)\phi(x + y)) \\
&= D^\alpha u_1 * u_2(\phi).
\end{aligned}
$$

Similarly, one shows that

$$
D^\alpha(u_1 * u_2) = u_1 * D^\alpha u_2.
$$

Example 6.2 For $(H(x) * \delta)'$, $x \in \mathbb{R}$ we have

$$
(H(x) * \delta(x))' = H'(x) * \delta(x) = \delta(x) * \delta(x) = \delta(x).
$$

As a consequence,

$$
H(x) * \delta'(x) = \delta(x).
$$

Exercise 6.3 Compute $(x^2 * \delta(x))'$, $x \in \mathbb{R}$.

Answer $2x$.

Exercise 6.4 Compute $(H(x) * P(x))'''$, $x \in \mathbb{R}$, where P is a polynomial of degree $n \in \mathbb{N}$.

Answer $P''(x)$.

5. Let $u_3 \in \mathscr{D}'(\mathbb{R}^n)$ and suppose $u_1 * u_2 * u_3$, $u_1 * u_2$, $u_2 * u_3$, $(u_1 * u_2) * u_3$ and $u_1 * (u_2 * u_3)$ exist. Then

$$
u_1 * u_2 * u_3 = (u_1 * u_2) * u_3 = u_1 * (u_2 * u_3). \tag{6.2}
$$

Exercise 6.5 Prove (6.2).

6.3 Existence

Let $X, B \subset \mathbb{R}^n$ be open sets and $A \subset \mathbb{R}^n$ a closed set. For a given $R > 0$, we set

$$T_R = \{(x, y) : x \in A, y \in B, |x + y| \leq R\}.$$

Theorem 6.1 *For $u_1 \in \mathscr{D}'(\mathbb{R}^n, A)$, $u_2 \in \mathscr{D}'(B)$, the convolution $u_1 * u_2$ exists and can be represented in the form*

$$u_1 * u_2(\phi) = u_1(x) \times u_2(y)(\xi(x)\eta(y)\phi(x + y)), \quad \phi \in \mathscr{C}_0^\infty(\mathbb{R}^n),$$

where $\xi, \eta \in \mathscr{C}_0^\infty(\mathbb{R}^n)$ and $\xi \equiv 1$ on A^ϵ, $\eta \equiv 1$ on B^ϵ, $\xi \equiv 0$ on $\mathbb{R}^n \setminus A^{2\epsilon}$, $\eta \equiv 0$ on $\mathbb{R}^n \setminus B^{2\epsilon}$.

Proof To prove this fact, we take $\phi \in \mathscr{C}_0^\infty(U_R)$ and set out to show that the limit

$$\lim_{k \to \infty} u_1(x) \times u_2(y)(\eta_k(x, y)\phi(x + y))$$

exists and does not depend on $\{\eta_k(x, y)\}_{k=1}^\infty$. Since $\mathrm{supp}(u_1 \times u_2) = \mathrm{supp}u_1 \times \mathrm{supp}u_2$, we have

$$\mathrm{supp}((u_1(x) \times u_2(y))\phi(x + y)) \subset T_R.$$

Because T_R is a bounded set, there exists $N = N(T_R) \in \mathbb{N}$ such that $\eta_k \equiv 1$ in T_R for every $k \geq N$. From here,

$$\lim_{k \to \infty} u_1(x) \times u_2(y)(\eta_k(x, y)\phi(x + y))$$
$$= u_1(x) \times u_2(y)(\phi(x + y)\eta_N(x, y))$$
$$= u_1(x) \times u_2(y)(\phi(x + y)), \quad x, y \in \mathbb{R}^n.$$

Consequently the limit $\lim_{k \to \infty} u_1(x) \times u_2(y)(\eta_k(x, y)\phi(x + y))$ exists and does not depend on $\{\eta_k(x, y)\}_{k=1}^\infty$. Therefore we can choose $\eta_k(x, y) = \xi_k(x)\tilde{\eta}_k(y)$, where $\xi_k, \tilde{\eta}_k \in \mathscr{C}_0^\infty(\mathbb{R}^n)$, $\xi_k \equiv 1$ on A^ϵ, $\tilde{\eta}_k \equiv 1$ on B^ϵ, $\xi_k \equiv 0$ on $\mathbb{R}^n \setminus A^{2\epsilon}$, $\tilde{\eta}_k \equiv 0$ on $\mathbb{R}^n \setminus B^{2\epsilon}$. We set $\xi = \xi_N$ and $\eta = \tilde{\eta}_N$. Thus,

$$\lim_{k \to \infty} u_1(x) \times u_2(y)(\eta_k(x, y)\phi(x + y)) = u_1(x) \times u_2(y)(\xi(x)\eta(y)\phi(x + y)), \quad x, y \in \mathbb{R}^n.$$

In addition, we have $u_1 * u_2 \in \mathscr{D}'(\overline{A + B})$ because $\mathrm{supp}(u_1 \times u_2) \subset \overline{A + B}$. This completes the proof.

Exercise 6.6 Let $u_1 \in \mathscr{D}'(\mathbb{R}^n, A)$, $u_2 \in \mathscr{D}'(B)$. Prove that

$$u_1 \mapsto u_1 * u_2$$

is a continuous operation from $\mathscr{D}'(\mathbb{R}^n, A)$ to $\mathscr{D}'(\overline{A + B})$.

Exercise 6.7 Let $u_1 \in \mathscr{D}'(\mathbb{R}^n)$, $u_2 \in \mathscr{E}'(\mathbb{R}^n)$. Prove that the convolution $u_1 * u_2$ exists and can be represented in the form

$$u_1 * u_2(\phi) = u_1(x) \times u_2(y)(\eta(y)\phi(x + y)), \quad \phi \in \mathscr{C}_0^\infty(\mathbb{R}^n),$$

where $\eta \in \mathscr{C}_0^\infty(\mathbb{R}^n)$, $\eta \equiv 1$ in $(\operatorname{supp} u_2)^\epsilon$ and $\eta \equiv 0$ in $\mathbb{R}^n \backslash (\operatorname{supp} u_2)^{2\epsilon}$.

If $u_1 \in \mathscr{C}^\infty(\mathbb{R}^n)$ and $u_2 \in \mathscr{E}'(\mathbb{R}^n)$, the convolution $u_1 * u_2 \in \mathscr{C}^\infty(\mathbb{R}^n)$ exists and can be represented as

$$u_1 * u_2(x) = \tilde{u}_2(y)(u_1(x - y)), \tag{6.3}$$

where $\tilde{u}_2$ is an extension of u_2 on $\mathscr{C}^\infty(\mathbb{R}^n)$.

If $u_1 \in \mathscr{C}^\infty(\mathbb{R}^n \backslash \{0\})$ and $u_2 \in \mathscr{E}'(\mathbb{R}^n)$, then the convolution $u_1 * u_2 \in \mathscr{C}^\infty(\mathbb{R}^n \backslash \operatorname{supp} u_2)$ exists and can be represented in the form (6.3).

6.4   The Convolution Algebras $\mathscr{D}'(\Gamma+)$ and $\mathscr{D}'(\Gamma)$

Let Γ be a closed cone.

Definition 6.2 We say that the set $A \subset \mathbb{R}^n$ is bounded on the side of the cone Γ if $A \subset \Gamma + K$ for some compact set $K \subset \mathbb{R}^n$.

A compact set A in $\mathbb{R}^n$ falls under this definition by taking $\Gamma = \{0\}$.
With $\mathscr{D}'(\Gamma+)$ we will indicate the space of distributions with supports bounded on the side of the cone Γ.

Definition 6.3 We say that the sequence $\{u_n\}_{n=1}^\infty$ in $\mathscr{D}'(\Gamma+)$ converges to 0 in $\mathscr{D}'(\Gamma+)$ if there exists a compact set $K \subset \mathbb{R}^n$ such that $\operatorname{supp} u_n \subset \Gamma + K$ and $u_n \to_{n\to\infty} 0$ in $\mathscr{D}'(\mathbb{R}^n)$.

If Γ is a closed, convex, acute cone, $C = \operatorname{int}\Gamma^*$, S is an $(n-1)$-dimensional C-like surface and we take $u_1 \in \mathscr{D}'(\Gamma+)$, $u_2 \in \mathscr{D}'(\overline{S}_+)$, then the convolution $u_1 * u_2$ exists in $\mathscr{D}'(\mathbb{R}^n)$ and can be written

$$u_1 * u_2(\phi) = u_1(x) \times u_2(y)(\xi(x)\eta(y)\phi(x + y)), \quad \phi \in \mathscr{C}_0^\infty(\mathbb{R}^n), \tag{6.4}$$

where $\xi, \eta \in \mathscr{C}_0^\infty(\mathbb{R}^n)$ and $\xi \equiv 1$ on $(\operatorname{supp} u_1)^\epsilon$, $\eta \equiv 1$ on $(\operatorname{supp} u_2)^\epsilon$, $\xi \equiv 0$ on $\mathbb{R}^n \backslash (\operatorname{supp} u_1)^{2\epsilon}$, $\eta \equiv 0$ on $\mathbb{R}^n \backslash (\operatorname{supp} u_2)^{2\epsilon}$. In addition, if $\operatorname{supp} u_1 \subset \Gamma + K$, where K is a compact set in $\mathbb{R}^n$, then $\operatorname{supp}(u_1 * u_2) \subset \overline{S}_+ + K$ and the operations

$$u_1 \mapsto u_1 * u_2 \quad \text{and} \quad u_2 \mapsto u_1 * u_2,$$

are continuous from $\mathscr{D}'(\Gamma+)$ and $\mathscr{D}'(\overline{S}_+)$ to $\mathscr{D}'(\overline{S}_+ + K)$, respectively.

If $u_1 \in \mathscr{D}'(\Gamma+)$, $u_2 \in \mathscr{D}'(\Gamma+)$, the convolution $u_1 * u_2$ exists in $\mathscr{D}'(\Gamma+)$ and can be represented as in (6.4). The operation

$$u_1 \mapsto u_1 * u_2$$

is a continuous map from $\mathscr{D}'(\Gamma+)$ to $\mathscr{D}'(\Gamma+)$. Taking the convolution as multiplication turns $\mathscr{D}'(\Gamma+)$ into a commutative and associative algebra.

The space of distributions in $\mathscr{D}'(\mathbb{R}^n)$ with support in a subset of Γ will be denoted with $\mathscr{D}'(\Gamma)$. If $u_1, u_2 \in \mathscr{D}'(\Gamma)$, then

$$\operatorname{supp}(u_1 * u_2) \subset \overline{\operatorname{supp} u_1 + \operatorname{supp} u_2} = \overline{\Gamma + \Gamma} = \Gamma.$$

Therefore $\mathscr{D}'(\Gamma)$ is a subalgebra of $\mathscr{D}'(\Gamma+)$.

Exercise 6.8 Let $u_1, u_2, u_3 \in \mathscr{D}'(\Gamma+)$. Prove that the convolution $u_1 * u_2 * u_3$ exists in $\mathscr{D}'(\Gamma+)$ and the operation

$$u_1 \mapsto u_1 * u_2 * u_3$$

from $\mathscr{D}'(\Gamma+)$ to $\mathscr{D}'(\Gamma+)$ is continuous.

Exercise 6.9 Generalize the previous exercise.

6.5 Regularization of Distributions

Let $X \subset \mathbb{R}^n$ be an open set and consider $u_1 \in \mathscr{D}'(X)$, $u_2 \in \mathscr{E}'(X)$ with $\operatorname{supp} u_2 \subset U_\epsilon \subset X$. The convolution

$$u_1 * u_2(\phi) = u_1(x) \times u_2(y)(\eta(y)\phi(x+y)), \quad \phi \in \mathscr{C}_0^\infty(\mathbb{R}^n), \tag{6.5}$$

where $\eta \in \mathscr{C}_0^\infty(X_\epsilon)$ is 1 on $(\operatorname{supp} u)_\epsilon$, is well defined, and does not depend on the choice of η. We know $u_1 * u_2 = u_2 * u_1$, $u_1 * \delta = u_1$ and that $u_1 \mapsto u_1 * u_2$ and $u_2 \mapsto u_1 * u_2$ are continuous. In particular, for $\alpha \in \mathscr{C}^\infty(X_\epsilon)$, the convolution

$$u_1 * \alpha(x) = u_{1y}(\alpha(x-y))$$

exists and belongs in $\mathscr{C}^\infty(X_\epsilon)$.

Definition 6.4 The distribution

$$u_{1\epsilon} = u_1 * \omega_\epsilon$$

is called the regularization of the distribution u_1.

Since $\omega_\epsilon \to_{\epsilon \to 0} \delta$ in $\mathscr{D}'(X)$, we have

$$u_{1\epsilon} = u_1 * \omega_\epsilon \to_{\epsilon \to 0} u_1 * \delta = u_1,$$

showing that any distribution can be considered as a weak limit of its regularization.

Theorem 6.2 *The space $\mathscr{C}_0^\infty(X)$ is dense in $\mathscr{D}'(X)$.*

Proof Take $u \in \mathscr{D}'(X)$ and let u_ϵ be its regularization. Define

$$X_1 \subset\subset X_2 \subset\subset X_3 \subset\subset \ldots \subset\subset X_k \subset\subset \ldots,$$

$$\cup_{k=1}^\infty X_k = X, \epsilon_k = \text{dist}(X_k, \partial X), \quad k = 1, 2, \ldots,$$

with $\eta_k \in \mathscr{C}_0^\infty(X_k)$ such that $\eta_k \equiv 1$ on $X_k, k = 1, 2, \ldots$. We consider the sequence $\{\eta_k u_{\epsilon_k}\}_{k=1}^\infty$, for which

$$\lim_{k\to\infty} \eta_k u_{\epsilon_k}(\phi) = \lim_{k\to\infty} u_{\epsilon_k}(\eta_k \phi)$$
$$= \lim_{k\to\infty} u_{\epsilon_k}(\phi) = u(\phi), \quad \phi \in \mathscr{C}_0^\infty(X).$$

Since $u \in \mathscr{D}'(X)$ is arbitrary, $\mathscr{C}_0^\infty(X)$ is dense in $\mathscr{D}'(X)$. This completes the proof.

6.6 Fractional Differentiation and Integration

The space $\mathscr{D}'(\overline{\mathbb{R}_+})$ will be denoted with $\mathscr{D}'_+$ for short.

Definition 6.5 Let $\alpha \in \mathbb{R}$. We define

$$f_\alpha(x) = \begin{cases} \dfrac{H(x)x^{\alpha-1}}{\Gamma(\alpha)} & \text{for } \alpha > 0, \quad x \in \mathbb{R}_+, \\ f_{\alpha+n}^{(n)}(x) & \text{for } \alpha \leq 0, \quad x \in \mathbb{R}_+, \quad n \in \mathbb{N}, \quad n+\alpha > 0. \end{cases}$$

Note that $f_\alpha \in \mathscr{D}'_+$ and

$$f_\alpha * f_\beta(x) = \frac{H(x)}{\Gamma(\alpha)\Gamma(\beta)} \int_0^x y^{\beta-1}(x-y)^{\alpha-1}dy, \quad x \in \mathbb{R}_+, \quad \alpha > 0, \quad \beta > 0.$$

$$(6.6)$$

Theorem 6.3 *We have $f_\alpha * f_\beta = f_{\alpha+\beta}$.*

Proof

Case 1. $\alpha > 0$, $\beta > 0$. We take $y = tx$ in (6.4) and we get

$$f_\alpha * f_\beta(x) = \frac{H(x)x^{\alpha+\beta-1}}{\Gamma(\alpha)\Gamma(\beta)} \int_0^1 t^{\beta-1}(1-t)^{\alpha-1} dt$$

$$= \frac{H(x)x^{\alpha+\beta-1}}{\Gamma(\alpha)\Gamma(\beta)} B(\alpha, \beta) = \frac{H(x)x^{\alpha+\beta-1}}{\Gamma(\alpha+\beta)} = f_{\alpha+\beta}(x), \quad x \in \overline{\mathbb{R}_+}.$$

Case 2. $\alpha \le 0$, $\beta > 0$. Then

$$f_\alpha(x) = f_{\alpha+n}^{(n)}(x), \quad \alpha + n > 0, \quad n \in \mathbb{N}, \quad x \in \overline{\mathbb{R}_+},$$

and we get

$$f_\alpha * f_\beta(x) = f_{\alpha+n}^{(n)} * f_\beta(x) = \left(f_{\alpha+n} * f_\beta\right)^{(n)}(x)$$

$$= \left(f_{\alpha+\beta+n}\right)^{(n)}(x) = \left(\frac{H(x)x^{\alpha+\beta+n-1}}{\Gamma(\alpha+\beta+n)}\right)^{(n)}$$

$$= \frac{H(x)(\alpha+\beta+n-1)\dots(\alpha+\beta)x^{\alpha+\beta-1}}{\Gamma(\alpha+\beta+n)}$$

$$= \frac{H(x)x^{\alpha+\beta-1}}{\Gamma(\alpha+\beta)} = f_{\alpha+\beta}(x), \quad x \in \overline{\mathbb{R}_+}.$$

Case 3. $\alpha > 0$, $\beta \le 0$. We omit the proof and leave it to the reader, since it merely reproduces that of case 2.

Case 4. $\alpha \le 0$, $\beta \le 0$. Let $n_1, n_2 \in \mathbb{N}$ be fixed so that

$$f_\alpha = f_{\alpha+n_1}^{(n_1)}, \quad f_\beta = f_{\beta+n_2}^{(n_2)}, \quad \alpha + n_1 > 0, \quad \beta + n_2 > 0.$$

Then

$$f_\alpha * f_\beta(x) = f_{\alpha+n_1}^{(n_1)} * f_{\beta+n_2}^{(n_2)}(x)$$

$$= \left(f_{\alpha+n_1} * f_{\beta+n_2}\right)^{(n_1+n_2)}(x)$$

$$= \left(f_{\alpha+\beta+n_1+n_2}\right)^{(n_1+n_2)}(x)$$

$$= \frac{1}{\Gamma(\alpha+\beta+n_1+n_2)}\left(H(x)x^{\alpha+\beta+n_1+n_2-1}\right)^{(n_1+n_2)}$$

$$= \frac{H(x)}{\Gamma(\alpha + \beta + n_1 + n_2)}(\alpha + \beta + n_1 + n_2 - 1)\ldots(\alpha + \beta)x^{\alpha+\beta-1}$$

$$= \frac{H(x)x^{\alpha+\beta-1}}{\Gamma(\alpha + \beta)} = f_{\alpha+\beta}(x), \quad x \in \overline{\mathbb{R}_+}.$$

This completes the proof.

Example 6.3 Let us consider f_0. We have

$$f_0(x) = f_1'(x) = \frac{H'(x)}{\Gamma(1)} = \delta(x).$$

Since $\mathscr{D}_+'$ is a commutative convolution algebra,

$$f_\alpha * f_{-\alpha}(x) = f_0(x) = \delta(x), \quad x \in \overline{\mathbb{R}_+},$$

for $\alpha \in \mathbb{R}$. Consequently $f_\alpha^{-1} = f_{-\alpha}$.
For $n \in \mathbb{Z}^-$, we have

$$f_n = f_0^{(-n)} = \delta^{(-n)}.$$

So,

$$f_n * u = \delta^{(-n)} * u = \delta * u^{(-n)} = u^{(-n)}$$

for $u \in \mathscr{D}_+'$. In the case when $n \in \mathbb{Z}^+$, we have $f_n(x) = \dfrac{H(x)x^{n-1}}{\Gamma(n)}, x \in \overline{\mathbb{R}_+}$.
Hence,

$$\left(f_1 * u\right)'(x) = f_1' * u(x)$$

$$= \frac{H'(x)}{\Gamma(1)} * u(x)$$

$$= \delta * u(x) = u(x), \quad x \in \overline{\mathbb{R}_+},$$

for $u \in \mathscr{D}_+'$.

Definition 6.6 When $\alpha < 0$ the operator $f_\alpha *$ is called fractional differentiation of order α. When $\alpha > 0$ it is known as fractional integration of order α.

Let $k \in (0, 1)$. Then

$$D^k u = D\left(D^{k-1}u\right) = D\left(f_{1-k} * u\right) \tag{6.7}$$

for $u \in \mathscr{D}_+'$.

Let $\phi \in \mathscr{C}_0^\infty(\overline{\mathbb{R}_+})$, $\xi, \eta \in \mathscr{D}_+'$ be chosen so that $\xi \equiv 1$ on $(\operatorname{supp} f_{1-k})^\epsilon$, $\eta \equiv 1$ on $(\operatorname{supp} u)^\epsilon$, $\xi \equiv 0$ on $\overline{\mathbb{R}_+} \backslash (\operatorname{supp} f_{1-k})^{2\epsilon}$, $\eta \equiv 0$ on $\overline{\mathbb{R}_+} \backslash (\operatorname{supp} f_{1-k})^{2\epsilon}$. Then

$$
f_{1-k} * u(\phi) = f_{1-k}(x) \times u(y)(\xi(x)\eta(y)\phi(x+y))
$$

$$
= \frac{H(x)x^{-k}}{\Gamma(1-k)}(\xi(x)u(y)(\eta(y)\phi(x+y)))
$$

$$
= \frac{H(x)x^{-k}}{\Gamma(1-k)}\left(\xi(x)\int_{-\infty}^{\infty} u(y)\phi(x+y)dy\right)
$$

$$
= \frac{1}{\Gamma(1-k)}\int_0^\infty x^{-k}\int_{-\infty}^{\infty} u(y)\phi(x+y)dydx
$$

$$
= \frac{1}{\Gamma(1-k)}\int_{-\infty}^{\infty}\int_0^\infty x^{-k}\phi(x+y)dxu(y)dy \qquad (x+y=z)
$$

$$
= \frac{1}{\Gamma(1-k)}\int_{-\infty}^{\infty} u(y)\int_y^\infty (z-y)^{-k}\phi(z)dzdy
$$

$$
= \frac{1}{\Gamma(1-k)}\int_{-\infty}^{\infty}\int_y^\infty u(y)(x-y)^{-k}\phi(x)dxdy
$$

$$
= \frac{1}{\Gamma(1-k)}\int_{-\infty}^{\infty}\int_0^x u(y)(x-y)^{-k}dy\phi(x)dx
$$

$$
= \frac{1}{\Gamma(1-k)}\int_0^x u(y)(x-y)^{-k}dy(\phi), \quad x \in \overline{\mathbb{R}_+}.
$$

As $\phi \in \mathscr{C}_0^\infty(\overline{\mathbb{R}_+})$ was chosen arbitrarily, we conclude

$$
f_{1-k} * u(x) = \frac{1}{\Gamma(1-k)}\int_0^x u(y)(x-y)^{-k}dy, \quad x \in \overline{\mathbb{R}_+}.
$$

The latter representation and (6.7) imply

$$
D^k u(x) = \frac{1}{\Gamma(1-k)}\frac{d}{dx}\int_0^x u(y)(x-y)^{-k}dy, \quad k \in (0,1), \quad x \in \overline{\mathbb{R}_+}.
$$

If $l \in \mathbb{N}$, then

$$D^{l+k}u = D^l(D^k u)$$

for $u \in \mathscr{D}'_+$.

Exercise 6.10 Compute $D^{\frac{1}{2}}H(x), x \in \overline{\mathbb{R}_+}$.

Answer $\dfrac{1}{\sqrt{\pi}} \dfrac{H(x)}{\sqrt{x}}$.

Exercise 6.11 Compute $D^{\frac{1}{3}}\delta(x), x \in \overline{\mathbb{R}_+}$.

Answer $\dfrac{1}{\Gamma(\frac{2}{3})} \dfrac{d}{dx}\left(H(x)x^{-\frac{1}{3}}\right)$.

Exercise 6.12 Let $k \in (0, 1)$ and $u \in \mathscr{D}'_+$, and call u_{-k} the kth primitive of u. Prove that

$$u_{-k} = f_k * u = \frac{1}{\Gamma(k)} \int_0^x (x - y)^{k-1}u(y)dy, \quad x \in \overline{\mathbb{R}_+}.$$

Exercise 6.13 Let $k \in (0, 1), l \in \mathbb{N}$ and $u \in \mathscr{D}'_+$. Prove that

$$u_{-l-k} = \frac{1}{\Gamma(k)} \int_0^x \int_0^{x_1} \cdots \int_0^{x_{l-1}} \int_0^{x_l} (x_l - y)^{k-1}u(y)dydx_l \ldots dx_1, \quad x \in \overline{\mathbb{R}_+}.$$

6.7 Advanced Practical Problems

Problem 6.1 Let $u(x, t) \in \mathscr{D}'(\mathbb{R}_x^n \times \mathbb{R}_t)$. Find

$$\left(D^\alpha \delta(x) \times \delta^{(\beta)}(t)\right) * u(x, t),$$

where $\alpha \in \mathbb{N}^n, \beta \in \mathbb{N}$.

Answer

$$D_x^\alpha D_t^\beta u(x, t).$$

Problem 6.2 Compute

1. $H(x) * H(x)x^2$,
2. $H(x) * H(x)\sin x$,
3. $H(x) * H(x)x^3$,

4. $H(x) * H(x)(x + \cos x)$,
5. $H(x) * H(x) f(x)$, $f \in \mathscr{C}^{\infty}(\mathbb{R})$

in $\mathscr{D}'(\mathbb{R})$.

1. *Solution* Fix $\phi \in \mathscr{C}_0^{\infty}(\mathbb{R})$ and choose $\xi, \eta \in \mathscr{C}_0^{\infty}(\mathbb{R})$ so that $\xi \equiv 1$ on $(\mathrm{supp}(H(x)x^2))^{\epsilon}$, $\eta \equiv 1$ on $(\mathrm{supp} H(x))^{\epsilon}$, $\xi \equiv 0$ on $\mathbb{R} \backslash (\mathrm{supp}(H(x)x^2))^{2\epsilon}$, $\eta \equiv 0$ on $\mathbb{R} \backslash (\mathrm{supp} H(x))^{2\epsilon}$. Then

$$H(x) * H(x)x^2(\phi) = H(x) \times H(y)y^2(\xi(x)\eta(y)\phi(x + y))$$

$$= \int_{-\infty}^{\infty} H(x) \int_{-\infty}^{\infty} H(y)y^2\phi(x + y)\xi(x)\eta(y)dydx$$

$$= \int_{-\infty}^{\infty} \int_{-\infty}^{\infty} H(x)H(y)y^2\phi(x + y)dydx \qquad (y + x = z)$$

$$= \int_{-\infty}^{\infty} \int_{-\infty}^{\infty} H(x)H(z - x)(z - x)^2\phi(z)dzdx$$

$$= \int_{-\infty}^{\infty} \phi(z) \int_{-\infty}^{\infty} H(x)H(z - x)(z - x)^2dxdz$$

$$= \int_{-\infty}^{\infty} H(z) \int_{0}^{z} (z - x)^2dx\phi(z)dz$$

$$= \int_{-\infty}^{\infty} H(z)\frac{z^3}{3}\phi(z)dz$$

$$= H(x)\frac{x^3}{3}(\phi).$$

Since $\phi \in \mathscr{C}_0^{\infty}(\mathbb{R})$ was arbitrarily chosen, we conclude that

$$H(x) * H(x)x^2 = H(x)\frac{x^3}{3}.$$

2. **Answer** $2H(x) \sin^2 \frac{x}{2}$,

3. **Answer** $H(x)\frac{x^4}{4}$,

4. **Answer** $H(x)\left(\dfrac{x^2}{2} + \sin x\right)$,

5. **Answer** $H(x)\displaystyle\int_0^x f(x-t)dt.$

Problem 6.3 Compute

1. $H(x)x * H(x)x^2$,
2. $H(x)x * H(x)\sin x$,
3. $H(x)\cos x * H(x)x^3$,
4. $H(x)x * H(x)e^{-x}$,
5. $H(x)f(x) * H(x)g(x),\ f, g \in \mathscr{C}^\infty(\mathbb{R})$

in $\mathscr{D}'(\mathbb{R})$.

1. *Solution* Let $\phi \in \mathscr{C}_0^\infty(\mathbb{R})$, $\xi, \eta \in \mathscr{C}_0^\infty(\mathbb{R})$ such that $\xi \equiv 1$ on $(\mathrm{supp}(H(x)x))^\epsilon$, $\eta \equiv 1$ on $(\mathrm{supp}(H(x)x^2))^\epsilon$, $\xi \equiv 0$ on $\mathbb{R}\backslash(\mathrm{supp}(H(x)x))^{2\epsilon}$, $\eta \equiv 0$ on $\mathbb{R}\backslash(\mathrm{supp}(H(x)x^2))^{2\epsilon}$. Then

$$H(x)x * H(x)x^2(\phi) = H(x)x \times H(y)y^2(\xi(x)\eta(y)\phi(x+y))$$

$$= \int_{-\infty}^\infty H(x)x \int_{-\infty}^\infty H(y)y^2\xi(x)\eta(y)\phi(x+y)dydx$$

$$= \int_{-\infty}^\infty H(x)x \int_{-\infty}^\infty H(y)y^2\phi(x+y)dydx$$

$$= \int_{-\infty}^\infty H(x)x \int_{-\infty}^\infty H(z-x)(z-x)^2\phi(z)dzdx$$

$$= \int_{-\infty}^\infty \phi(z) \int_{-\infty}^\infty H(x)H(z-x)x(z-x)^2dxdz$$

$$= \int_{-\infty}^\infty \phi(z)H(z) \int_0^z x(z-x)^2dxdz$$

$$= \int_{-\infty}^\infty H(z)\frac{z^4}{12}\phi(z)dz$$

$$= H(x)\frac{x^4}{12}(\phi).$$

Therefore

$$H(x)x * H(x)x^2 = H(x)\frac{x^4}{4}.$$

2. **Answer** $H(x)(x - \sin x)$,

3. **Answer** $H(x)(3x^2 + 6\cos x - 6)$,

4. **Answer** $H(x)(x - 1 + e^{-x})$,

5. **Answer** $H(x)\displaystyle\int_0^x f(y)g(x - y)dy.$

Problem 6.4 Prove

1. $\delta(x - a) * \delta(x - b) = \delta(x - a - b)$, $x \in \mathbb{R}$, $a, b = \text{const}$,
2. $\delta^{(m)}(x - a) * \left(\delta^{(k)}(x - b) * u(x)\right) = u^{(k+m)}(x - a - b)$, $x \in \mathbb{R}$,

$u \in \mathscr{D}'(\mathbb{R}), a, b = \text{const}, k, m \in \mathbb{N}.$

Problem 6.5 In $\mathscr{D}'(\mathbb{R}^2)$ compute

1. $H(at - |x|) * \left(H(t) \times \delta(x)\right), a > 0,$
2. $H(at - |x|) * \left(\delta(t) \times \delta(x)\right), a > 0,$
3. $H(at - |x|) * \left(H(t) \sin t \times \delta(x)\right), a > 0,$
4. $H(at - |x|) * \left(H(t)(t^2 + t + 1) \times \delta(x)\right), a > 0,$
5. $H(at - |x|) * \left(H(t)(1 + \cos t) \times \delta(x)\right), a > 0,$
6. $H(at - |x|) * \left(f(t) \times \delta(x)\right), a > 0,$

where $f \in \mathscr{C}(t \geq 0)$ and $f \equiv 0$ for $t < 0.$

Answer

1. $H(t)\left(t - \dfrac{|x|}{a}\right),$
2. $H(at - |x|),$
3. $H(at - |x|)2\sin^2\left(\dfrac{t}{2} - \dfrac{|x|}{2a}\right),$
4. $H(t)\left(\dfrac{1}{3}\left(t - \dfrac{|x|}{a}\right)^3 + \dfrac{1}{2}\left(t - \dfrac{|x|}{a}\right)^2 + t - \dfrac{|x|}{a}\right),$
5. $H(t)\left(t - \dfrac{|x|}{a} + \sin\left(t - \dfrac{|x|}{a}\right)\right),$
6. $H(at - |x|)\displaystyle\int_0^{t - \frac{|x|}{a}} f(\tau)d\tau.$

Problem 6.6 Let $X_1 \subset \mathbb{R}^{n_1}$, $X_2 \subset \mathbb{R}^{n_2}$ be open sets and $K \in \mathscr{D}'(X_1 \times X_2)$. Define the map $\mathscr{K} : \mathscr{C}_0^\infty(X_2) \mapsto \mathscr{D}'(X_1)$ by

$$\mathscr{K}(\phi)(x_1) = \int_{X_2} K(x_1, x_2)\phi(x_2)dx_2, \quad \phi \in \mathscr{C}_0^\infty(X_2).$$

Prove that

1. $\mathscr{K}$ is continuous if and only if $\mathscr{K}\phi_j \to_{j\to\infty} 0$ in $\mathscr{D}'(X_1)$ as $\phi_j \to_{j\to\infty} 0$ in $\mathscr{C}_0^\infty(X_2)$.
2.

$$\mathscr{K}\phi(\psi) = K(\psi \times \phi), \quad \phi \in \mathscr{C}_0^\infty(X_2), \psi \in \mathscr{C}_0^\infty(X_1).$$

Hint Use the definition of distribution and direct product by $\mathscr{C}_0^\infty$ functions.

Definition 6.7 The distribution $K(x_1, x_2)$ is called the kernel of the map $\mathscr{K}$.

Problem 6.7 Let $X_1 \subset \mathbb{R}^{n_1}$, $X_2 \subset \mathbb{R}^{n_2}$ be open sets, $K \in \mathscr{C}^\infty(X_1 \times X_2)$, and define $\mathscr{K} : \mathscr{C}^\infty(X_2) \mapsto \mathscr{C}^\infty(X_1)$ by

$$\mathscr{K}\phi(x_1) = \int_{X_2} K(x_1, x_2)\phi(x_2)dx_2, \quad \phi \in \mathscr{C}_0^\infty(X_2).$$

Prove that $\mathscr{K}$ can be extended to a map from $\mathscr{E}'(X_2)$ to $\mathscr{C}^\infty(X_1)$

$$\mathscr{K}u(x_1) = u\Big(K(x_1, \cdot)\Big), \quad u \in \mathscr{E}'(X_2), \quad x_1 \in X_1.$$

Problem 6.8 Let $u_1 \in \mathscr{D}'^k(\mathbb{R}^n)$, $u_2 \in \mathscr{C}_0^\infty(\mathbb{R}^n)$. Prove that $u_1 * u_2$ defines a continuous function

$$x \mapsto u_1(u_2(x - \cdot)).$$

Problem 6.9 Take $u_1, u_2 \in \mathscr{D}'(X)$, u_2 with compact support. Prove that

$$\mathrm{singsupp}(u_1 * u_2) \subset \mathrm{singsupp}u_1 + \mathrm{singsupp}u_2.$$

Solution Let $\psi \in \mathscr{C}_0^\infty(\mathbb{R}^n)$ be such that $\psi \equiv 1$ on a neighbourhood of $\mathrm{singsupp}u_2$. Then

$$u_2 = (1 - \psi)u_2 + \psi u_2.$$

By the definition of ψ it follows that $(1 - \psi)u_2 \in \mathscr{C}_0^\infty(\mathbb{R}^n)$. Therefore $u_1 * ((1 - \psi)u_2)$ is a $\mathscr{C}^\infty$ function on

$$\left\{ x : \{x\} - \mathrm{supp}(\psi u_2) \subset \overline{\mathrm{singsupp} u_1} \right\}.$$

Consequently

$$\mathrm{singsupp}(u_1 * u_2) = \mathrm{singsupp}(u_1 * (\psi u_2)) \subset \mathrm{singsupp} u_1 + \mathrm{singsupp}(\psi u_2).$$

We also have

$$\mathrm{singsupp}(\psi u_2) \subset \mathrm{singsupp} u_2,$$

and the claim follows.

Problem 6.10 Let P be a differential operator with constant coefficients

$$P = \sum_\alpha a_\alpha D^\alpha.$$

Prove

1. $P(u) = P(\delta) * u$ for $u \in \mathscr{D}'(\mathbb{R}^n)$,
2. $P(u_1 * u_2) = P(u_1) * u_2 = u_1 * P(u_2)$

for $u_1, u_2 \in \mathscr{D}'(X)$, where u_2 has compact support.

Solution

1. We have

$$P(u) = \sum_\alpha a_\alpha \left(D^\alpha \delta * u \right) = \sum_\alpha \left(a_\alpha D^\alpha \delta * u \right) = \left(\sum_\alpha a_\alpha D^\alpha \delta \right) * u = P(\delta) * u.$$

2. We have

$$P(u_1 * u_2) = \sum_\alpha a_\alpha D^\alpha (u_1 * u_2) = \sum_\alpha a_\alpha D^\alpha u_1 * u_2 = \left(\sum_\alpha a_\alpha D^\alpha u_1 \right) * u_2 = P(u_1) * u_2.$$

At the same time

$$P(u_1 * u_2) = \sum_\alpha a_\alpha D^\alpha (u_1 * u_2) = \sum_\alpha a_\alpha u_1 * D^\alpha u_2 = u_1 * \left(\sum_\alpha a_\alpha D^\alpha u_2 \right) = u_1 * P(u_2).$$

Problem 6.11 Let $u_1, u_2 \in \mathcal{D}'(\mathbb{R}^n)$, u_2 with compact support. Suppose that for every $y \in \mathrm{supp} u_2$ we can find an integer $j \geq 0$ and an open neighbourhood V_y of y for which

1. $u_1 \in \mathcal{D}'(\{x\} - V_y)$,
2. $u_2 \in \mathcal{C}^{k+j}(V_y)$

or

1. $u_1 \in \mathcal{C}^{k+j}(\{x\} - V_y)$,
2. $u_2 \in \mathcal{D}'^j(V_y)$,

where $x' \in \mathbb{R}^n$. Prove that $u_1 * u_2 \in \mathcal{C}^k$ on a neighbourhood of x.

Problem 6.12 Given $f \in \mathcal{C}(\mathbb{R}^n)$, compute $f * \mu \delta_S$.

Problem 6.13 Let $\dfrac{\partial}{\partial n}(\nu \delta_S) \in \mathcal{D}'(\mathbb{R}^n)$, $f \in \mathcal{C}^1(\mathbb{R}^n)$. Find

$$f * \frac{\partial}{\partial n}(\nu \delta_S),$$

where ν is a piecewise-continuous function.

Problem 6.14 Let $\mu \in \mathcal{C}(\mathbb{R}^n)$. Find

1. $|x|^{2-n} * \mu \delta_S, n \geq 3$,
2. $\log |x| * \mu \delta_S, n = 2$.

Problem 6.15 Let $\mathscr{E}_n(x) = |x|^{2-n}$, $n \geq 3$ and $g \in L^1(\mathbb{R}^n)$. Compute

1. $V_n = \mathscr{E}_n * g$,
2. $\Delta_n(\mathscr{E}_n * g)$.

Problem 6.16 Compute

1. $|x|^2 * \delta_{S_R}$,
2. $\sin |x|^2 * \delta_{S_R}$,
3. $e^{|x|^2} * \delta_{S_R}$,
4. $|x| * \delta_{S_R}$,
5. $f(|x|) * \delta_{S_R}$,

where $f(x) \in \mathcal{C}([0, \infty))$.

Problem 6.17 Let $w(t)$ be a continuous function on $t \geq 0$ and $w(t) = 0$ for $t < 0$. Define $\mathscr{E}_3(x, t) = \dfrac{H(t)}{4\pi t} \delta_{S_t}(x)$. Find $\mathscr{E}_3(t) * w(t)$.

Problem 6.18 Let

$$\mathcal{E}_1(x, t) = \frac{1}{2} H(t - |x|),$$

$$\mathcal{E}_2(x, t) = \frac{H(t - |x|)}{2\pi \sqrt{t^2 - |x|^2}},$$

$$\mathcal{E}_3(x, t) = \frac{H(t)}{4\pi t} \delta_{S(t)}(x).$$

Also let $\tilde{u}(x) \in \mathscr{C}(\mathbb{R}^i)$, $i = 1, 2, 3$. Find $\mathcal{E}_i(x, t) * \tilde{u}(x)$, $i = 1, 2, 3$.

Problem 6.19 Let $f(x, t) \in \mathscr{D}'(\mathbb{R}_x^i \times \mathbb{R}_t^1)$, $i = 1, 2, 3$, be a distribution for which $\text{supp} f \subset \{(x, t) : t \geq 0\}$. Find $\mathcal{E}_i * f$, $i = 1, 2, 3$.

Problem 6.20 Let $f(\lambda) \in \mathscr{C}^1(\lambda \geq 0)$, $f'(0) = 0$. Find

$$-f(|x|) * \frac{\partial}{\partial n} \delta_{S_R}.$$

Problem 6.21 Let $u_1, u_2, \ldots, u_n \in \mathscr{C}_0$. Prove that

$$|u_1 * u_2 * \cdots * u_n(0)| \leq ||u_1||_{p_1} \cdots ||u_n||_{p_n},$$

where

$$\frac{1}{p_1} + \frac{1}{p_2} + \cdots + \frac{1}{p_n} = n - 1, \quad 1 \leq p_j \leq \infty, \quad j = 1, 2, \cdots, n.$$

Hint Distinguish two cases: $k = 2$ and $k > 2$.

Problem 6.22 Let $1 \leq p_j \leq \infty$, $j = 1, 2, \cdots, n$ and

$$\frac{1}{p_1} + \cdots + \frac{1}{p_n} = n - 1 + \frac{1}{q}, \quad 1 \leq q \leq \infty.$$

Take $u_i \in \mathscr{C}_0$, $i = 1, 2, \cdots, n$, and prove

$$||u_1 * u_2 * \cdots * u_n||_q \leq ||u_1||_{p_1} \cdots ||u_n||_{p_n}.$$

Problem 6.23 Let $k_a(y) = |y|^{-\frac{n}{a}}$ and $\frac{1}{a'} + \frac{1}{a} = 1$, $\frac{1}{p} + \frac{1}{p'} = 1$, $1 \leq p < a'$, $u \in L^\infty \cap L^p$. Prove

$$||k_a * u||_\infty \leq C_{p,a} ||u||_p^{\frac{p'}{a}} ||u||_\infty^{1 - \frac{p}{a'}}.$$

Solution Let $R > 0$ be fixed. Then

$$|k_a * u(x)| = \left| \int_{\mathbb{R}^n} k_a(x-y)u(y)dy \right| = \left| \int_{\mathbb{R}^n} k_a(y)u(x-y)dy \right|$$

$$\leq \int_{\mathbb{R}^n} |y|^{-\frac{n}{a}}|u(x-y)|dy = \int_{|y|\leq R} |y|^{-\frac{n}{a}}|u(x-y)|dy + \int_{|y|\geq R} |y|^{-\frac{n}{a}}|u(x-y)|dy.$$

For

$$\int_{|y|\leq R} |y|^{-\frac{n}{a}}|u(x-y)|dy,$$

we have the following estimate

$$\int_{|y|\leq R} |y|^{-\frac{n}{a}}|u(x-y)|dy \leq c_1\|u\|_\infty \int_0^R \rho^{-\frac{n}{a}}\rho^{n-1}d\rho = \|u\|_\infty c_2 R^{n-\frac{n}{a}}. \qquad (6.8)$$

For

$$\int_{|y|\geq R} |y|^{-\frac{n}{a}}|u(x-y)|dy$$

we have the following estimate

$$\int_{|y|\geq R} |y|^{-\frac{n}{a}}|u(x-y)|dy \leq \|u\|_p \left(\int_{|y|\geq R} |y|^{-\frac{n}{a}p'}dy \right)^{\frac{1}{p'}}$$

$$= \|u\|_p \left(c_3 \int_R^\infty \rho^{n-1-\frac{n}{a}p'}d\rho \right)^{\frac{1}{p'}} = c_4\|u\|_p R^{\frac{n}{p'}-\frac{n}{a}}. \qquad (6.9)$$

Combining (6.8) and (6.9),

$$|k_a * u(x)| \leq C_{p,a}\left(R^{n-\frac{n}{a}}\|u\|_\infty + R^{\frac{n}{p'}-\frac{n}{a}}\|u\|_p \right).$$

We choose R so that

$$R^{\frac{n}{p}} = \|u\|_p \frac{1}{\|u\|_\infty}.$$

Then

$$R^{n-\frac{n}{a}} = ||u||_p^{\frac{p}{a'}} ||u||_\infty^{-\frac{p}{a'}},$$
$$R^{n-\frac{n}{a}}||u||_\infty = ||u||_p^{\frac{p}{a'}} ||u||_\infty^{1-\frac{p}{a'}},$$
$$R^{\frac{n}{p'}-\frac{n}{a}}||u||_p = ||u||_p^{\frac{p}{a'}} ||u||_\infty^{1-\frac{p}{a'}}.$$

Consequently

$$|k_a * u(x)| \le C_{p,a}||u||_p^{\frac{p}{a'}} ||u||_\infty^{1-\frac{p}{a'}}.$$

Here c_1, c_2, c_3, c_4 and $C_{p,a}$ are nonnegative constants.

Problem 6.24 Let $u \in L^1(\mathbb{R}^n)$ and s be a positive number. Prove that

1. u can be written as

$$u = v + \sum_{k=1}^{\infty} w_k,$$

where

$$||v||_1 + \sum_{k=1}^{\infty} ||w_k||_1 \le 3||u||_1, \quad |v(x)| \le 2^n s, \quad x \in \mathbb{R}^n.$$

2. for every sequence of pairwise disjoint cubes I_k we have $w_k(x) = 0$ for $x \notin I_k$,

$$\int_{\mathbb{R}^n} w_k(x)dx = 0, s \sum_{k=1}^{\infty} \mu(I_k) \le ||u||_1.$$

Solution Let us subdivide $\mathbb{R}^n$ into cubes I_n such that $\mu(I_n) > \dfrac{1}{s}\displaystyle\int_{\mathbb{R}^n} |u(x)|dx$. Then

$$s\mu(I_n) > \int_{\mathbb{R}^n} |u(x)|dx = \sum_i \int_{I_i} |u(x)|dx > \int_{I_n} |u(x)|dx,$$

i.e.,

$$\frac{1}{\mu(I_n)} \int_{I_n} |u(x)|dx < s.$$

Now, we divide I_1 into 2^n equal parts so that the average of $|u|$ on each one is greater than or equal to s. Therefore

$$s\mu(I_{1k}) \le \int_{I_{1k}} |u(x)|dx \le \int_{I_1} |u(x)|dx \le s\mu(I_1) = 2^n s\mu(I_{1k}).$$

Let

$$v(x) = \frac{1}{\mu(I_{1k})} \int_{I_{1k}} u(y)dy, \quad x \in I_{1k}, \tag{6.10}$$

and

$$w_{1k}(x) = \begin{cases} u(x) - v(x) & \text{for} \quad x \in I_{1k}, \\ 0 & \text{for} \quad x \notin I_{1k}. \end{cases} \tag{6.11}$$

Now, we divide I_2 into 2^n equal parts so that the average of $|u|$ on each is greater than or equal to s. We extend the definitions (6.10) and (6.11) to these cubes. Continuing in this way we produce a sequence of functions w_{jk} and a sequence of cubes I_{jk}. We complete the definition of v by setting

$$v(x) = u(x) \quad \text{for} \quad x \notin O = \cup_k I_k.$$

Then

$$u = v + \sum_{k=1}^{\infty} w_k.$$

We also have

$$\int_{I_k} \left(|v(x)| + |w_k(x)| \right) dx \leq 3 \int_{I_k} |u(x)| dx.$$

Therefore

$$\sum_{k=1}^{\infty} \int_{I_k} |v(x)|dx + \sum_{k=1}^{\infty} \int_{I_k} |w_k(x)|dx \leq 3 \sum_{k=1}^{\infty} \int_{I_k} |u(x)|dx,$$

$$\int_{\mathbb{R}^n} |v(x)|dx + \sum_{k=1}^{\infty} ||w_k||_1 \leq 3||u||_1.$$

Since $I_k \cap I_l = \emptyset, k \neq l$, we have $w_k(x) = 0$ for $x \notin I_k$, $v(x) = u(x)$ for $x \notin O$ and

$$|v(x)| \leq 2^n s \quad \text{for} \quad x \in O.$$

If $x \notin O$, there exist sufficiently small cubes containing x on each of which the average of $|u|$ is less than s. Consequently

$$|u(x)| \leq s, \quad x \notin O.$$

From the inequality

$$s\mu(I_k) \le \int_{I_k} |u(x)|dx,$$

we obtain

$$s \sum_{k=1}^{\infty} \mu(I_k) \le ||u||_1.$$

Problem 6.25 Let I be a cube with centre at the origin, I^* a cube with the same centre and twice the edge. Take $w \in L^1(\mathbb{R}^n)$, supp$w \subset I$, $\int_{\mathbb{R}^n} w(x)dx = 0$. Prove

$$\left(\int_{CI^*} |k_a * w(x)|^a dx \right)^{\frac{1}{a}} \le C_a ||w||_1. \tag{6.12}$$

Here CI^* is the complement of I^*.

Solution Let the side of the cube I is L. By the mean value theorem, we have

$$|k_a * w(x)| = \left| \int_{\mathbb{R}^n} k_a(x-y)w(y)dy \right| \le \int_{\mathbb{R}^n} |(k_a(x-y) - k_a(x))||w(y)|dy \le CL|x|^{-1-\frac{n}{a}}||w||_1$$

when $x \notin I^*$. Here C is a constant. Hence,

$$\left(\int_{CI^*} |k_a * w(x)|^a dx \right)^{\frac{1}{a}} \le CL||w||_1 \left(\int_{CI^*} |x|^{-a-n}dx \right)^{\frac{1}{a}}.$$

Note that

$$\left(\int_{|x| \ge L} |x|^{-a-n}dx \right)^{\frac{1}{a}} = C_1 \left(\int_L^{\infty} r^{-a-1}dr \right)^{\frac{1}{a}}$$

$$= \frac{C_1}{L},$$

where C_1 is a constant. Therefore there exists a constant C_a so that (6.12) holds.

Problem 6.26 Prove

$$\mu \left\{ x : |k_a * u(x)| > t \right\} t^a \le C_a ||u||_1^a,$$

for $t > 0$, $a > 0$.

Solution Let us suppose $||u||_1 = 1$ (otherwise, we may consider $\dfrac{u}{||u||_1}$). Then u can be represented as

$$u = v + \sum_{k=1}^{\infty} w_k.$$

We also have (when $p = 1$)

$$|k_a * v(x)| \le c||v||_1^{\frac{1}{a}} \le c_1 s^{\frac{1}{a}}, \quad x \in \mathbb{R}^n.$$

Let s satisfies

$$c_1 s^{\frac{1}{a}} = \frac{t}{2}.$$

Then

$$|k_a * u(x)| > t, \quad x \in \mathbb{R}^n,$$

implies

$$\sum_{k=1}^{\infty} |k_a * w_k(x)| > \frac{t}{2}, \quad x \in \mathbb{R}^n.$$

Let

$$O = \cup_k I_k^*,$$

where I_k^* is the cube with twice the edge of I_k. We have

$$\mu(O) < \frac{2^n}{s}, \quad \int_{CO} \left(\sum_{k=1}^{\infty} |k_a * w_k(x)| \right)^a dx \le c_1,$$

$$\mu\left\{ x : \sum_{k=1}^{\infty} |k_a * w_k(x)| > \frac{t}{2} \right\} \le \frac{2^n}{s} + c\left(\frac{t}{2} \right)^{-a} \le c_1 t^{-a}.$$

Problem 6.27 Let $1 < a < \infty$, $1 < p < q < \infty$, $\dfrac{1}{p} + \dfrac{1}{a} = 1 + \dfrac{1}{q}$. Prove

$$||k_a * u||_q \le C_{p,a} ||u||_p$$

for $u \in \mathscr{C}_0(\mathbb{R}^n)$.

Solution For convenience we will suppose that

$$||u||_p = 1.$$

Let

$$\mu(t) = \mu\left\{x : |k_a * w(x)| > t\right\}.$$

Then

$$||k_a * u||_q^q = \int_{\mathbb{R}^n} |k_a * u(x)|^q dx = -\int_0^\infty t^q d\mu(t) = q \int_0^\infty t^{q-1}\mu(t)dt.$$

Set

$$u = v + w,$$

where

$$u = \begin{cases} v & \text{for} \quad |u| \le s, \\ w & \text{for} \quad |u| > s. \end{cases}$$

Then

$$||k_a * v||_\infty \le cs^{1-\frac{p}{a'}} = cs^{\frac{p}{q}}.$$

Now, we choose s so that

$$cs^{\frac{p}{q}} = \frac{t}{2}.$$

If

$$|k_a * u(x)| > t, \quad x \in \mathbb{R}^n,$$

then we have

$$|k_a * w(x)| > \frac{t}{2}, \quad x \in \mathbb{R}^n.$$

Consequently

$$\mu(t) \leq \mu\left\{x : |k_a * w(x)| > \frac{t}{2}\right\} \leq c' t^{-a} \|w\|_1^q,$$

$$\|k_a * u\|_q^q \leq c'' \int_0^\infty t^{q-1-a}\left(\int_{|u(x)|>s} |u(x)|dx\right)^a dt$$

$$\leq c''\left(\int_0^\infty \left(\int_{|u(x)|>s} t^{q-1-a}dt\right)^{\frac{1}{a}} |u(x)|dx\right)^a.$$

We note that

$$\int_{s<|u(x)|} t^{q-1-a}dt \approx t^{q-a}.$$

When $s = |u(x)|$, we have

$$\int_{s=|u(x)|} t^{q-1-a}dt \approx |u(x)|^{\frac{(q-a)p}{q}} = |u(x)|^{\frac{ap}{p'}}.$$

Consequently

$$\|k_a * u\|_q^q \leq c\left(\int_{\mathbb{R}^n} |u(x)|^{1+\frac{p}{p'}}dx\right)^a = c\left(\int_{\mathbb{R}^n} |u(x)|^p dx\right)^a$$

and then there exists a $C_{p,a}$ such that

$$\|k_a * u\|_q \leq C_{p,a}\|u\|_p.$$

Problem 6.28 Let $k \in \mathscr{C}^1(\mathbb{R}^n \setminus \{0\})$ be a homogeneous function of degree $-\dfrac{n}{a}$, $1 \leq p \leq \infty$ and $0 < \gamma = n\left(1 - \dfrac{1}{a} - \dfrac{1}{p}\right) < 1$. Prove

$$\sup_{\substack{x \neq y \\ x,y \in \mathbb{R}^n}} |k * u(x) - k * u(y)||x - y|^{-\gamma} \leq c\|u\|_p$$

for $u \in L^p(\mathbb{R}^n) \cap \mathscr{E}'(\mathbb{R}^n)$.

Solution Let $x \in \mathbb{R}^n$ and $h = |x|$. We have

$$k * u(x) - k * u(0) = \int_{\mathbb{R}^n} \Big(k(x - y) - k(-y)\Big)u(y)dy$$

$$= \int_{|y|\leq 2h} \Big(k(x - y) - k(-y)\Big)u(y)dy + \int_{|y|\geq 2h} \Big(k(x - y) - k(-y)\Big)u(y)dy.$$

Now, we consider

$$\int_{|y|\leq 2h} \Big(k(x - y) - k(-y)\Big)u(y)dy.$$

Then

$$\Big| \int_{|y|\leq 2h} \Big(k(x - y) - k(-y)\Big)u(y)dy\Big|$$

$$\leq \Big(\int_{|y|\leq 2h} \big|k(x - y) - k(-y)\big|^{p'} dy\Big)^{\frac{1}{p'}}\Big(\int_{|y|\leq 2h} |u(y)|^p dy\Big)^{\frac{1}{p}}$$

$$\leq c\|u\|_p\Big(\int_{|y|\leq 2h} |k(y)|^{p'} dy\Big)^{\frac{1}{p'}}$$ (6.13)

$$\leq c_1\|u\|_p h^{\left(n - \frac{np'}{a}\right)\frac{1}{p'}}$$

$$= c_1 h^\gamma \|u\|_p,$$

where c and c_1 are positive constants. As we saw earlier, using the mean value theorem we obtain

$$\Big| \int_{|y|\geq 2h} \Big(k(x - y) - k(-y)\Big)u(y)dy\Big| \leq c_2 h^\gamma \|u\|_p,$$

where c_2 is a positive constant. From here and (6.13), we find

$$\sup_{x\neq 0, x\in\mathbb{R}^n} |k * u(x) - k * u(0)| \leq c_3 h^\gamma \|u\|_p$$

for some positive constant c_3.

Problem 6.29 Let $u \in \mathscr{D}'(X)$, $p > n$, $D_j u \in L^p_{\text{loc}}$, $j = 1, 2, \cdots, n$. Prove

$$\sup_{x\neq y; x, y\in K} \frac{|u(x) - u(y)|}{|x - y|^\gamma} < \infty, \quad \gamma = 1 - \frac{n}{p}.$$

Problem 6.30 Let $u \in \mathscr{D}'(X)$, $1 < p < \infty$, $m \in \mathbb{N}$. Let $D^{\alpha}u \in L_{\text{loc}}^{p}(X)$ for $|\alpha| = m$. Prove that for $|\alpha| < m$ we have

1. $D^{\alpha}u \in L_{\text{loc}}^{q}(X)$ if $q < \infty$, $\dfrac{1}{p} \le \dfrac{1}{q} + \dfrac{m - |\alpha|}{n}$,

2. $D^{\alpha}u$ is Hölder continuous of order γ, where $0 < \gamma < 1$ and $\dfrac{1}{p} \le (m - |\alpha| - \gamma)\dfrac{1}{n}$.

Problem 6.31 In $\mathscr{D}'(R^{1})$ compute

1. $\left(\dfrac{d}{dx}\right)^{\frac{1}{4}}(H(x) * \delta(x))$,

2. $\left(\dfrac{d}{dx}\right)^{\frac{1}{3}} H(x)$.

Answer

1. $\dfrac{H(x)}{\Gamma\left(\frac{3}{4}\right)x^{\frac{1}{4}}}$,

2. $\dfrac{H(x)}{\Gamma\left(\frac{2}{3}\right)x^{\frac{1}{3}}}$.

Problem 6.32 In $\mathscr{D}'(\mathbb{R})$ compute

1. $\lim\limits_{k \to \infty} \delta(x + k)$,
2. $\lim\limits_{k \to \infty} \delta(x - k)$,
3. $\delta(x + k) * \delta(x - k)$, $k \in \mathbb{R}$.

Answer

1. 0,
2. 0,
3. $\delta(x)$.

Problem 6.33 Let

$$f_{\alpha}(x) = \frac{1}{\sqrt{2\pi}\alpha}e^{-\frac{x^{2}}{2\alpha^{2}}}, \quad x \in \mathbb{R}, \quad \alpha > 0.$$

Prove that $f_{\alpha} \in \mathscr{D}'(\mathbb{R})$ and

$$f_{\alpha} * f_{\beta} = f_{\sqrt{\alpha^{2} - \beta^{2}}}.$$

Problem 6.34 Let

$$f_{\alpha}(x) = \frac{1}{\pi}\frac{\alpha}{\alpha^{2} + x^{2}}, \quad x \in \mathbb{R}, \quad \alpha > 0.$$

Prove that $f_\alpha \in \mathscr{D}'(\mathbb{R})$ and

$$f_\alpha * f_\beta = f_{\alpha+\beta}.$$

Problem 6.35 Prove that the function

$$u(x) = \frac{\sin \pi \alpha}{\pi} \int_0^x \frac{g'(\xi)}{(x - \xi)^{1-\alpha}} d\xi$$

solves

$$\int_0^x \frac{u(\xi)}{(x - \xi)^\alpha} d\xi = g(x), \quad g(0) = 0, \quad g \in \mathscr{C}^1(x \geq 0), \quad 0 < \alpha < 1.$$

Problem 6.36 Prove

$$e^{ax} f * e^{ax} g = e^{ax}(f * g), \quad f, g \in \mathscr{D}'_+.$$

Problem 6.37 Let $f \in \mathscr{D}'(\mathbb{R}^n)$. Prove that the convolution $f * 1$ exists and is constant.

Problem 6.38 Let $u \in \mathscr{D}'(\mathbb{R}^n)$, $\phi \in \mathscr{C}_0^\infty(\mathbb{R}^n)$. Prove

1. $u * \phi \in \mathscr{C}^\infty(\mathbb{R}^n)$,
2. $\operatorname{supp}(u * \phi) \subset \operatorname{supp} u + \operatorname{supp} \phi$,
3. $D^\alpha(u * \phi) = D^\alpha u * \phi = u * D^\alpha \phi$

for every $\alpha \in \mathbb{N}^n$.

2. *Solution* Let $u * \phi(x) \neq 0$. Then $x - y \in \operatorname{supp} \phi$, so $x \in \operatorname{supp} u + \operatorname{supp} \phi$. Since x is arbitrary in $\operatorname{supp}(u * \phi)$, we conclude that

$$\operatorname{supp}(u * \phi) \subset \operatorname{supp} u + \operatorname{supp} \phi.$$

3. *Solution* From the definition of $D^\alpha u$, it follows that

$$D^\alpha(u * \phi) = D^\alpha u * \phi = u * D^\alpha \phi.$$

Problem 6.39 Let $u \in \mathscr{D}'(\mathbb{R}^n)$, $\phi, \psi \in \mathscr{C}_0^\infty(\mathbb{R}^n)$. Prove

$$u * (\phi * \psi) = (u * \phi) * \psi.$$

Solution

$$u * (\phi * \psi)(x) = \lim_{h \to 0} u\left(\sum_{k \in \mathbb{Z}^n} \phi(x - \cdot - kh)h^n \psi(kh) \right)$$

$$= \lim_{h \to 0} \sum_{k \in \mathbb{Z}^n} (u * \phi)(x - kh)\psi(kh)h^n$$

$$= \int_{\mathbb{R}^n} (u * \phi)(x - y)\psi(y)dy = (u * \phi) * \psi.$$

Problem 6.40 Let $u, v \in \mathscr{D}'(X)$, where X is a real open interval. Prove

1. $u' \geq 0$ if and only if u defines a nondecreasing function,
2. $v'' \geq 0$ if and only if v defines a convex function.

6.8   Notes and References

In this chapter we define convolution of distributions. We deduct some of its basic properties. We prove that the convolution of distributions exists and it is well defined. We introduce a regularization of distributions. As applications of the convolution of distributions, we introduce fractional differentiation and integration. In the chapter are investigated the convolution algebras $\mathscr{D}'(\Gamma+)$ and $\mathscr{D}'(\Gamma)$. Additional materials can be found in [7, 16, 17, 20, 21, 24, 25] and references therein.

Chapter 7
Tempered Distributions

7.1 Definition

Definition 7.1 A linear continuous functional on $\mathscr{S}(\mathbb{R}^n)$ is called a tempered distribution. The space of tempered distributions is indicated by $\mathscr{S}'(\mathbb{R}^n)$.

Definition 7.2 A sequence $\{u_n\}_{n=1}^{\infty}$ in $\mathscr{S}'(\mathbb{R}^n)$ is said to converge in $\mathscr{S}'(\mathbb{R}^n)$ to $u \in \mathscr{S}'(\mathbb{R}^n)$ if $u_n(\phi) \to_{n\to\infty} u(\phi)$ for every $\phi \in \mathscr{S}(\mathbb{R}^n)$.

Note that the convergence in $\mathscr{S}'(\mathbb{R}^n)$ implies convergence in $\mathscr{D}'(\mathbb{R}^n)$.

Definition 7.3 A set $M' \subset \mathscr{S}'(\mathbb{R}^n)$ is called weakly bounded if for every $\phi \in \mathscr{S}(\mathbb{R}^n)$ there is a constant C_ϕ such that $|u(\phi)| \leq C_\phi$ for every $u \in M'$.

Theorem 7.1 *If $M' \subset \mathscr{S}'(\mathbb{R}^n)$ is a weakly bounded set, there exist constants $K > 0$ and $m \in \mathbb{N}$ such that*

$$|u(\phi)| \leq K\|\phi\|_{\mathscr{S},m}, \quad u \in M', \phi \in \mathscr{S}(\mathbb{R}^n).$$

Proof In order to show this let us suppose that the assertion is false, i.e., there exist sequences $\{u_k\}_{k=1}^{\infty}$ in M' and $\{\phi_k\}_{k=1}^{\infty}$ in $\mathscr{S}(\mathbb{R}^n)$ such that

$$|u_k(\phi_k)| > k\|\phi_k\|_{\mathscr{S},k}, \quad k \in \mathbb{N}. \tag{7.1}$$

We define the functions

$$\psi_k(x) = \frac{1}{\sqrt{k}}\frac{\phi_k(x)}{\|\phi_k\|_{\mathscr{S},k}}, \quad k \in \mathbb{N}.$$

© The Author(s), under exclusive license to Springer Nature Switzerland AG 2021
S. G. Georgiev, *Theory of Distributions*,
https://doi.org/10.1007/978-3-030-81265-2_7

So, $\psi_k \in \mathscr{S}(\mathbb{R}^n)$, $k \in \mathbb{N}$, and

$$\|\psi_k\|_{\mathscr{S},p} = \frac{1}{\sqrt{k}} \frac{\|\phi_k\|_{\mathscr{S},p}}{\|\phi_k\|_{\mathscr{S},k}}, \qquad k, p \in \mathbb{N}.$$

Moreover,

$$\|\phi_k\|_{\mathscr{S},p} \le \|\phi_k\|_{\mathscr{S},k}$$

for every $k \ge p$. Hence,

$$\|\psi_k\|_{\mathscr{S},p} \le \frac{1}{\sqrt{k}} \to_{k\to\infty} 0.$$

Since $p \in \mathbb{N}$ was arbitrary and $\mathscr{S} = \bigcap_{p\in\mathbb{N}\cup\{0\}} \mathscr{S}_p$, we have $\psi_k \to_{k\to\infty} 0$ in $\mathscr{S}(\mathbb{R}^n)$. Using techniques of the sort of (2.9)–(2.14), we conclude that

$$u_k(\psi_k) \to_{k\to\infty} 0. \tag{7.2}$$

On the other hand, using (7.1), we have

$$|u_k(\phi_k)| > \sqrt{k}\sqrt{k}\|\phi_k\|_{\mathscr{S},k}, \qquad k \in \mathbb{N},$$

from which

$$|u_k(\psi_k)| > \sqrt{k} \to_{k\to\infty} \infty.$$

This contradicts with (7.2). This completes the proof.

From this we also deduce that any tempered distribution u has finite order m. It can be extended to a linear continuous functional from the smallest dual space $\mathscr{S}'_m$, and

$$|u(\phi)| \le \|u\|_{\mathscr{S},-m}\|\phi\|_{\mathscr{S},m},$$

where $\|u\|_{\mathscr{S},-m}$ is a norm in $\mathscr{S}'_m$.

Example 7.1 Let u be defined on $\mathbb{R}^n$ and suppose

$$\int_{\mathbb{R}^n} \frac{|u(x)|}{(1+|x|)^m} dx < \infty$$

for some $m \ge 0$. Define the functional on $\mathscr{S}(\mathbb{R}^n)$

$$u(\phi) = \int_{\mathbb{R}^n} u(x)\phi(x)dx, \qquad \phi \in \mathscr{S}(\mathbb{R}^n). \tag{7.3}$$

This is well defined. In fact, let $\phi \in \mathscr{S}(\mathbb{R}^n)$ and C be a positive constant such that

$$\sup_{x \in \mathbb{R}^n} \left((1 + |x|)^m |\phi(x)|\right) \leq C.$$

Then

$$|u(\phi)| = \left| \int_{\mathbb{R}^n} u(x)\phi(x)dx \right| \leq \int_{\mathbb{R}^n} |u(x)||\phi(x)|dx$$

$$= \int_{\mathbb{R}^n} \frac{|u(x)|}{(1 + |x|)^m} (1 + |x|)^m |\phi(x)|dx \leq C \int_{\mathbb{R}^n} \frac{|u(x)|}{(1 + |x|)^m}dx < \infty.$$

It is a linear and continuous functional on $\mathscr{S}(\mathbb{R}^n)$, so $u \in \mathscr{S}'(\mathbb{R}^n)$.

Exercise 7.1 Prove that $e^x \notin \mathscr{S}'(\mathbb{R})$.

Exercise 7.2 Prove that $\cos\left(e^x\right)$ belongs to $\mathscr{S}'(\mathbb{R})$ but not to $\mathscr{S}(\mathbb{R})$.

Exercise 7.3 Show that $\mathscr{S}'(\mathbb{R}^n)$ is a $\mathbb{C}$-vector space.

7.2   Direct Product

We remark that the function $\psi(x) = u_1(y)(\phi(x, y))$, where $\phi \in \mathscr{S}(\mathbb{R}^{n+m})$, $u_1 \in \mathscr{S}'(\mathbb{R}^m)$, satisfies

$$D^\alpha \psi(x) = u_1(y)(D_x^\alpha \phi(x, y))$$

for every $\alpha \in \mathbb{N}^n \cup \{0\}$. Since $u_1 \in \mathscr{S}'(\mathbb{R}^m)$, there exist $q \in \mathbb{N}$ and a positive constant C_{u_1} such that

$$|D^\alpha \psi(x)| \leq C_{u_1} \sup_{y \in \mathbb{R}^m, |\beta| \leq q} (1 + |y|^2)^{\frac{q}{2}} |D_x^\alpha D_y^\beta \phi(x, y)|.$$

Therefore

$$||\psi||_{\mathscr{S},p} = \sup_{x \in \mathbb{R}^n, |\alpha| \leq p} (1 + |x|^2)^{\frac{p}{2}} |D^\alpha \psi(x)|$$

$$\leq C_{u_1} \sup_{\substack{(x,y) \in \mathbb{R}^{n+m} \\ |\alpha| \leq p, |\beta| \leq q}} (1 + |x|^2)^{\frac{p}{2}} (1 + |y|^2)^{\frac{q}{2}} |D_x^\alpha D_y^\beta \phi(x, y)|$$

$$\leq C_{u_1} ||\phi||_{\mathscr{S},p+q}, \quad \phi \in \mathscr{S}(\mathbb{R}^{n+m}),$$

for $p, q \in \mathbb{N}$. Let $u_1 \in \mathscr{S}'(\mathbb{R}^n)$, $u_2 \in \mathscr{S}'(\mathbb{R}^m)$. Then the functional

$$u_1(\psi) = u_1(x)(u_2(y)(\phi(x, y))), \quad \phi \in \mathscr{S}(\mathbb{R}^{n+m}),$$

is linear and continuous on $\mathscr{S}(\mathbb{R}^{n+m})$.

Definition 7.4 The direct product of u_1 and u_2 is

$$u_1(x) \times u_2(y)(\phi) = u_1(x)(u_2(y)(\phi(x, y))), \quad \phi \in \mathscr{S}(\mathbb{R}^{n+m}).$$

Notice that $u_1(x) \times u_2(y) \in \mathscr{S}'(\mathbb{R}^{n+m})$.

Since $\mathscr{C}_0^\infty(\mathbb{R}^n)$ is dense in $\mathscr{S}(\mathbb{R}^n)$, all properties of direct products in $\mathscr{D}'(\mathbb{R}^n)$ hold true for the space $\mathscr{S}'(\mathbb{R}^n)$.

Exercise 7.4 Let $u_1 \in \mathscr{S}'(\mathbb{R}^n)$, $u_2 \in \mathscr{S}'(\mathbb{R}^m)$. Prove that the operation

$$u_1(x) \mapsto u_1(x) \times u_2(y)$$

from $\mathscr{S}'(\mathbb{R}^n)$ to $\mathscr{S}'(\mathbb{R}^{n+m})$ is linear and continuous.

7.3   Convolution

Take $u_1, u_2 \in \mathscr{S}'(\mathbb{R}^n)$ so that the convolution $u_1 * u_2$ exists in $\mathscr{D}'(\mathbb{R}^n)$.

1. Let $u_1 \in \mathscr{S}'(\mathbb{R}^n)$, $u_2 \in \mathscr{E}'(\mathbb{R}^n)$. Since $u_1 * u_2$ exists in $\mathscr{D}'(\mathbb{R}^n)$, we have

$$u_1 * u_2(\phi) = u_1(x) \times u_2(y)(\eta(y)\phi(x + y)), \quad \phi \in \mathscr{S}(\mathbb{R}^n),$$

where $\eta \in \mathscr{C}_0^\infty(\mathbb{R}^n)$ and $\eta \equiv 1$ on $(\mathrm{supp}\, u_2)^\epsilon$. As $u_1 \in \mathscr{S}'(\mathbb{R}^n)$, $u_2 \in \mathscr{E}'(\mathbb{R}^n)$, the direct product $u_1(x) \times u_2(y)$ exists in $\mathscr{S}'(\mathbb{R}^{2n})$. Then the convolution $u_1 * u_2$ exists in $\mathscr{S}'(\mathbb{R}^n)$. We claim that $\phi \mapsto \eta(y)\phi(x+y)$ is a continuous operation on $\mathscr{S}(\mathbb{R}^n)$. In fact, we have

$$\|\eta(y)\phi(x + y)\|_{\mathscr{S},p} \le \sup_{(x,y)\in\mathbb{R}^{2n}, |\alpha|\le p} (1 + |x|^2 + |y|^2)^{\frac{p}{2}} |D^\alpha(\eta(y)\phi(x + y))|$$

$$\le C_\alpha \sup_{(x,y)\in\mathbb{R}^{2n}, |\alpha|\le p} (1 + |x|^2 + |y|^2)^{\frac{p}{2}} |D^\alpha \phi(x + y)|$$

$$= C_\alpha \|\phi\|_{\mathscr{S},p}, \quad C_\alpha = \text{const.}$$

Therefore the map $\phi \mapsto \eta(y)\phi(x + y)$ from $\mathscr{S}(\mathbb{R}^n)$ to itself is continuous, and $u_1 \mapsto u_1 * u_2$ is continuous from $\mathscr{S}'(\mathbb{R}^n)$ to $\mathscr{S}'(\mathbb{R}^n)$.

2. Let Γ be a closed, convex, acute cone in $\mathbb{R}^n$ with vertex at 0, $C = \int \Gamma^*$, S is a strictly C-like surface. Now, we suppose $u_1 \in \mathscr{S}'(\Gamma+)$ and $u_2 \in \mathscr{S}'(\overline{S_+})$. The convolution $u_1 * u_2$ exists in $\mathscr{S}'(\mathbb{R}^n)$ and can be represented as

$$u_1 * u_2(\phi) = u_1(x) \times u_2(y)(\xi(x)\eta(y)\phi(x+y)), \quad \phi \in \mathscr{S}(\mathbb{R}^n),$$

where $\xi, \eta \in \mathscr{C}_0^\infty(\mathbb{R}^n)$, $\xi \equiv 1$ on $(\mathrm{supp} u_1)^\epsilon$, $\eta \equiv 1$ on $(\mathrm{supp} u_2)^\epsilon$ and $\xi \equiv 0$ on $\mathbb{R}^n \backslash (\mathrm{supp} u_1)^{2\epsilon}$, $\eta \equiv 0$ on $\mathbb{R}^n \backslash (\mathrm{supp} u_2)^{2\epsilon}$. If K is a compact set in $\mathbb{R}^n$ and $\mathrm{supp} u_1 \subset \Gamma + K$, the map $u_1 \mapsto u_1 * u_2$ is continuous from $\mathscr{S}'(\Gamma + K)$ to $\mathscr{S}'(\overline{S_+} + K)$.
   The set $\mathscr{S}'(\Gamma+)$ is a convolution subalgebra of $\mathscr{D}'(\Gamma+)$ and $\mathscr{S}'(\Gamma)$ is a convolution subalgebra of $\mathscr{S}'(\Gamma+)$.
3. Let $u \in \mathscr{S}'(\mathbb{R}^n)$ and $\eta \in \mathscr{S}(\mathbb{R}^n)$. Then the convolution $u_1 * \eta$ exists in Θ_M. It can be represented in the form

$$u * \eta(\phi) = u(\eta * \phi(-x)), \quad \phi \in \mathscr{S}(\mathbb{R}^n).$$

We note that there exists a natural number m such that

$$|D^\alpha (u * \eta)(x)| \leq C_u (1 + |x|^2)^{\frac{m}{2}} \|\eta\|_{\mathscr{S}, m+|\alpha|}, \quad x \in \mathbb{R}^n.$$

Here $C_u = \mathrm{const}$. In fact, let $\{\eta_k(x, y)\}_{k=1}^\infty$ be a sequence in $\mathscr{C}_0^\infty(\mathbb{R}^{2n})$ such that $\eta_k \to_{k\to\infty} 1$ in $\mathbb{R}^{2n}$ and $\phi \in \mathscr{S}(\mathbb{R}^n)$. Then

$$\int_{\mathbb{R}^n} \eta(y)\eta_k(x, y)\phi(x+y)dy \to_{k\to\infty} \int_{\mathbb{R}^n} \eta(y)\phi(x+y)dy$$

in $\mathscr{S}(\mathbb{R}^n)$. Since $u * \eta$ exists in $\mathscr{D}'(\mathbb{R}^n)$, we have

$$u * \eta(\phi) = \lim_{k\to\infty} u(x) \times \eta(y)(\eta_k(x, y)\phi(x+y))$$
$$= \lim_{k\to\infty} u(x)\left(\int_{\mathbb{R}^n} \eta(y)\eta_k(x, y)\phi(x+y)dy \right)$$
$$= u(x)\left(\int_{\mathbb{R}^n} \eta(y)\phi(x+y)dy \right) = u(x)\left(\int_{\mathbb{R}^n} \phi(\xi)\eta(\xi - x)d\xi \right)$$
$$= u(x)(\eta * \phi(-x)).$$

We note that $\phi\eta \in \mathscr{S}(\mathbb{R}^n)$ and

$$u * \eta(\phi) = \int_{\mathbb{R}^n} u(x)(\eta(\xi - x))\phi(\xi)d\xi.$$

But $\phi \in \mathscr{S}(\mathbb{R}^n)$ was chosen arbitrarily, so

$$u * \eta(x) = u(y)(\eta(x - y)).$$

If m is the order of u, then

$$
\begin{aligned}
|D^\alpha(u * \eta)(x)| &\leq C_u \|D_x^\alpha \eta(x - y)\|_{\mathscr{S},m} \\
&= C_u \sup_{y \in \mathbb{R}^n, |\beta| \leq m} (1 + |y|^2)^{\frac{m}{2}} |D_x^\alpha D_y^\beta \eta(x - y)| \\
&= C_u \sup_{\xi \in \mathbb{R}^n, |\beta| \leq m} (1 + |x - \xi|^2)^{\frac{m}{2}} |D^{\alpha+\beta}\eta(\xi)| \\
&\leq C_u (1 + |x|^2)^{\frac{m}{2}} \sup_{\xi \in \mathbb{R}^n, |\beta| \leq m} (1 + |\xi|^2)^{\frac{m}{2}} |D^{\alpha+\beta}\eta(\xi)| \\
&\leq C_u (1 + |x|^2)^{\frac{m}{2}} \|\eta\|_{\mathscr{S},m+|\alpha|}.
\end{aligned}
$$

7.4 Advanced Practical Problems

Problem 7.1 Prove that for every distribution $u \in \mathscr{S}'(\mathbb{R}^n)$ there exist constants $K \geq 0$ and $m \in \mathbb{N}$ such that

$$|u(\phi)| \leq K \|\phi\|_m, \quad \phi \in \mathscr{S}(\mathbb{R}^n).$$

Problem 7.2 Prove that any tempered distribution has finite order.

Problem 7.3 Prove

$$\mathscr{S}_0' \subset \mathscr{S}_1' \subset \cdots \subset \mathscr{S}_m' \subset \cdots \subset \mathscr{S}' = \bigcup_{p \in \mathbb{N} \cup \{0\}} \mathscr{S}_p'.$$

Problem 7.4 Prove that the embedding $\mathscr{S}_p' \subset \mathscr{S}_{p+1}'$ is continuous for any $p \in \mathbb{N}$.

Problem 7.5 Prove that every weakly convergent sequence in $\mathscr{S}_p'$, $p \in \mathbb{N}$, converges in the norm of $\mathscr{S}_{p+1}'$.

Problem 7.6 Prove that $\mathscr{S}_p'$, $p \in \mathbb{N}$, is a weakly complete space.

Problem 7.7 Show that $\mathscr{S}'(\mathbb{R}^n)$ is a complete space.

Problem 7.8 Let $u \in \mathscr{E}'(\mathbb{R}^n)$. Prove that $u \in \mathscr{S}'(\mathbb{R}^n)$ and $u(\phi) = u(\eta\phi)$ for every $\phi \in \mathscr{S}(\mathbb{R}^n)$, where $\eta \in \mathscr{C}_0^\infty(\mathbb{R}^n)$ and $\eta \equiv 1$ on a neighbourhood of $\operatorname{supp} u$.

Definition 7.5 A measure μ on $\mathbb{R}^n$ is called tempered if

$$\int_{\mathbb{R}^n} (1 + |x|)^{-m} \mu(dx) < \infty$$

for some $m \geq 0$.

Problem 7.9 Let μ be a tempered measure on $\mathbb{R}^n$ and define the functional

$$\mu(\phi) = \int_{\mathbb{R}^n} \phi(x)\mu(dx), \quad \phi \in \mathscr{S}(\mathbb{R}^n).$$

Prove that $\mu \in \mathscr{S}'(\mathbb{R}^n)$.

Problem 7.10 Let $u \in \mathscr{S}'(\mathbb{R}^n)$. Prove

1. $D^\alpha u \in \mathscr{S}'(\mathbb{R}^n)$ for every $\alpha \in \mathbb{N}^n \cup \{0\}$,
2. the map $u \mapsto D^\alpha u$ is a linear continuous operation on $\mathscr{S}'(\mathbb{R}^n)$.

Problem 7.11 Let $u \in \mathscr{S}'(\mathbb{R}^n)$, A is an invertible $n \times n$ matrix. Prove that $u(Ax + b) \in \mathscr{S}'(\mathbb{R}^n)$, where $b = (b_1, b_2, \ldots, b_n)$, $b_l = $ const, $l = 1, 2, \ldots, n$, and the map $u(x) \mapsto u(Ax + b)$ is a linear and continuous operation on $\mathscr{S}'(\mathbb{R}^n)$.

Problem 7.12 Let $u \in \mathscr{S}'(\mathbb{R}^n)$, $a \in \Theta_M$. Prove that $au \in \mathscr{S}'(\mathbb{R}^n)$ and that $u \mapsto au$ is a linear and continuous operation from $\mathscr{S}'(\mathbb{R}^n)$ to $\mathscr{S}'(\mathbb{R}^n)$.

Problem 7.13 Let $a_n \in \mathbb{C}$, $|a_n| \leq C(1 + |n|)^N$ for some constants $C > 0$ and $N \geq 0$, $n = 1, 2, \ldots$. Prove that

$$\sum_{k=1}^{\infty} a_k \delta(x - k) \in \mathscr{S}'(\mathbb{R}^n).$$

Problem 7.14 Let $u \in \mathscr{S}'(\mathbb{R}^n)$. Prove that there exists a tempered function g in $\mathbb{R}^n$ and a constant $m \in \mathbb{N}$ such that

$$u(x) = D_1^m D_2^m \ldots D_n^m g(x), \quad x \in \mathbb{R}^n.$$

Solution Since $u \in \mathscr{S}'(\mathbb{R}^n)$, there exist $p \in \mathbb{N}$ and a positive constant C_u such that

$$
\begin{aligned}
|u(\phi)| \leq C_u \|\phi\|_p &= C_u \sup_{x \in \mathbb{R}^n, |\alpha| \leq p} (1 + |x|^2)^{\frac{p}{2}} |D^\alpha \phi(x)| \\
&\leq C_u \max_{|\alpha| \leq p} \int_{\mathbb{R}^n} \left| D_1 D_2 \ldots D_n \left((1 + |x|^2)^{\frac{p}{2}} D^\alpha \phi(x) \right) \right| dx, \quad \phi \in \mathscr{S}(\mathbb{R}^n).
\end{aligned}
\tag{7.4}
$$

We define the functions

$$\psi_\alpha(x) = D_1 D_2 \ldots D_n \left((1 + |x|^2)^{\frac{p}{2}} D^\alpha \phi(x) \right), \quad \phi \in \mathscr{S}(\mathbb{R}^n).$$

In this way we have a one-to-one mapping $\phi \mapsto \{\psi_\alpha\}$ from $\mathscr{S}(\mathbb{R}^n)$ to the direct sum $\bigoplus_{|\alpha| \leq p} L^1(\mathbb{R}^n)$ equipped with the norm

$$\|\{f_\alpha\}\| = \max_{|\alpha| \leq p} \|f_\alpha\|_1.$$

Call

$$M = \left\{ \{\psi_\alpha\}, \phi \in \mathscr{S}(\mathbb{R}^n) \right\}.$$

Then M is a subset of $\bigoplus_{|\alpha| \leq p} L^1(\mathbb{R}^n)$, on which we define the functional

$$u^*\left(\{\psi_\alpha\} \right) = u(\phi), \qquad \{\psi_\alpha\} \in M.$$

Using (7.4), we get

$$\left| u^*(\{\psi_\alpha\}) \right| = |u(\phi)| \leq C_u \left\| D_1 D_2 \ldots D_n \left((1 + |x|^2)^{\frac{p}{2}} D^\alpha \phi(x) \right) \right\|_1 = C_u \left\| \{\psi_\alpha\} \right\|.$$

We conclude that u^* is continuous. We also recall that $L^\infty(\mathbb{R}^n)$ is the dual space to $L^1(\mathbb{R}^n)$. By the Hahn–Banach and Riesz theorems, there exists a vector-valued map $\{\chi_\alpha\} \in \bigoplus_{|\alpha| \leq p} L^\infty(\mathbb{R}^n)$ such that

$$u^*\left(\{\psi_\alpha\} \right) = \sum_{|\alpha| \leq p} \int_{\mathbb{R}^n} \chi_\alpha(x) \psi_\alpha(x) dx.$$

Hence,

$$u(\phi) = \sum_{|\alpha| \leq p} \int_{\mathbb{R}^n} \chi_\alpha(x) D_1 D_2 \ldots D_n \left((1 + |x|^2)^{\frac{p}{2}} D^\alpha \phi(x) \right) dx$$

for $\phi \in \mathscr{S}(\mathbb{R}^n)$. Integrating by parts, we infer the existence of functions g_α, $|\alpha| \leq p + 2$, $g_\alpha \in \mathscr{S}(\mathbb{R}^n)$, such that

$$u(\phi) = (-1)^{pn} \int_{\mathbb{R}^n} \sum_{|\alpha| \leq (p+2)n} g_\alpha(x) D_1^{p+2} D_2^{p+2} \ldots D_n^{p+2} \phi(x) dx.$$

Since $\phi \in \mathscr{S}(\mathbb{R}^n)$, we conclude that

$$u(x) = \sum_{|\alpha| \leq (p+2)n} D_1^{p+2} D_2^{p+2} \ldots D_n^{p+2} g_\alpha(x).$$

Problem 7.15 Let $u \in \mathscr{S}'(\mathbb{R}^n)$. Prove that there exists $p \in \mathbb{N} \cup \{0\}$ such that for every positive ϵ there are functions $g_{\alpha,\epsilon} \in \mathscr{S}(\mathbb{R}^n)$, $|\alpha| \leq p$, such that $g_{\alpha,\epsilon} \equiv 0$ on

$\mathbb{R}^n\backslash(\operatorname{supp}u)^\epsilon$ and

$$u(x) = \sum_{|\alpha|\le p} D^\alpha g_{\alpha,\epsilon}(x). \tag{7.5}$$

Solution Let $\epsilon > 0$ and $\eta \in \Theta_M$, $\eta \equiv 1$ on $(\operatorname{supp}u)^{\frac{\epsilon}{3}}$, $\eta \equiv 0$ on $\mathbb{R}^n\backslash(\operatorname{supp}u)^\epsilon$. Using the previous problem, there exist $m \in \mathbb{N} \cup \{0\}$ and $g \in \mathscr{S}(\mathbb{R}^n)$ such that

$$u(x) = D_1^m D_2^m \ldots D_n^m g(x).$$

Since $u(x) = \eta(x)u(x)$,

$$u(x) = \eta(x)D_1^m D_2^m \ldots D_n^m g(x) = D_1^m D_2^m \ldots D_n^m (\eta(x)g(x)) - \sum_{|\alpha|\le mn-1} \eta_\alpha(x)D^\alpha g(x),$$

where $\eta_\alpha \in \Theta_M$, $\eta_\alpha(x) = 0$ for $x \notin (\operatorname{supp}u)^\epsilon$. The function $\eta_\alpha(x)D^\alpha g(x)$ we represent in the form

$$\eta_\alpha(x)D^\alpha g(x) = D^\alpha(\eta_\alpha(x)g(x)) - F(x),$$

and so forth. Note that we obtain (7.5) for $p = mn$ and $g_{\alpha,\epsilon} = \chi_\alpha g$, where $\chi_\alpha \in \Theta_M$ and $\operatorname{supp}\chi_\alpha \subset \overline{(\operatorname{supp}u)^\epsilon}$.

Problem 7.16 Let $u_1 \in \mathscr{S}'(\mathbb{R}^m)$. Prove that $D^\alpha \psi(x) = u_1(y)(D_x^\alpha \phi(x, y))$ for every $\phi \in \mathscr{S}(\mathbb{R}^{n+m})$ and $\alpha \in \mathbb{N}^n \cup \{0\}$.

Problem 7.17 Prove that $\mathscr{S}(\mathbb{R}^n)$ is dense in $\mathscr{S}'(\mathbb{R}^n)$.

Solution Let $u \in \mathscr{S}'(\mathbb{R}^n)$ and consider $u_\epsilon = u*\omega_\epsilon$. Then $u_\epsilon \in \Theta_M$ and $u_\epsilon \to_{\epsilon\to 0} u$ in $\mathscr{S}'(\mathbb{R}^n)$. Since the space $\mathscr{S}(\mathbb{R}^n)$ is dense in Θ_M, because if $a \in \Theta_M$ we have $ae^{-\epsilon|x|^2} \in \mathscr{S}(\mathbb{R}^n)$, $\epsilon > 0$, and $ae^{-\epsilon|x|^2} \to_{\epsilon\to 0} a$ in $\mathscr{S}'(\mathbb{R}^n)$. The claim follows.

7.5 Notes and References

In this chapter we introduce tempered distributions. They are investigated the direct product of tempered distributions and convolution of tempered distributions. In the chapter are given some applications of tempered distributions. Additional materials can be found in [7, 16, 17, 20, 21, 24, 25] and references therein.

Chapter 8
Integral Transforms

8.1 The Fourier Transform in $\mathscr{S}(\mathbb{R}^n)$

Definition 8.1 The Fourier transform of $\phi \in \mathscr{S}(\mathbb{R}^n)$ is the integral

$$\mathscr{F}(\phi)(x) = \int_{\mathbb{R}^n} e^{-i(x,\xi)}\phi(\xi)d\xi, \quad x \in \mathbb{R}^n.$$

Note that $\mathscr{F}(\phi)$ is bounded and continuous on $\mathbb{R}^n$. Furthermore, $\mathscr{F}(\phi) \in \mathscr{C}^\infty(\mathbb{R}^n)$ and

$$D^\alpha \mathscr{F}(\phi)(x) = \int_{\mathbb{R}^n} (-i\xi)^\alpha e^{-i(x,\xi)}\phi(\xi)d\xi = \mathscr{F}((-i\xi)^\alpha \phi)(x),$$

$$\mathscr{F}(D^\alpha \phi)(x) = \int_{\mathbb{R}^n} D^\alpha \phi(\xi)e^{-i(x,\xi)}d\xi = (ix)^\alpha \mathscr{F}(\phi)(x), \quad x \in \mathbb{R}^n,$$

for every $\alpha \in \mathbb{N}^n \cup \{0\}$, where

$$(-i\xi)^\alpha = (-i)^{|\alpha|}\xi_1^{\alpha_1}\xi_2^{\alpha_2}\ldots\xi_n^{\alpha_n}, \quad \xi = (\xi_1, \xi_2, \ldots, \xi_n),$$

$$(ix)^\alpha = i^{|\alpha|}x_1^{\alpha_1}x_2^{\alpha_2}\ldots x_n^{\alpha_n}, \quad x = (x_1, x_2, \ldots, x_n).$$

In particular, $\mathscr{F}(\phi)$ is an integrable function on $\mathbb{R}^n$. Observe that every function $\phi \in \mathscr{S}(\mathbb{R}^n)$ can be represented by means of its Fourier transform $\mathscr{F}(\phi)$ and inverse Fourier transform

$$\mathscr{F}^{-1}(\phi)(\xi) = \frac{1}{(2\pi)^n}\int_{\mathbb{R}^n} e^{i(x,\xi)}\phi(x)dx, \quad \xi \in \mathbb{R}^n,$$

© The Author(s), under exclusive license to Springer Nature Switzerland AG 2021
S. G. Georgiev, *Theory of Distributions*,
https://doi.org/10.1007/978-3-030-81265-2_8

as follows

$$\phi = \mathscr{F}^{-1}(\mathscr{F}(\phi)) = \mathscr{F}(\mathscr{F}^{-1}(\phi)).$$

Explicitly,

$$\phi(x) = \frac{1}{(2\pi)^n} \int_{\mathbb{R}^n} e^{i(x,\xi)} \mathscr{F}(\phi)(\xi) d\xi, \quad x \in \mathbb{R}^n.$$

As

$$(1 + |\xi|^2)^{\frac{p}{2}} |D^\alpha \mathscr{F}(\phi)(\xi)| \le (1 + |\xi|^2)^{\left[\frac{p+1}{2}\right]} |D^\alpha \mathscr{F}(\phi)(\xi)|$$

$$\le \left| \int_{\mathbb{R}^n} (I - \Delta)^{\left[\frac{p+1}{2}\right]} \left((-ix)^\alpha \phi(x) \right) e^{-i(x,\xi)} dx \right|$$

$$\le C \sup_{x \in \mathbb{R}^n} (1 + |x|^2)^{\frac{n+1}{2}} \left| (I - \Delta)^{\left[\frac{p+1}{2}\right]} (x^\alpha \phi(x)) \right|, \quad \xi \in \mathbb{R}^n,$$

for $p \in \mathbb{N}$, it follows that

$$\|\mathscr{F}(\phi)\|_{\mathscr{S},p} \le C_p \|\phi\|_{\mathscr{S},p+n+1},$$

where C_p is a constant independent of ϕ. By the last estimate, we conclude that $\phi \mapsto \mathscr{F}(\phi)$ is a linear and continuous map on $\mathscr{S}(\mathbb{R}^n)$. Every element $\phi \in \mathscr{S}(\mathbb{R}^n)$ can be represented as the Fourier transform of the function $\psi = \mathscr{F}^{-1}(\phi) \in \mathscr{S}(\mathbb{R}^n)$, where $\phi = \mathscr{F}(\psi)$. If $\mathscr{F}(\phi) = 0$, then $\phi = 0$. Therefore the map $\phi \mapsto \mathscr{F}(\phi)$ is one-to-one on $\mathscr{S}(\mathbb{R}^n)$.

Exercise 8.1 Compute $\mathscr{F}\left(e^{-ax^2}\right), x \in \mathbb{R}, a = \text{const} > 0$.

Answer $\sqrt{\dfrac{\pi}{a}} e^{-\frac{\xi^2}{4a}}, \xi \in \mathbb{R}$.

Exercise 8.2 Let A be a positive definite $n \times n$ matrix. Prove that

$$\mathscr{F}\left(e^{i(Ax,x)}\right) = \frac{\pi^{\frac{n}{2}}}{\sqrt{\det A}} e^{-i\frac{\pi n}{4} - \frac{i}{4}(A^{-1}\xi,\xi)}, \quad x, \xi \in \mathbb{R}^n.$$

8.2 The Fourier Transform in $\mathscr{S}'(\mathbb{R}^n)$

Definition 8.2 The Fourier transform of a distribution $u \in \mathscr{S}'(\mathbb{R}^n)$ is defined as follows

$$\mathscr{F}(u)(\phi) = u(\mathscr{F}(\phi)) \quad \text{for} \quad \phi \in \mathscr{S}(\mathbb{R}^n).$$

Since the map $\phi \mapsto \mathscr{F}(\phi) : \mathscr{S}(\mathbb{R}^n) \mapsto \mathscr{S}(\mathbb{R}^n)$ is linear and continuous, the operation $u \mapsto \mathscr{F}(u)$ is linear and continuous from $\mathscr{S}'(\mathbb{R}^n)$ to itself.

For $u \in \mathscr{S}'(\mathbb{R}^n)$, we define the operator $\mathscr{F}^{-1}$ in the following manner

$$\mathscr{F}^{-1}(u)(x) = \frac{1}{(2\pi)^n} \mathscr{F}(u)(-x).$$

As $\phi \mapsto \mathscr{F}(\phi)$ is continuous and goes from $\mathscr{S}(\mathbb{R}^n)$ to $\mathscr{S}(\mathbb{R}^n)$, and $\mathscr{S}(\mathbb{R}^n)$ is dense in $\mathscr{S}'(\mathbb{R}^n)$, we conclude that

$$\mathscr{F}^{-1}(\mathscr{F}(u)) = \mathscr{F}(\mathscr{F}^{-1}(u)) = u, \quad u \in \mathscr{S}'(\mathbb{R}^n).$$

It follows that for every distribution $u \in \mathscr{S}'(\mathbb{R}^n)$ there exists a distribution $v \in \mathscr{S}'(\mathbb{R}^n)$ such that $v = \mathscr{F}^{-1}(u)$ and $u = \mathscr{F}(v)$. If $\mathscr{F}(u) = 0$, then $u = 0$.

Example 8.1 Let us compute $\mathscr{F}(\delta)$. Take $\phi \in \mathscr{S}(\mathbb{R}^n)$, so

$$\mathscr{F}(\delta)(\phi) = \delta(\mathscr{F}(\phi)) = \mathscr{F}(\phi)(0) = \int_{\mathbb{R}^n} \phi(x)dx.$$

This implies $\mathscr{F}(\delta) = 1$.

Exercise 8.3 Compute $\mathscr{F}(H(x)e^{-x})$, $x \in \mathbb{R}$.

Answer $\dfrac{1}{1 + i\xi}$.

Let $u(x, y) \in \mathscr{S}'(\mathbb{R}^{n+m})$, $x \in \mathbb{R}^n$, $y \in \mathbb{R}^m$. We introduce the Fourier transform $\mathscr{F}_x(u)$ with respect to the variable $x = (x_1, x_2, \ldots, x_n)$ by

$$\mathscr{F}_x(u)(\phi) = u(\mathscr{F}_\xi(\phi)), \quad \phi \in \mathscr{S}(\mathbb{R}^{n+m}),$$

where

$$\mathscr{F}_\xi(\phi)(x, y) = \int_{\mathbb{R}^n} e^{-i(\xi, x)} \phi(\xi, y)d\xi.$$

The map $\phi(\xi, y) \mapsto \mathscr{F}_\xi(\phi)$ is an isomorphism from $\mathscr{S}(\mathbb{R}^{n+m})$ to itself, and $\mathscr{F}_x(u) \in \mathscr{S}'(\mathbb{R}^{n+m})$ for $u \in \mathscr{S}'(\mathbb{R}^{n+m})$. The inverse Fourier transform $\mathscr{F}_\xi^{-1}$ is defined by

$$\mathscr{F}_\xi^{-1}(u) = \frac{1}{(2\pi)^n} \mathscr{F}_\xi(u(-\xi, y))(x, y).$$

The map $u \mapsto \mathscr{F}_x(u)$ is an isomorphism of $\mathscr{S}'(\mathbb{R}^{n+m})$ onto $\mathscr{S}'(\mathbb{R}^{n+m})$.

Example 8.2 Let $a = \text{const} \in \mathbb{R}$. Then

$$\mathscr{F}(\delta(x-a))(\phi) = \delta(x-a)(\mathscr{F}(\phi)) = \mathscr{F}(\phi)(a) = \int_{\mathbb{R}^n} e^{-i\xi a}\phi(\xi)d\xi, \quad \phi \in \mathscr{S}(\mathbb{R}).$$

Since $\phi \in \mathscr{S}(\mathbb{R})$ is arbitrary, $\mathscr{F}(\delta(x-a)) = e^{-i\xi a}$.

Exercise 8.4 Prove that

$$\mathscr{F}\left(\frac{\delta(x-a) + \delta(x+a)}{2}\right) = \cos(a\xi), \quad a = \text{const},$$

in $\mathscr{S}'(\mathbb{R})$.

8.3   Properties of the Fourier Transform in $\mathscr{S}'(\mathbb{R}^n)$

1. For any $u \in \mathscr{S}'(\mathbb{R}^n), \alpha \in \mathbb{N}^n$, we have

$$D^\alpha \mathscr{F}(u) = \mathscr{F}((-ix)^\alpha u).$$

Example 8.3 Let $\phi \in \mathscr{S}(\mathbb{R}^n)$. Then

$$\mathscr{F}(x^\alpha)(\phi) = (i)^{|\alpha|}\mathscr{F}((-ix)^\alpha 1)(\phi) = (i)^{|\alpha|}D^\alpha \mathscr{F}(1)(\phi),$$

$\alpha \in \mathbb{N}^n \cup \{0\}$. On the other hand,

$$\mathscr{F}(\delta)(\phi) = 1(\phi).$$

Using the inverse Fourier transform, we get

$$\delta(\phi) = \mathscr{F}^{-1}(1)(\phi) = \frac{1}{(2\pi)^n}\mathscr{F}(1)(\phi), \quad \text{i.e.,} \quad \mathscr{F}(1)(\phi) = (2\pi)^n\delta(\phi).$$

Therefore

$$\mathscr{F}(x^\alpha)(\phi) = (2\pi)^n(i)^{|\alpha|}D^\alpha\delta(\phi),$$

in other words

$$\mathscr{F}(x^\alpha) = (2\pi)^n(i)^{|\alpha|}D^\alpha\delta(\xi).$$

2. For any $u \in \mathscr{S}'(\mathbb{R}^n), \alpha \in \mathbb{N}^n \cup \{0\}$, we have

$$\mathscr{F}(D^\alpha u) = (i\xi)^\alpha \mathscr{F}(u).$$

Example 8.4 Let us find $\mathscr{F}(\delta'')$ in $\mathscr{S}'(\mathbb{R})$. Take $\phi \in \mathscr{S}(\mathbb{R})$ and compute

$$\mathscr{F}(\delta''(\phi))(\xi) = (i\xi)^2 \mathscr{F}(\delta(\phi)) = (i\xi)^2 \delta(\mathscr{F}(\phi))$$
$$= (i\xi)^2 \delta\left(\int_{-\infty}^{\infty} e^{-ix\xi} \phi(x)dx\right) = (i\xi)^2 \int_{-\infty}^{\infty} \phi(x)dx,$$

so

$$\mathscr{F}(\delta'')(x) = (ix)^2.$$

3. For any $u \in \mathscr{S}'(\mathbb{R}^n)$, we have

$$\mathscr{F}(u(x - x_0)) = e^{-i(\xi, x_0)} \mathscr{F}(u).$$

4. For any $u \in \mathscr{S}'(\mathbb{R}^n)$, we have

$$\mathscr{F}(u)(\xi + \xi_0) = \mathscr{F}(e^{-i(\xi_0, x)} u)(\xi).$$

5. For any nonsingular $n \times n$ matrix A, we have

$$\mathscr{F}(u(Ax))(\xi) = \frac{1}{|\det A|} \mathscr{F}(u)(A^{-1T}\xi), \quad u \in \mathscr{S}'(\mathbb{R}^n).$$

6. For any $u \in \mathscr{S}'(\mathbb{R}^n)$, $v \in \mathscr{S}'(\mathbb{R}^m)$, we have

$$\mathscr{F}(u(x) \times v(y))(\xi, \eta) = \mathscr{F}_x(u(x) \times \mathscr{F}(v)(\eta))$$
$$= \mathscr{F}_y(\mathscr{F}(u)(\xi) \times v(y)) = \mathscr{F}u(\xi) \times \mathscr{F}(v)(\eta).$$

7. For any $u \in \mathscr{S}'(\mathbb{R}^{n+m})$, $\alpha \in \mathbb{N}^n \cup \{0\}$, $\beta \in \mathbb{N}^m \cup \{0\}$, we have

$$D_x^\alpha D_y^\beta \mathscr{F}_x(u) = \mathscr{F}_x((-ix)^\alpha D_y^\beta u),$$
$$\mathscr{F}_x(D_x^\alpha D_y^\beta u) = (i\xi)^\alpha D_y^\beta \mathscr{F}_x(u).$$

The proofs of the above properties are left to the reader.

8.4 The Fourier Transform of Distributions with Compact Support

If we take $u \in \mathscr{E}'(\mathbb{R}^n)$ we know already that $u \in \mathscr{S}'(\mathbb{R}^n)$, so it admits a Fourier transform in $\mathscr{S}'(\mathbb{R}^n)$. What is more, the Fourier transform exists in Θ_M and can be represented in the form

$$\mathscr{F}(u)(\xi) = u(x)(\eta(x)e^{-i(\xi, x)}), \tag{8.1}$$

where $\eta \in \mathscr{C}_0^\infty(\mathbb{R}^n)$ and $\eta \equiv 1$ on a neighbourhood of supp$u$. We claim that there are constants $C_\alpha > 0$ and $m \in \mathbb{N}$ such that

$$|D^\alpha \mathscr{F}(u)(\xi)| \leq ||u||_{\mathscr{S},-m} C_\alpha (1 + |\xi|^2)^{\frac{m}{2}}, \quad \xi \in \mathbb{R}^n, \alpha \in \mathbb{N}^n \cup \{0\}.$$

Indeed, let $\phi \in \mathscr{S}(\mathbb{R}^n)$ be arbitrary. So,

$$D^\alpha \mathscr{F}(u)(\phi) = (-1)^{|\alpha|} \mathscr{F}(u)(D^\alpha \phi) = (-1)^{|\alpha|} u(\mathscr{F}(D^\alpha \phi))$$
$$= (-1)^{|\alpha|} u(\eta(x)(ix)^\alpha \mathscr{F}(\phi)) = u(x)\left(\int_{\mathbb{R}^n} \eta(x)(-ix)^\alpha \phi(\xi) e^{-i(x,\xi)} d\xi\right)$$
$$= \int_{\mathbb{R}^n} u(x)\left(\eta(x)(-ix)^\alpha e^{-i(x,\xi)}\right) \phi(\xi) d\xi$$

and therefore

$$D^\alpha \mathscr{F}(u)(\xi) = u(x)(\eta(x)(-ix)^\alpha e^{-i(x,\xi)}), \quad \alpha \in \mathbb{N}^n \cup \{0\}, \tag{8.2}$$

From here, we obtain (8.1) for $\alpha = 0$. But (8.2) implies

$$|D^\alpha \mathscr{F}(u)(\xi)| = |u(x)(\eta(x)(-ix)^\alpha e^{-i(x,\xi)})|$$
$$\leq ||u||_{\mathscr{S},-m} ||\eta(x)(-ix)^\alpha e^{-ix\xi}||_{\mathscr{S},m}$$
$$= ||u||_{\mathscr{S},-m} \sup_{\substack{x \in \mathbb{R}^n \\ |\beta| \leq m}} (1 + |x|^2)^{\frac{m}{2}} |D_x^\beta (\eta(x)(-ix)^\alpha e^{-i(x,\xi)})|$$
$$\leq ||u||_{\mathscr{S},-m} C_\alpha (1 + |\xi|^2)^{\frac{m}{2}}.$$

Therefore $\mathscr{F}(u) \in \Theta_M$.

8.5   The Fourier Transform of Convolutions

Let $u \in \mathscr{S}'(\mathbb{R}^n)$ and $v \in \mathscr{E}'(\mathbb{R}^n)$. So, the convolution $u * v$ is defined in $\mathscr{S}'(\mathbb{R}^n)$. Choose $\phi \in \mathscr{S}(\mathbb{R}^n)$ and $\eta \in \mathscr{C}_0^\infty(\mathbb{R}^n)$ so that $\eta \equiv 1$ on a neighbourhood of supp$v$. Then

$$u * v(\phi) = u(x) \times v(y)(\eta(y)\phi(x + y)),$$

and

$$\mathscr{F}(u * v)(\phi) = u(x)(v(y)(\eta(y)\mathscr{F}(\phi(x + y))))$$
$$= u(x)\left(v(y)\left(\eta(y)\int_{\mathbb{R}^n} \phi(\xi)e^{-i((x+y),\xi)}d\xi\right)\right)$$
$$= u(x)\left(\int_{\mathbb{R}^n} v(y)(\eta(y)e^{-i(y,\xi)})e^{-i(x,\xi)}\phi(\xi)d\xi\right)$$
$$= u(x)\left(\int_{\mathbb{R}^n} \mathscr{F}(v)(\xi)e^{-i(x,\xi)}\phi(\xi)d\xi\right)$$
$$= u(\mathscr{F}(\mathscr{F}(v)\phi) = \mathscr{F}(u)(\mathscr{F}(v)\phi) = \mathscr{F}(u).\mathscr{F}(v)(\phi).$$

Consequently

$$\mathscr{F}(u * v) = \mathscr{F}(u).\mathscr{F}(v).$$

8.6 The Laplace Transform

8.6.1 Definition

Definition 8.3 Let Γ be a closed, convex, acute cone in $\mathbb{R}^n$ with vertex at 0, and set $C = \text{int}\Gamma^*$, so $C \neq \emptyset$ is an open, convex cone. Define

$$T^C = \mathbb{R}^n + iC = \{z = x + iy : x \in \mathbb{R}^n, y \in C\}.$$

The Laplace transform of $u \in \mathscr{S}'(\Gamma+)$ is defined by

$$L(u)(z) = \mathscr{F}(u(\xi)e^{(y,\xi)})(x). \tag{8.3}$$

This is well defined. Indeed, pick $\eta \in \mathscr{C}^\infty(\mathbb{R}^n)$ with the following properties: $|D^\alpha\eta(\xi)| \leq c_\alpha, \eta \equiv 1$ on $(\text{supp}u)^\epsilon$ and $\eta \equiv 0$ on $\mathbb{R}^n \backslash (\text{supp}u)^{2\epsilon}$, and $\epsilon > 0$ arbitrary. Since $\eta(\xi)e^{(y,\xi)} \in \mathscr{S}(\mathbb{R}^n)$ for every $y \in C$, we have

$$u(\xi)e^{(y,\xi)} = u(\xi)\eta(\xi)e^{(y,\xi)} \in \mathscr{S}'(\Gamma+)$$

for any $y \in C$. We conclude that (8.3) is well defined. It has the following representation

$$L(u)(z) = u(\xi)(\eta(\xi)e^{-i(z,\xi)}). \tag{8.4}$$

Observe that the Laplace transform does not depend on the choice of η. Indeed, let $\phi \in \mathscr{S}(\Gamma+)$ be arbitrary. Then $\eta(\xi)e^{\langle y,\xi\rangle}\phi(x)e^{-i\langle x,\xi\rangle} \in \mathscr{S}(\mathbb{R}^{2n})$ and

$$
\begin{aligned}
L(u)(z) &= \mathscr{F}(u(\xi)e^{\langle y,\xi\rangle})(\phi) = u(\xi)e^{\langle y,\xi\rangle}\mathscr{F}(\phi) \\
&= u(\xi)e^{\langle y,\xi\rangle}\left(\int_{\mathbb{R}^n} \phi(x)e^{-i\langle x,\xi\rangle}dx\right) = u(\xi)\left(\eta(\xi)e^{\langle y,\xi\rangle}\int_{\mathbb{R}^n}\phi(x)e^{-i\langle x,\xi\rangle}dx\right) \\
&= u(\xi)\left(\int_{\mathbb{R}^n}\phi(x)e^{-i\langle z,\xi\rangle}\eta(\xi)dx\right) = \int_{\mathbb{R}^n} u(\xi)\left(e^{-i\langle z,\xi\rangle}\eta(\xi)\right)\phi(x)dx.
\end{aligned}
$$

Equation (8.4) now follows.

Example 8.5 Let us compute $L(\delta(\xi - \xi_0))$. With $\phi \in \mathscr{S}(\Gamma+)$,

$$
\begin{aligned}
L(\delta(\xi - \xi_0))(\phi) &= \mathscr{F}(\delta(\xi - \xi_0)e^{\langle y,\xi\rangle})(\phi) = \delta(\xi - \xi_0)(e^{\langle y,\xi\rangle}\mathscr{F}(\phi)) \\
&= \delta(\xi - \xi_0)\left(e^{\langle y,\xi\rangle}\int_{\mathbb{R}^n}e^{-i\langle x,\xi\rangle}\phi(x)dx\right) = \delta(\xi - \xi_0)\left(\int_{\mathbb{R}^n}e^{-i\langle z,\xi\rangle}\phi(x)dx\right) \\
&= \int_{\mathbb{R}^n}e^{-i\langle z,\xi_0\rangle}\phi(x)dx.
\end{aligned}
$$

Since $\phi \in \mathscr{S}(\Gamma+)$ is arbitrary, we find

$$
L(\delta(\xi - \xi_0))(z) = e^{-i\langle z,\xi_0\rangle}.
$$

8.6.2 Properties

Let us write $v(z) = L(u)(z)$. Since $\eta(\xi)e^{-i\langle z,\xi\rangle}$ is a continuous function in the variable $z \in T^C$ in $\mathscr{S}(\Gamma+)$, for $z, z_0 \in T^C$, we have

$$
\eta(\xi)e^{-i\langle z,\xi\rangle} \to_{z\to z_0} \eta(\xi)e^{-i\langle z_0,\xi\rangle}
$$

in $\mathscr{S}(\Gamma+)$. Hence,

$$
v(z) = u(\xi)(\eta(\xi)e^{-i\langle z,\xi\rangle}) \to_{z\to z_0} u(\xi)(\eta(\xi)e^{-i\langle z_0,\xi\rangle}) = v(z_0)
$$

and v is continuous in T^C.

Take $e_1 = (1, 0, \ldots, 0)$ and $z \in T^C$ and consider

$$
\chi_h(\xi) = \frac{1}{h}\left(\eta(\xi)e^{-i\langle z+he_1,\xi\rangle} - \eta(\xi)e^{-i\langle z,\xi\rangle}\right) \to_{h\to 0} \eta(\xi)(-i\xi_1)e^{-i\langle z,\xi\rangle}
$$

in $\mathscr{S}(\Gamma+)$. Then

$$\frac{v(z+he_1)-v(z)}{h} = \frac{1}{h}\left(u(\xi)(\eta(\xi)e^{-i(z+he_1,\xi)}) - u(\xi)(\eta(\xi)e^{-i(z,\xi)})\right)$$

$$= u(\xi)(\chi_h(\xi)) \to_{h\to 0} u(\xi)(\eta(\xi)(-i\xi_1)e^{-i(z,\xi)})$$

$$= (-i\xi_1)u(\xi)(\eta(\xi)e^{-i(z,\xi)}),$$

so

$$\frac{\partial v}{\partial z_1} = (-i\xi_1)u(\xi)(\eta(\xi)e^{-i(z,\xi)})$$

and finally

$$D^\alpha L(u) = L((-i\xi)^\alpha u), \quad \forall \alpha \in \mathbb{N}^n \cup \{0\}.$$

Definition 8.4 The distribution $u \in \mathscr{S}'(\Gamma+)$ for which $v = L(u)$ is called a spectral function of v.

If a spectral function u exists, then it must be unique, and we have a representation

$$u(\xi) = e^{-(y,\xi)}\mathscr{F}_x^{-1}(v(x+iy))(\xi).$$

Using the features of the Fourier transform one can easily deduce the following properties for the Laplace transform.

1. $L(D^\alpha u) = (iz)^\alpha L(u) \quad$ for any $u \in \mathscr{S}'(\Gamma+), \alpha \in \mathbb{N}^n \cup \{0\}$.

   Example 8.6

   $$L(D^\alpha \delta(\xi - \xi_0)) = (iz)^\alpha e^{-i(z,\xi_0)}.$$

2. $L(u(\xi)e^{-i(a,\xi)}) = L(u)(z+a) \quad$ for any $u \in \mathscr{S}'(\Gamma+), \operatorname{Im} a \in C$.
3. $L(u(\xi + \xi_0)) = e^{i(z,\xi_0)}L(u)(z)$.
4. $L(u(A\xi)) = \dfrac{1}{|\det A|}L(u)(A^{-1T}z) \quad$ for $z \in T^{A^{T}C}$, where A is invertible of order n.
5. $L(u_1 \times u_2)(z,\zeta) = L(u_1)(z)L(u_2)(\zeta) \quad$ for any $u_1 \in \mathscr{S}'(\Gamma_1+), u_2 \in \mathscr{S}'(\Gamma_2+), (z,\zeta) \in T^{C_1 \times C_2}$.
6. $L(u_1 * u_2) = L(u_1)L(u_2) \quad$ for any $u_1, u_2 \in \mathscr{S}'(\Gamma+)$.

Example 8.7 Let us compute $L(H(\xi)\sin(\omega\xi))$ in $\mathscr{S}'(\Gamma+)$, $\omega \in \mathbb{C}$, $n = 1$. Let $\phi \in \mathscr{S}(\Gamma+)$. Then

$$L(H(\xi)\sin(\omega\xi))(\phi) = \mathscr{F}(H(\xi)\sin(\omega\xi)e^{y\xi})(\phi)$$

$$= H(\xi)\sin(\omega\xi)e^{y\xi}(\mathscr{F}(\phi)) = H(\xi)\sin(\omega\xi)e^{y\xi}\int_{-\infty}^{\infty} e^{-ix\xi}\phi(x)dx$$

$$= \int_{0}^{\infty}\int_{-\infty}^{\infty} \sin(\omega\xi)e^{y\xi}e^{-ix\xi}\phi(x)dxd\xi = \int_{-\infty}^{\infty}\phi(x)\int_{0}^{\infty}\sin(\omega\xi)e^{-iz\xi}d\xi dx$$

$$= \int_{-\infty}^{\infty} \frac{\omega}{\omega^2 - z^2}\phi(x)dx.$$

This proves that

$$L(H(\xi)\sin(\omega\xi)) = \frac{\omega}{\omega^2 - z^2}.$$

Exercise 8.5 Compute $L(H(\xi)\cos(\omega\xi))$ in $\mathscr{S}'(\Gamma+)$, $n = 1$, $\omega \in \mathbb{C}$.

Answer $\dfrac{iz}{\omega^2 - z^2}$.

Exercise 8.6 Compute $L(H(\xi)e^{i\omega\xi})$ in $\mathscr{S}'(\Gamma+)$, $n = 1$, $\omega \in \mathbb{C}$.

Answer $\dfrac{i}{w - z}$.

Exercise 8.7 Compute $L(H(\xi)e^{-i\omega\xi})$ in $\mathscr{S}'(\Gamma+)$, $n = 1$, $\omega \in \mathbb{C}$.

Answer $-\dfrac{i}{w + z}$.

8.7 Advanced Practical Problems

Problem 8.1 Compute in $\mathscr{S}(\mathbb{R})$

$$\mathscr{F}\left(e^{-\frac{x^2}{4}}\cos(\alpha x)\right), \qquad \alpha = \text{const}.$$

Problem 8.2 Let $u, v \in \mathscr{S}(\mathbb{R}^n)$. Prove

$$\int_{\mathbb{R}^n} \mathscr{F}(u)(x)v(x)dx = \int_{\mathbb{R}^n} u(x)\mathscr{F}(v)(x)dx.$$

Solution We have

$$\int_{\mathbb{R}^n} \mathscr{F}(u)(x)v(x)dx = \int_{\mathbb{R}^n}\int_{\mathbb{R}^n} e^{-i(x,\xi)}u(\xi)d\xi\, v(x)dx = \int_{\mathbb{R}^n}\int_{\mathbb{R}^n} e^{-i(x,\xi)}u(\xi)v(x)d\xi dx$$

$$= \int_{\mathbb{R}^n} u(\xi) \int_{\mathbb{R}^n} e^{-i(x,\xi)}v(x)dx d\xi = \int_{\mathbb{R}^n} u(\xi)\mathscr{F}(v)(\xi)d\xi.$$

Problem 8.3 Prove that

$$\mathscr{F}\mathscr{F}^{-1} = \mathscr{F}^{-1}\mathscr{F} = I \quad \text{in} \quad \mathscr{S}(\mathbb{R}^n).$$

Problem 8.4 Show

$$\mathscr{F}(f * g) = \mathscr{F}(f)\mathscr{F}(g)$$

for $f, g \in \mathscr{S}(\mathbb{R}^n)$.

Problem 8.5 Prove

$$\lim_{|\lambda|\to\infty} \mathscr{F}(f)(\lambda) = 0$$

for $f \in \mathscr{S}(\mathbb{R}^n)$.

Problem 8.6 Let $f \in \mathscr{S}(\mathbb{R})$. Prove

1. $\mathscr{F}(f(-x)) = \mathscr{F}(f)(-\xi)$,
2. $\mathscr{F}(f(ax + b)) = \dfrac{1}{a}e^{i\frac{b\xi}{a}}\mathscr{F}(f)\left(\dfrac{\xi}{a}\right), a, b, c = \text{const}, a > 0$,
3. $\mathscr{F}(e^{iax} f(x)) = \mathscr{F}(f)(\xi - a), a = \text{const} \neq 0$,
4. $\mathscr{F}(e^{iax} f(bx + c)) = \dfrac{1}{b}e^{i\frac{c}{b}(\xi-a)}\mathscr{F}(f)\left(\dfrac{\xi - a}{b}\right), a, b, c = \text{const}, b > 0$.

Problem 8.7 In $\mathscr{S}'(\mathbb{R})$ compute $\mathscr{F}(u)$ when

1. $u = H(1 - |x|)$,
2. $u = e^{-4x^2}$,
3. $u = e^{ix^2}$,
4. $u = e^{-ix^2}$,
5. $u = H(x)e^{-3x}$,
6. $u = H(-x)e^{4x}$,
7. $u = e^{-2|x|}$,
8. $u = \dfrac{2}{1 + x^2}$,
9. $u = H(x)e^{-2x}\dfrac{x^{\alpha-1}}{\Gamma(\alpha)}$.

1. *Solution* Fix an arbitrary $\phi \in \mathscr{S}(\mathbb{R})$, so

$$\mathscr{F}(H(1 - |x|))(\phi) = H(1 - |x|)(\mathscr{F}(\phi))$$

$$= H(1 - |x|)\left(\int_{-\infty}^{\infty} e^{-ix\xi}\phi(\xi)d\xi\right) = \int_{-1}^{1}\int_{-\infty}^{\infty} e^{-ix\xi}\phi(\xi)d\xi dx$$

$$= \int_{-\infty}^{\infty}\phi(\xi)\int_{-1}^{1} e^{-ix\xi}dxd\xi = \int_{-\infty}^{\infty} 2\frac{\sin\xi}{\xi}\phi(\xi)d\xi = 2\frac{\sin\xi}{\xi}(\phi).$$

Hence,

$$\mathscr{F}(H(1 - |x|)) = 2\frac{\sin\xi}{\xi}.$$

Problem 8.8 In $\mathscr{S}'(\mathbb{R})$ compute $\mathscr{F}(u)$ when

1. $u = H(x - a), a = \text{const}$,
2. $u = \text{sign}x$,
3. $u = P\dfrac{1}{x}$,
4. $u = \dfrac{1}{x \pm i0}$,
5. $u = |x|$,
6. $u = H(x)x^k, k \in \mathbb{N}$,
7. $u = |x|^k, k \in \mathbb{N}, k \geq 2$,
8. $u = x^k P\dfrac{1}{x}, k \in \mathbb{N}$,
9. $u = P\dfrac{1}{x^2}$,
10. $u = x^k \delta(x), k \in \mathbb{N}$,
11. $u = x^k \delta^{(m)}(x), k, m \in \mathbb{N}, m \geq k$,
12. $u = P\dfrac{1}{x^3}$,
13. $u = H^{(1/2)}(x)$,
14. $u = \displaystyle\sum_{k=-\infty}^{\infty} a_k \delta(x - k), a_k = \text{const}, |a_k| \leq C(1 + |k|)^m, C = \text{const} > 0$, for
    some $m \geq 2$,
15. $u = H(\pm x)$,
16. $u = P\dfrac{1}{x}$.

Problem 8.9 Find

1. $\mathscr{F}\left(P\dfrac{1}{|x|^2}\right)$,

2. $\mathscr{F}\left(\dfrac{H(1-|x|)}{\sqrt{1-|x|^2}}\right)$

in $\mathscr{S}'(\mathbb{R}^2)$.

Problem 8.10 Find $\mathscr{F}\left(\dfrac{1}{|x|^k}\right)$, $0 < k < n$, in $\mathscr{S}'(\mathbb{R}^n)$.

Problem 8.11 Find

1. $\mathscr{F}_x(\delta(x,t))$,
2. $\mathscr{F}_x(H(t-|x|))$, $n = 1$,

in $\mathscr{S}'(\mathbb{R}^{n+1}(x,t))$, $(x,t) = (x_1, x_2, \ldots, x_n, t)$.

Problem 8.12 Find

1. $\mathscr{F}_\xi^{-1}\left(H(t)e^{-\xi^2 t}\right)$,

2. $\mathscr{F}_\xi^{-1}\left(H(t)\dfrac{\sin(\xi t)}{\xi}\right)$

in $\mathscr{S}'(\mathbb{R}^n)$.

Problem 8.13 Find $\mathscr{F}_\xi^{-1}\left(H(t)\dfrac{\sin(|\xi|t)}{|\xi|}\right)$ in $\mathscr{S}'(\mathbb{R}^3)$.

Problem 8.14 Find $L(H(\xi)J_0(\xi))$ in $\mathscr{S}'(\Gamma+)$, $n = 1$.

Problem 8.15 Using the Laplace transform, solve the following Cauchy problems in $\mathscr{S}'(\Gamma+)$, $n = 1$:

1. $u'(t) + 3u(t) = e^{-2t}$, $u(0) = 0$,
2. $u''(t) + 5u'(t) + 6u(t) = 12$, $u(0) = 2$, $u'(0) = 0$,
3.

$$\begin{cases} u'(t) + 5u(t) + 2v(t) = e^{-t}, \\ v'(t) + 2v(t) + 2u(t) = 0, \\ u(0) = 1, \ v(0) = 0. \end{cases}$$

Problem 8.16 Using the Laplace transform solve the following equations in $\mathscr{S}'(\Gamma+)$, $n = 1$:

1. $(H(t)\sin t) * u(t) = \delta(t)$,
2. $(H(t)\cos t) * u(t) = \delta(t)$,
3. $u(t) + 2(H(t)\cos t) * u(t) = \delta(t)$,
4.

$$\begin{cases} H(t) * u_1(t) + \delta'(t) * u_2(t) = \delta(t) \\ \delta(t) * u_1(t) + \delta'(t) * u_2(t) = 0. \end{cases}$$

8.8   Notes and References

In this chapter we introduce the Fourier transform for $\mathscr{S}$ functions and the Fourier transform for tempered distributions. They are deducted some of their basic properties. It is investigated the Fourier transform of convolution of tempered distributions. It is defined the Laplace transform for distributions and they are deduct some of its properties. Additional materials can be found in [7, 16, 17, 20, 21, 24, 25] and references therein.

Chapter 9
Fundamental Solutions

9.1 Definition and Properties

Let us write

$$P(D) = \sum_{|\alpha| \leq m} a_\alpha D^\alpha, \quad a_\alpha = \text{const}, \sum_{|\alpha|=m} |a_\alpha| \neq 0.$$

Definition 9.1 Given P as above, the distribution $u \in \mathscr{D}'(\mathbb{R}^n)$ is called fundamental solution if

$$P(D)u = \delta.$$

Consider the polynomial

$$P(\xi) = \sum_{|\alpha| \leq m} a_\alpha \xi^\alpha.$$

There exists a transformation

$$\xi = A\xi', \quad \text{with } \det A \neq 0, \quad A = (a_{kj}),$$

under which P reads

$$P(\xi') = a{\xi'_1}^m + \sum_{0 \leq k \leq m-1} P_k(\xi'_2, \ldots, \xi'_n){\xi'_1}^k, \quad a = \text{const} \neq 0. \tag{9.1}$$

© The Author(s), under exclusive license to Springer Nature Switzerland AG 2021
S. G. Georgiev, *Theory of Distributions*,
https://doi.org/10.1007/978-3-030-81265-2_9

Note that there exists a constant $\kappa = \kappa(m)$ such that for every point $\xi \in \mathbb{R}^n$ there is a $k \in \mathbb{N} \cup \{0\}$ for which

$$|P(\xi_1 + i\tau \frac{k}{m}, \xi_2, \ldots, \xi_n)| \geq a\kappa, \quad |\tau| = 1. \tag{9.2}$$

Before to prove the classical Malgrange theorem, we will formulate the Paley-Wiener-Schwartz theorem. For its proof we refer the reader to [35].

Theorem 9.1 (Paley-Wiener-Schwartz Theorem) *For a function f to be integral and to satisfy the conditions of growth: for any $\epsilon > 0$ there is a constant $M = M(\epsilon) > 0$ such that*

$$|f(z)| \leq M(\epsilon)e^{(a+\epsilon)|y|}(1 + |z|^2)^{\frac{\alpha}{2}}, \quad z \in \mathbb{C},$$

for $a = a(f) \geq 0, \alpha = \alpha(f) \geq 0, z = x + iy$, it is necessary and sufficient that its spectral function $g \in \mathscr{E}'(\overline{U_a})$.

Theorem 9.2 (Malgrange-Ehrenpreis Theorem) *Every differential operator with constant coefficients has a fundamental solution in $\mathscr{D}'(\mathbb{R}^n)$.*

Proof Without loss of generality, we suppose

$$P(\xi) = a\xi_1^m + \sum_{k=0}^{m-1} P_k(\xi_2, \ldots, \xi_n)\xi_1^k, \quad a = \text{const} > 0.$$

We will prove the Malgrange-Ehrenpreis theorem for the polynomial $P(i\xi)$. Let $\phi_0, \phi_1, \phi_2, \ldots, \phi_m \in \mathscr{C}_0^\infty(\mathbb{R}^n)$ be chosen so that $\sum_{k=0}^{m} \phi_k(\xi) = 1, \phi_k(\xi) \geq 0$ for $\xi \in \mathbb{R}^n$, and $\phi_k(\xi) = 0$ for those $\xi \in \mathbb{R}^n$ for which

$$\min_{|\tau|=1} \left| P\left(i\xi_1 - \tau\frac{k}{m}, i\xi_2, \ldots, i\xi_n\right) \right| < a\kappa.$$

If $L(\phi)$ denotes the Laplace transform of $\phi \in \mathscr{C}_0^\infty(\mathbb{R}^n)$, we set

$$u(\phi) = \frac{1}{(2\pi)^n} \sum_{k=0}^{m} \int_{\mathbb{R}^n} \phi_k(\xi) \frac{1}{2\pi i} \int_{|\tau|=1} \frac{L(\phi)(\xi_1 + i\tau\frac{k}{m}, \xi_2, \ldots, \xi_n)\, d\tau}{P(\xi_1 - \tau\frac{k}{m}, i\xi_2, \ldots, i\xi_n)} \frac{d\tau}{\tau} d\xi. \tag{9.3}$$

We fix $\phi \in \mathscr{C}_0^\infty(\mathbb{R}^n)$ and choose $R > 0$ so that $\mathrm{supp}\phi \subset \overline{U}_R$. Since $L(\phi)$ is an entire function, by the Paley-Wiener-Schwartz theorem we have

$$
\left| L(\phi)\left(\xi_1 + i\tau\frac{k}{m}, \xi_2, \ldots, \xi_n \right) \right| \leq
$$
$$
\left(1 + \left| \xi_1 + i\tau\frac{k}{m} \right|^2 + |\xi_2|^2 + \cdots + |\xi_n|^2 \right)^{-N} \max_{|\tau|=1} e^{|\mathrm{Re}\tau|\frac{k}{m}} \int_{|x|<R} |(1 - \Delta)^N \phi(x)| dx
$$

$$(9.4)$$

for every $N \geq 0$. Fixing $N > \dfrac{n}{2}$ ensures that

$$
\int_{\mathbb{R}^n} \left(1 + \left| \xi_1 + i\tau\frac{k}{m} \right|^2 + |\xi_2|^2 + \cdots + |\xi_n|^2 \right)^{-N} d\xi < \infty.
$$

We note that

$$
\min_{|\tau|=1} \left| P\left(i\xi_1 - \tau\frac{k}{m}, i\xi_2, \ldots, i\xi_n \right) \right| \geq a\kappa
$$

for $\xi \in \mathbb{R}^n$ with $\phi_k(\xi) \neq 0$ and for every $k = 0, 1, 2, \ldots, m$. Then, using (9.4),

$$
|u(\phi)| \leq \frac{1}{(2\pi)^n} \sum_{k=0}^n \int_{\mathbb{R}^n} \phi_k(\xi) \frac{\max_{|\tau|=1} |L(\phi)(\xi_1 + i\tau\frac{k}{m}, \xi_2, \ldots, \xi_n)|}{\min_{|\tau|=1} |P(i\xi_1 - \tau\frac{k}{m}, i\xi_2, \ldots, i\xi_n)|} d\xi
$$

$$
\leq \frac{1}{(2\pi)^n a\kappa} \sum_{k=0}^m \max_{|\tau|=1} e^{|\mathrm{Re}\tau|\frac{k}{m}} \int_{\mathbb{R}^n} \left(1 + \left| \xi_1 + i\tau\frac{k}{m} \right|^2 + \xi_2^2 + \cdots + \xi_n^2 \right)^{-N} d\xi
$$

$$
\times \int_{|x|<R} |(1 - \Delta)^N \phi(x)| dx.
$$

Let

$$
K_N = \frac{1}{(2\pi)^n a\kappa} \sum_{k=0}^m \max_{|\tau|=1} e^{|\mathrm{Re}\tau|\frac{k}{m}} \int_{\mathbb{R}^n} \left(1 + \left| \xi_1 + i\tau\frac{k}{m} \right|^2 + \xi_2^2 + \cdots + \xi_n^2 \right)^{-N} d\xi,
$$

so

$$
|u(\phi)| \leq K_N \int_{|x|<R} |(1 - \Delta)^N \phi(x)| dx
$$

for every $\phi \in \mathscr{C}_0^\infty(\overline{U}_R)$. Therefore u is a linear and continuous functional on $\mathscr{C}_0^\infty(\mathbb{R}^n)$.

Moreover,

$$P(D)u(\phi) = u(P(-D)\phi)$$

$$= \frac{1}{(2\pi)^n} \sum_{k=0}^{m} \int_{\mathbb{R}^n} \phi_k(\xi) \frac{1}{2\pi i} \int_{|\tau|=1} \frac{L(P(-D)\phi)(\xi_1 + i\tau\frac{k}{m}, \xi_2, \ldots, \xi_n)\, d\tau}{P(i\xi_1 - \tau\frac{k}{m}, i\xi_2, \ldots, i\xi_n)} \frac{d\tau}{\tau} d\xi$$

$$\left(L(P(-D)\phi)\left(\xi_1 + i\tau\frac{k}{m}, \xi_2, \ldots, \xi_n\right) = P\left(i\xi_1 - \tau\frac{k}{m}, \xi_2, \ldots, \xi_n\right) L(\phi)\right)$$

$$= \frac{1}{(2\pi)^n} \sum_{k=0}^{m} \int_{\mathbb{R}^n} \phi_k(\xi) \frac{1}{2\pi i} \int_{|\tau|=1} L(\phi)\left(\xi_1 + i\tau\frac{k}{m}, \xi_2, \ldots, \xi_n\right) \frac{d\tau}{\tau} d\xi$$

$$= \frac{1}{(2\pi)^n} \sum_{k=0}^{m} \int_{\mathbb{R}^n} \phi_k(\xi) \mathscr{F}(\phi)(\xi) d\xi = \frac{1}{(2\pi)^n} \int_{\mathbb{R}^n} \mathscr{F}(\phi)(\xi) d\xi$$

$$= \phi(0) = \delta(\phi),$$

so finally

$$P(D)u(\phi) = \delta(\phi).$$

Since $\phi \in \mathscr{C}_0^\infty(\mathbb{R}^n)$ was chosen arbitrarily,

$$P(D)u = \delta$$

follows, showing that u is a fundamental solution for $P(D)$. This completes the proof.

Example 9.1 The distribution $u(t) = H(t)e^{-at} \in \mathscr{D}'(\mathbb{R})$, $a = \text{const} > 0$, is a fundamental solution for the operator $\dfrac{d}{dt} + a$.

Indeed, let $\phi \in \mathscr{C}_0^\infty(\mathbb{R})$ be arbitrary but fixed. Then

$$\left(\frac{d}{dt} + a\right)u(\phi) = \left(\frac{d}{dt} + a\right)(H(t)e^{-at})(\phi)$$

$$= \frac{d}{dt}(H(t)e^{-at})(\phi) + aH(t)e^{-at}(\phi)$$

$$= -H(t)e^{-at}(\phi') + a\int_0^\infty e^{-at}\phi(t)dt$$

$$= -\int_0^\infty e^{-at}\phi'(t)dt + a\int_0^\infty e^{-at}\phi(t)dt$$

$$= \phi(0) = \delta(\phi).$$

Hence, $\left(\dfrac{d}{dt} + a\right)u(t) = \delta(t)$.

Exercise 9.1 Check if $H(x)\dfrac{\sin(ax)}{a} \in \mathcal{D}'(\mathbb{R})$ is a fundamental solution for $\dfrac{d^2}{dx^2} + a^2$, $a = \text{const} \neq 0$.

9.2 Fundamental Solutions of Ordinary Differential Operators

Let $(a, b) \subseteq \mathbb{R}, 0 \in (a, b)$. Consider the differential operator

$$Pu = u^{(n)} + a_1(t)u^{(n-1)} + \cdots + a_n(t)u, \tag{9.5}$$

where $a_j \in \mathscr{C}^\infty(a, b)$, $j = 1, 2, \ldots, n$. It is known that there exists a fundamental system $\{u_j\}_{j=1}^n$, $u_j \in \mathscr{C}^\infty(a, b)$, $j = 1, 2, \ldots, n$, of solutions of the equation $Pu = 0$. Moreover, if the coefficients of (9.5) are analytical in (a, b), then u_j, $j = 1, 2, \ldots, n$, are analytical in (a, b). Now, consider the equation

$$Pu = \delta. \tag{9.6}$$

We will search a solution of the equation (9.6) in the following form

$$u = \sum_{j=1}^n c_j u_j,$$

where $c_j \in \mathcal{D}'(a, b)$, $j = 1, 2, \ldots, n$, will be determined below using the Lagrange method. We have the system

$$
\begin{aligned}
\sum_{j=1}^n c_j' u_j^{(k)} &= 0, \quad k = 0, 1, \ldots, n - 2, \\
\sum_{j=1}^n c_j' u_j^{(n-1)} &= \delta \quad \text{in} \quad (a, b).
\end{aligned}
\tag{9.7}
$$

The last system is reduced in $(a, b) \backslash \{0\}$ to the system

$$\sum_{j=1}^n c_j' u_j^{(k)} = 0, \quad k = 0, 1, \ldots, n - 1. \tag{9.8}$$

Let

$$W(t) = \det \begin{pmatrix} u_1(t) & u_2(t) & \ldots & u_n(t) \\ u_1'(t) & u_2'(t) & \ldots & u_n'(t) \\ \ldots & \ldots & \ldots & \ldots \\ u_1^{(n-1)}(t) & u_2^{(n-1)}(t) & \ldots & u_n^{(n-1)}(t) \end{pmatrix}, \quad t \in (a,b)\setminus\{0\}.$$

With $W_{ij}, i, j = 1, 2, \ldots, n$, we denote the minor of the element $u_j^{(i-1)}, i, j = 1, 2, \ldots, n$. We multiply the kth equation of the system (9.8) with $W_{kj}, k, j = 1, 2, \ldots, n$, then we add the obtained equations and we get

$$c_j' W = 0 \quad \text{in} \quad (a,b)\setminus\{0\}, \quad j = 1, 2, \ldots, n.$$

Take $\phi \in \mathscr{C}_0^\infty((a,b)\setminus\{0\})$ and set $\psi = \dfrac{\phi}{W}$. Then

$$0 = \left(c_j' W\right)(\psi) = \left(c_j' W\right)\left(\frac{\phi}{W}\right) = c_j'(\phi), \quad j = 1, 2, \ldots, n.$$

Therefore $c_j' = 0, j = 1, 2, \ldots, n$, in $(a,b)\setminus\{0\}$. Let

$$h(t) = \begin{cases} \alpha, & 0 < t < b, \\ \beta, & a < t < 0, \end{cases}$$

where α and β are constants. Take $\phi \in \mathscr{C}_0^\infty(a,b)$ arbitrarily. Then

$$h'(\phi) = -h\left(\phi'\right)$$

$$= -\beta \int_a^0 \phi'(t)dt - \alpha \int_0^b \phi'(t)dt$$

$$= -\beta\phi(0) + \alpha\phi(0)$$

$$= (\alpha - \beta)\phi(0)$$

$$= (\alpha - \beta)\delta(\phi).$$

Therefore $h' = (\alpha - \beta)\delta$. From here, we conclude that $c_j' = a_j\delta, j = 1, 2, \ldots, n$, for some constants $a_j, j = 1, 2, \ldots, n$. Hence, the system (9.7) we can rewrite in the form

$$\left(\sum_{j=1}^{n} a_j u_j^{(k)}\right)\delta = 0, \quad k = 0, 1, \ldots, n-2,$$

$$\left(\sum_{j=1}^{n} a_j u_j^{(n-1)}\right)\delta = \delta,$$

whereupon

$$\sum_{j=1}^{n} a_j u_j^{(k)}(0) = 0, \quad k = 0, 1, \ldots, n-2,$$

$$\sum_{j=1}^{n} a_j u_j^{(n-1)}(0) = 1. \tag{9.9}$$

Consequently, for any choice of the functions

$$c_j(t) = \begin{cases} \alpha_j, & t \in (0, b), \\ \beta_j, & t \in (a, 0), \end{cases} \quad j = 1, 2, \ldots, n,$$

the linear combination

$$U(t) = \sum_{j=1}^{n} c_j(t) u_j(t)$$

is a fundamental solution of (9.5). In particular, if $\{u_j\}_{j=1}^{n}$ is a fundamental system of solutions of (9.5) such that

$$u_j^{(k)}(0) = \begin{cases} 0, & k \neq j-1, \\ 1, & k = j-1, \end{cases} \quad j = 1, 2, \ldots, n,$$

then, using (9.9), we find

$$a_j = 0, \quad j = 1, 2, \ldots, n-1, \quad a_n \cdot 1 = 1,$$

and the functions

$$X_1(t) = H(t)u_n(t), \quad X_2(t) = -H(-t)u_n(t)$$

form a fundamental system of (9.5).

Exercise 9.2 Prove that

$$H(x)e^{\pm ax}\frac{x^{m-1}}{(m-1)!}, \quad m = 2, 3, \ldots, a = \text{const},$$

is a fundamental solution for the operator

$$\left(\frac{d}{dx} \mp a\right)^m.$$

9.3   Fundamental Solution of the Heat Operator

Consider the heat operator

$$P(D) = D_0 - \sum_{j=1}^{n} D_j^2,$$

where $D_0 = \dfrac{\partial}{\partial t}$, and the equation

$$P(D)E = \delta(x, t).$$

The Fourier transform on the variable x yields

$$\mathscr{F}_x(D_0 E) - \mathscr{F}_x\left(\sum_{j=1}^{n} D_j^2 E\right) = \mathscr{F}_x(\delta),$$

whereupon

$$D_0\mathscr{F}_x(E)(\xi, t) + |\xi|^2 \mathscr{F}_x(E)(\xi, t) = 1(\xi) \cdot \delta(t).$$

Therefore

$$\mathscr{F}_x(E)(\xi, t) = (1(\xi) \cdot H(t))\, e^{-|\xi|^2 t}$$

and hence,

$$E(x, t) = \mathscr{F}_\xi^{-1}\left(H(t)e^{-|\xi|^2 t}\right) = \frac{H(t)}{(2\sqrt{\pi t})^n}e^{-\frac{|x|^2}{4t}}, \quad x \in \mathbb{R}^n, \quad t > 0.$$

9.4 Fundamental Solution of the Laplace Operator

Consider the differential operator

$$P(D) = \sum_{j=1}^{n} D_j^2$$

and the equation

$$P(D)E = \delta.$$

Take the Fourier transform and we get

$$-|\xi|^2 \mathscr{F}(E)(\xi) = 1(\xi)$$

or

$$\mathscr{F}(E)(\xi) = -\frac{1}{|\xi|^2}.$$

Then

$$E(x) = -\mathscr{F}^{-1}\left(\frac{1}{|\xi|^2}\right)(x), \quad x \in \mathbb{R}^n.$$

Hence, for $n = 2$, we have

$$E(x) = \frac{1}{2\pi} \log |x|, \quad x \in \mathbb{R}^2,$$

and for $n \geq 3$, we have

$$E(x) = -\frac{1}{(n-2)\sigma_n} \frac{1}{|x|^{n-2}}, \quad x \in \mathbb{R}^n,$$

where σ_n is the "area" of the boundary of the n-dimensional unit ball.

9.5 Advanced Practical Problems

Problem 9.1 Using (9.3) find a fundamental solution for the following operators

1. $\dfrac{d^2}{dx^2} + 4\dfrac{d}{dx}$,

2. $\dfrac{d^2}{dx^2} - 4\dfrac{d}{dx} + 1,$

3. $\dfrac{d^2}{dx^2} + 3\dfrac{d}{dx} + 2,$

4. $\dfrac{d^2}{dx^2} - 4\dfrac{d}{dx} + 5,$

5. $\dfrac{d^3}{dx^3} - 1,$

6. $\dfrac{d^3}{dx^3} - 3\dfrac{d^2}{dx^2} + 2\dfrac{d}{dx},$

7. $\dfrac{d^4}{dx^4} - 1,$

8. $\dfrac{d^4}{dx^4} - 2\dfrac{d^2}{dx^2} + 1.$

Answer

1. $H(x)\dfrac{1 - e^{-4x}}{4},$

2. $H(x)xe^x,$

3. $H(x)(e^{-x} - e^{-2x}),$

4. $H(x)e^{2x}\sin x,$

5. $\dfrac{H(x)}{3}\left(e^x - e^{-\frac{x}{2}}\left(\cos\dfrac{\sqrt{3}}{2}x + \sqrt{3}\sin\dfrac{\sqrt{3}}{2}x\right)\right),$

6. $\dfrac{H(x)}{2}(1 - e^x)^2,$

7. $\dfrac{H(x)}{2}(\operatorname{sh}x - \sin x),$

8. $\dfrac{H(x)}{2}(x\operatorname{ch}x - \operatorname{sh}x).$

Problem 9.2 Prove that

$$u(x, t) = \frac{H(t)}{2\sqrt{\pi t}}e^{t - \frac{(x+t)^2}{4t}}$$

is a fundamental solution for the operator

$$\frac{\partial}{\partial t} - \frac{\partial^2}{\partial x^2} - \frac{\partial}{\partial x} - 1.$$

Problem 9.3 Prove that

$$-H(t)H(-x)e^{t+x}$$

is a fundamental solution for the operator

$$\frac{\partial^2}{\partial x \partial t} - \frac{\partial}{\partial x} - \frac{\partial}{\partial t} + 1.$$

Problem 9.4 Let $n = 3$. Prove that

$$u(x) = -\frac{e^{\pm ik|x|}}{4\pi |x|}$$

satisfies the equation

$$\Delta u + k^2 u = \delta(x).$$

Problem 9.5 Prove that

$$u(x, t) = \frac{1}{2a} H(at - |x|)$$

satisfies the equation

$$u_{tt} - a^2 u_{xx} = \delta(x, t), \quad x \in \mathbb{R}, \quad t \in \mathbb{R},$$

and

$$u(x, t) \to_{t \to +0} 0, \quad \frac{\partial u(x, t)}{\partial t} \to_{t \to +0} \delta(x), \quad \frac{\partial^2 u(x, t)}{\partial t^2} \to_{t \to +0} 0$$

in $\mathscr{D}'(\mathbb{R})$.

Problem 9.6 Prove that

$$u(x, y) = \frac{1}{\pi (x + iy)}$$

satisfies the equation

$$\frac{1}{2}\left(\frac{\partial u}{\partial x} + i\frac{\partial u}{\partial y}\right) = \delta(x, y).$$

Problem 9.7 Prove that

$$u(x, t) = \frac{i H(t)}{(2\sqrt{\pi t})^n} e^{-i\frac{\pi n}{4}} e^{i\frac{|x|^2}{4t}}$$

solves

$$\frac{1}{i}u_t - \Delta u = \delta(x,t), \quad x \in \mathbb{R}^n, \quad t \in \mathbb{R}.$$

Problem 9.8 Define

$$u(x,t) = \frac{H(t)}{(2\sqrt{\pi t})^n} e^{-\frac{|x|^2}{4t}}$$

and suppose

$$F(x) = \int_0^\infty u(x,t)dt,$$

exists for any $t \geq 0$ and almost every $x \in \mathbb{R}^n$. Prove

$$\Delta F = -\delta(x,t).$$

9.6 Notes and References

In this chapter it is given a definition for a fundamental solution of a differential operator. It is proved the classical Malgrange-Eherenpreis theorem. As applications, they are deducted the fundamental solutions for ordinary differential operators, the heat operator and the Laplace operator. Additional materials can be found in [7, 16, 17, 20, 21, 24, 25] and references therein.

Chapter 10
Sobolev Spaces

10.1 Definitions

Definition 10.1 Let A be an open set in $\mathbb{R}^n$, $m \in \mathbb{N}$ and $1 \le p \le \infty$. The Sobolev space $W^{m,p}(A)$ consists of functions in $L^p(A)$ whose partial derivatives up to order m, in the sense of distributions, can be identified with functions in $L^p(A)$.

Equivalently,

$$W^{m,p}(A) = \{u \in L^p(A) : D^\alpha u \in L^p(A) \text{ for any } \alpha \in \mathbb{N}^n \cup \{0\}, |\alpha| \le m\}.$$

Notice that clearly $W^{0,q}(A) = L^q(A)$.

For $p = 2$, the symbol $W^{m,2}(A)$ is generally replaced by $H^m(A)$, and in the case $A = \mathbb{R}^n$, we can use the Fourier transform $\xi \mapsto \mathscr{F}(u)(\xi)$ of $u \in L^2(A)$ to give the following characterisation

$$W^{m,2}(\mathbb{R}^n) = H^m(\mathbb{R}^n) = \{u \in L^2(\mathbb{R}^n) : \xi \mapsto (1 + |\xi|^2)^{\frac{m}{2}} \mathscr{F}(u)(\xi) \in L^2(\mathbb{R}^n)\}.$$

Example 10.1 Let $n = 2$. We seek conditions on $\beta > 0$ so that the function $u(x, y) = x(x^2 + y^2)^{-\beta}$, $x, y \in U_1$, is, away from the origin, an element of $H^1(U_1)$. In polar coordinates $x = r \cos\phi$, $y = r \sin\phi$, $0 < r < 1$, $\phi \in [0, 2\pi]$, we have

$$u(x, y) = x(x^2 + y^2)^{-\beta} = r^{1-2\beta} \cos\phi,$$

$$|u(x, y)|^2 = r^{2-4\beta} \cos^2\phi,$$

$$u_x(x, y) = (x^2 + y^2)^{-\beta} - 2\beta x^2 (x^2 + y^2)^{-\beta-1} = r^{-2\beta}(1 - 2\beta \cos^2\phi),$$

$$|u_x(x, y)|^2 = r^{-4\beta}(1 - 2\beta \cos^2\phi)^2,$$

$$u_y(x, y) = -2\beta xy(x^2 + y^2)^{-\beta-1} = -2\beta r^{-2\beta} \cos\phi \sin\phi,$$

$$|u_y(x, y)|^2 = 4\beta^2 r^{-4\beta} \cos^2\phi \sin^2\phi, \quad x, y \in U_1,$$

© The Author(s), under exclusive license to Springer Nature Switzerland AG 2021
S. G. Georgiev, *Theory of Distributions*,
https://doi.org/10.1007/978-3-030-81265-2_10

substituting which produces

$$\int_{U_1} |u(x, y)|^2 dxdy = \int_0^1 \int_0^{2\pi} r^{3-4\beta} \cos^2 \phi d\phi dr < \infty \Longleftrightarrow \beta < 1,$$

$$\int_{U_1} |u_x(x, y)|^2 dxdy = \int_0^1 \int_0^{2\pi} r^{1-4\beta} (1 - 2\beta \cos^2 \phi)^2 d\phi dr < \infty \Longleftrightarrow \beta < \frac{1}{2},$$

$$\int_{U_1} |u_y(x, y)|^2 dxdy = 4\beta^2 \int_0^1 \int_0^{2\pi} r^{1-4\beta} \cos^2 \phi \sin^2 \phi d\phi dr < \infty \Longleftrightarrow \beta < \frac{1}{2}.$$

Consequently $u \in H^1(U_1)$ if $0 < \beta < \dfrac{1}{2}$.

Exercise 10.1 Let $n = 2$ and $u(x, y) = xy(x^2 + y^2)^{-\beta}$, $(x, y) \in U_1 \setminus \{(0, 0)\}$, $\beta > 0$. Find conditions on the parameter β so that $u \in H^1(U_1)$.

Exercise 10.2 Define u on U_1 by

$$u(x) = (1 - |x|)^\beta (-\log(1 - |x|))^\alpha, \quad x \in U_1,$$

where $\alpha, \beta > 0$ are real. Find conditions for α and β so that $u \in W^{1,p}(U_1)$, $1 \le p \le \infty$.

10.2   Elementary Properties

One can endow the Sobolev spaces with a norm

$$\|u\|_{W^{m,p}(A)} = \begin{cases} \left(\displaystyle\sum_{0 \le |\alpha| \le p} \|D^\alpha u\|_{L^p(A)}^p \right)^{\frac{1}{p}} & \text{if } 1 \le p < \infty, \\ \displaystyle\max_{0 \le |\alpha| \le m} \|D^\alpha u\|_{L^\infty(A)} & \text{if } p = \infty. \end{cases}$$

In particular, the space $H^m(A)$ admits an inner product

$$(u, v) = \sum_{0 \le |\alpha| \le m} (D^\alpha u, D^\alpha v).$$

Exercise 10.3 Check that $\| \cdot \|_{W^{m,p}(A)}$ fulfils the axioms for being a norm.

Exercise 10.4 Check that (u, v) is indeed an inner product.

Definition 10.2 A sequence $\{u_k\}_{k=1}^{\infty}$ in $W^{m,p}(A)$ converges to $u \in W^{m,p}(A)$ if

$$||u_k - u||_{W^{m,p}(A)} \to_{k\to\infty} 0.$$

Definition 10.3 A sequence $\{u_k\}_{k=1}^{\infty}$ in $W^{m,p}(A)$ converges to $u \in W^{m,p}(A)$ in $W^{m,p}_{loc}(A)$ if $u_k \to_{k\to\infty} u$ in $W^{m,p}(V)$ for every $V \subset\subset A$.

Definition 10.4 We denote by $W^{m,p}_0(A)$ the closure of $C_0^{\infty}(A)$ in $W^{m,p}(U)$.

In particular, $H_0^m(A) = W_0^{m,2}(A)$.

Exercise 10.5 Prove that $u \in W_0^{m,p}(A)$ if and only if there exist functions $u_k \in C_0^{\infty}(A)$ such that $u_k \to_{k\to\infty} u$ in $W^{m,p}(A)$.

Now, we set out to prove a number of elementary, but important properties of the Sobolev spaces.

Theorem 10.1 *Let $u \in W^{m,p}(A)$. For any $0 \le |\alpha| \le m$, we have $D^\alpha u \in W^{m-|\alpha|,p}(A)$ and*

$$D^\beta(D^\alpha u) = D^\alpha(D^\beta u) = D^{\alpha+\beta} u$$

for $0 \le |\alpha| + |\beta| \le m$.

Proof To prove this property, we observe that $u \in W^{m,p}(A)$ implies that $D^\alpha u$ is a well-defined distribution, and $D^\alpha u \in L^p(A)$ for $0 \le |\alpha| \le m$. Moreover, $D^\alpha u \in W^{m-|\alpha|,p}(A)$ if $D^\beta(D^\alpha u)$ exists and belongs to $L^p(A)$ for $0 \le |\beta| \le m - |\alpha|$.

Pick $\phi \in \mathscr{C}_0^{\infty}(A)$. Then, for $0 \le |\alpha| + |\beta| \le m$, we have

$$\int_A D^\alpha u D^\beta \phi \, dx = (-)^{|\alpha|} \int_A u D^\alpha(D^\beta \phi) dx$$

$$= (-1)^{|\alpha|} \int_A u D^{\alpha+\beta} \phi \, dx$$

$$= (-1)^{|\beta|} \int_A D^{\alpha+\beta} u \phi \, dx.$$

On the other hand,

$$\int_A D^\alpha u D^\beta \phi \, dx = (-1)^{|\beta|} \int_A D^\beta(D^\alpha u) \phi \, dx,$$

so overall

$$D^\beta(D^\alpha u) = D^{\alpha+\beta}u.$$

This completes the proof.

Theorem 10.2 *For any $u, v \in W^{m,p}(A)$ and constants $\lambda, \mu, \lambda u + \mu v \in W^{m,p}(A)$.*

Proof For this, let $\phi \in \mathscr{C}_0^\infty(A)$ be arbitrary. As $u, v \in W^{m,p}(A)$, for $0 \le |\alpha| \le m$ we have $D^\alpha u, D^\alpha v \in L^p(A)$ and

$$\int_A D^\alpha(\lambda u + \mu v)\phi dx = (-1)^{|\alpha|} \int_A (\lambda u + \mu v)D^\alpha \phi dx$$

$$= \lambda(-1)^{|\alpha|} \int_A u D^\alpha \phi dx + \mu(-1)^{|\alpha|} \int_A v D^\alpha \phi dx$$

$$= \lambda \int_A D^\alpha u \phi dx + \mu \int_A D^\alpha v \phi dx$$

$$= \int_A (\lambda D^\alpha u + \mu D^\alpha v)\phi dx.$$

This completes the proof.

Exercise 10.6 If $u \in W^{m,p}(A)$ and V is open in A, prove that $u \in W^{m,p}(V)$.

Theorem 10.3 (Leibniz Formula) *If $\zeta \in \mathscr{C}_0^\infty(A)$ and $u \in W^{m,p}(A)$, then $\zeta u \in W^{m,p}(A)$ and*

$$D^\alpha(\zeta u) = \sum_{\beta \le \alpha} \binom{\alpha}{\beta} D^\beta \zeta D^{\alpha-\beta}u.$$

Proof First of all, fix $\phi \in \mathscr{C}_0^\infty(A)$. Then, for $|\alpha| = 1$, we get

$$\int_A D^\alpha(u\zeta)\phi dx = -\int_A u\zeta D^\alpha \phi dx$$

$$= -\int_A u(D^\alpha(\zeta\phi) - D^\alpha \zeta\phi)dx$$

$$= \int_A (D^\alpha u\zeta + u D^\alpha \zeta)\phi dx.$$

Let $l < m$. We suppose that the assertion holds for $|\alpha| \le l$ and prove it for $|\alpha| = l + 1$. Let $\alpha = \beta + \gamma$, where $|\beta| = l$ and $|\gamma| = 1$. Then

$$\int_A u\zeta \, D^\alpha \phi \, dx = \int_A u\zeta \, D^\beta (D^\gamma \phi) dx = (-1)^{|\beta|} \int_A D^\beta (u\zeta) D^\gamma \phi \, dx$$

so by induction hypothesis

$$= (-1)^{|\beta|} \int_A \sum_{\sigma \le \beta} \binom{\beta}{\sigma} D^\sigma \zeta \, D^{\beta-\sigma} u \, D^\gamma \phi \, dx$$

$$= (-1)^{|\beta|+|\gamma|} \int_A \sum_{\sigma \le \beta} \binom{\beta}{\sigma} D^\gamma (D^\sigma \zeta \, D^{\beta-\sigma} u) \, \phi \, dx$$

$$= (-1)^{|\alpha|} \int_A \sum_{\sigma \le \beta} \binom{\beta}{\sigma} \left(D^{\gamma+\sigma} \zeta \, D^{\beta-\sigma} u + D^\sigma \zeta \, D^{\beta+\gamma-\sigma} u \right) \phi \, dx$$

$$= (-1)^{|\alpha|} \int_A \sum_{\sigma \le \alpha} \binom{\alpha}{\sigma} D^\sigma \zeta \, D^{\alpha-\sigma} u \phi \, dx,$$

proving the assertion.

Theorem 10.4 *For every $m = 1, 2, \ldots, 1 \le p \le \infty$, $W^{m,p}(A)$ is a Banach space.*

Proof We shall prove the statement for $p < \infty$, and leave the reader to see to $p = \infty$. Let $\{u_l\}_{l=1}^\infty \subset W^{m,p}(A)$ be a Cauchy sequence. Then, for every $l \in \mathbb{N}$ and $0 \le |\alpha| \le m$, we have

$$\|D^\alpha u_{l+q} - D^\alpha u_l\|_{L^p(A)} \to_{q\to\infty} 0.$$

Therefore the sequence $\{D^\alpha u_l\}_{l=1}^\infty$ is fundamental in $L^p(A)$, for any α such that $0 \le |\alpha| \le m$. Since $L^p(A)$ is a Banach space, this sequence converges in $L^p(A)$ to some u_α, for any $0 \le |\alpha| \le m$. Let $u_0 = u$. We claim that $D^\alpha u = u_\alpha$ for any $0 \le |\alpha| \le m$. In fact,

$$\int_A u D^\alpha \phi \, dx = \lim_{l\to\infty} \int_A u_l D^\alpha \phi \, dx = (-1)^{|\alpha|} \lim_{l\to\infty} \int_A D^\alpha u_l \phi \, dx = (-1)^{|\alpha|} \int_A u_\alpha \phi \, dx,$$

so $D^\alpha u = u_\alpha, 0 \le |\alpha| \le m$. As

$$\|D^\alpha u_l - D^\alpha u\|_{L^p(A)} \to_{l\to\infty} 0$$

for every $0 \leq |\alpha| \leq m$, we conclude that

$$||u_l - u||_{W^{m,p}(A)} \to_{l \to \infty} 0.$$

This completes the proof.

Exercise 10.7 Prove that $H^m(A)$ is a Hilbert space for any $m \in \mathbb{N}$.

Exercise 10.8 Prove that if $u \in W^{1,p}(0, 1)$ for some $1 < p < 1$, then

$$|u(x) - u(y)| \leq |x - y|^{1-\frac{1}{p}} \left(\int_0^1 |u'(t)|^p dt \right)^{\frac{1}{p}}$$

for a.e. $x, y \in [0, 1]$.

10.3   Approximation by Smooth Functions

Let $\epsilon > 0$ and $A \subset \mathbb{R}^n$ be open. We define

$$u_\epsilon(x) = \omega_\epsilon \star u(x)$$

for $u \in W^{m,p}(A)$ and $x \in A_\epsilon$.

Theorem 10.5 *If $u \in W^{m,p}(A)$, then $u_\epsilon \in \mathscr{C}^\infty(A_\epsilon)$.*

Proof To see this, let α be an arbitrary multi-index. Then

$$D^\alpha u_\epsilon(x) = D^\alpha \int_A \omega_\epsilon(x - y)u(y)dy$$

$$= \int_A D_x^\alpha \omega_\epsilon(x - y)u(y)dy$$

$$= (-1)^{|\alpha|} \int_A D_y^\alpha \omega_\epsilon(x - y)u(y)dy, \quad x \in A,$$

from which

$$D^\alpha u_\epsilon(x) = \int_A \omega_\epsilon(x - y)D^\alpha u(y)dy = \omega_\epsilon \star D^\alpha u(x), \quad x \in A,$$

and

$$|D^\alpha u_\epsilon(x)| = \left| \int_A D^\alpha_x \omega_\epsilon(x-y)u(y)dy \right|$$

$$\leq \int_A |D^\alpha_x \omega_\epsilon(x-y)||u(y)|dy$$

$$\leq \left(\int_A |D^\alpha_x \omega_\epsilon(x-y)|^q dy \right)^{\frac{1}{q}} \left(\int_A |u(y)|^p dy \right)^{\frac{1}{p}}$$

$$\leq C\|u\|_{W^{m,p}(A)}, \quad x \in A.$$

This completes the proof.

Theorem 10.6 *If $u \in W^{m,p}_{loc}(A)$, then*

$$u_\epsilon \to_{\epsilon \to 0} u$$

in $W^{m,p}_{loc}(A)$.

Proof Let α be given. From the previous property $D^\alpha u_\epsilon$ exists, and if $V \subset\subset A$, then

$$\|u_\epsilon - u\|_{W^{m,p}(V)} = \left(\sum_{|\alpha| \leq m} \int_V |D^\alpha u_\epsilon - D^\alpha u|^p dx \right)^{\frac{1}{p}}$$

$$\leq 2^{\frac{1}{p}} \sum_{|\alpha| \leq m} \left(\int_V |D^\alpha u_\epsilon - D^\alpha u|^p dx \right)^{\frac{1}{p}} \tag{10.1}$$

$$= 2^{\frac{1}{p}} \sum_{|\alpha| \leq m} \|D^\alpha u_\epsilon - D^\alpha u\|_{L^p(V)}.$$

Since $(D^\alpha u)_\epsilon = D^\alpha u_\epsilon$ and $D^\alpha u \in L^p(V)$, $0 \leq |\alpha| \leq m$, using the properties of the convolution, we see that

$$\|D^\alpha u_\epsilon - D^\alpha u\|_{L^p(V)} \to_{\epsilon \to 0} 0.$$

Hence, $u_\epsilon \to_{\epsilon \to 0} u$ in $W^{m,p}_{loc}(A)$. This completes the proof.

Theorem 10.7 *If A is a bounded set in $\mathbb{R}^n$ and $u \in W^{m,p}(A)$, there exists a sequence $\{u_l\}_{l=1}^{\infty}$ in $\mathscr{C}^{\infty}(A) \cap W^{m,p}(A)$ such that*

$$u_l \to_{l \to \infty} u$$

in $W^{m,p}(A)$.

Proof We claim $A = \cup_{i=1}^{\infty} A_i$, where $A_i = \{x \in A : \text{dist}(x, \partial A) > \frac{1}{i}\}$, $i \in \mathbb{N}$.

Let $x \in A$ and $x \notin \partial A$. Therefore there exists $i \in \mathbb{N}$ such that $\text{dist}(x, \partial A) > \frac{1}{i}$, so

$x \in A_i$ and $x \in \cup_{i=1}^{\infty} A_i$. Consequently $A \subset \cup_{i=1}^{\infty} A_i$. Conversely, if $x \in \cup_{i=1}^{\infty} A_i$ there exists $j \in \mathbb{N}$ such that $x \in A_j$. Therefore $x \in A$ and $\mathrm{dist}(x, \partial A) > \dfrac{1}{j}$, so we conclude $\cup_{i=1}^{\infty} A_i \subset A$. Let $V_i = A_{i+3} \backslash \bar{A}_i$, $i = 1, 2, \ldots$, and choose $V_0 \subset\subset A$. Then, as above, $A = \cup_{i=0}^{\infty} V_i$. Let $\{\zeta_i\}_{i=1}^{\infty}$ be a sequence of smooth functions such that

$$\zeta_i \in \mathscr{C}_0^{\infty}(V_i), \quad 0 \le \zeta_i \le 1, \quad \sum_{i=0}^{\infty} \zeta_i = 1.$$

We define $\epsilon_i > 0$ and $\delta > 0$, $u^i = (\zeta_i u) \star \omega_{\epsilon_i}$ so that

$$||u^i - \zeta_i u||_{W^{m,p}(A)} < \frac{\delta}{2^{i+1}},$$
$$\mathrm{supp}\, u^i \subset W_i = A_{i+4} \backslash \bar{A}_i \supset V_i.$$

Note that $\zeta_i u \in W^{m,p}(A)$. From the properties of the convolution, we know

$$u^i \to_{\epsilon_i \to 0} \zeta_i u$$

in $W^{m,p}(A)$. Let

$$v = \sum_{i=0}^{\infty} u^i.$$

Since $\sum_{i=0}^{\infty} \zeta_i = 1$, we have that $u = \sum_{i=0}^{\infty} \zeta_i u$. Let $V \subset\subset A$. Only a finite number of elements are different from zero in V, so $v \in \mathscr{C}^{\infty}(A)$. Therefore

$$||v - u||_{W^{m,p}(A)} = \left|\left| \sum_{i=0}^{\infty} u^i - \sum_{i=0}^{\infty} \zeta_i u \right|\right|_{W^{m,p}(V)}$$

$$= \left|\left| \sum_{i=0}^{\infty} (u^i - \zeta_i u) \right|\right|_{W^{m,p}(V)}$$

$$\le \sum_{i=0}^{\infty} ||u^i - \zeta_i u||_{W^{m,p}(V)}$$

$$< \delta \sum_{i=0}^{\infty} \frac{1}{2^{i+1}}$$

$$= \delta,$$

and consequently

$$\sup_{V \subset\subset U} \|v - u\|_{W^{m,p}(V)} \le \delta.$$

Let v_l be the function which corresponds to the space $W^{m,p}(V_l)$. Then

$$\|v_l - u\|_{W^{m,p}(A)} \to_{l\to\infty} 0.$$

This completes the proof.

Exercise 10.9 Let A be a bounded set with $\mathscr{C}^1$ boundary ∂A. Prove that if $u \in W^{m,p}(A)$, $1 \le p < \infty$, there exists a sequence $\{u_l\}_{l=1}^\infty$ in $\mathscr{C}^\infty(\overline{A})$ such that

$$\|u_l - u\|_{W^{m,p}(A)} \to_{l\to\infty} 0.$$

Solution Let $x^0 \in \partial A$. Since ∂A is $\mathscr{C}^1$, there exists a radius $r > 0$ and a $\mathscr{C}^1$ function $\gamma : \mathbb{R}^{n-1} \to \mathbb{R}$ such that

$$A \cap U(x^0, r) = \{x \in U(x^0, r) : x_n > \gamma(x_1, x_2, \dots, x_{n-1})\}.$$

Let $V = A \cap U(x^0, \frac{r}{2})$ and define

$$x^\epsilon = x + \lambda\epsilon e_n, \quad x \in V, \epsilon > 0, \lambda > 0, e_n = (0, 0, \dots, n).$$

There exists a large enough $\lambda > 0$ and a small enough $\epsilon > 0$ such that $U(x^\epsilon, \epsilon)$ lies in $A \cap U(x^0, r)$ for every $x \in V$. Let $u^\epsilon(x) = u(x^\epsilon)$ and define

$$v_\epsilon = \omega_\epsilon \star u_\epsilon.$$

Take $|\alpha| \le m$, so

$$\|D^\alpha v_\epsilon - D^\alpha u\|_{L^p(V)} = \|D^\alpha v_\epsilon - D^\alpha u_\epsilon + D^\alpha u_\epsilon - D^\alpha u\|_{L^p(V)}$$
$$\le \|D^\alpha v_\epsilon - D^\alpha u_\epsilon\|_{L^p(V)} + \|D^\alpha u_\epsilon - D^\alpha u\|_{L^p(V)}. \tag{10.2}$$

We have

$$|D^\alpha v_\epsilon - D^\alpha u_\epsilon\|_{L^p(V)} \to_{\epsilon\to 0} 0 \tag{10.3}$$

and

$$\|D^\alpha u_\epsilon - D^\alpha u\|_{L^p(V)} \to_{\epsilon\to 0} 0. \tag{10.4}$$

From (10.2)–(10.4), it follows

$$||D^\alpha v_\epsilon - D^\alpha u||_{L^p(V)} \to_{\epsilon \to 0} 0 \quad \text{for} \quad 0 \le |\alpha| \le m.$$

Consequently

$$||v_\epsilon - u||_{W^{m,p}(V)} \to_{\epsilon \to 0} 0.$$

Since ∂A is compact, there exist finitely many points x_i^0, radii r_i and functions $v_i \in \mathscr{C}^\infty(\overline{V}_i)(i = 1, 2, \dots, N)$ for which

$$\partial A \subset \cup_{i=1}^N U(x_i^0, \frac{r_i}{2}), \qquad ||v_i - u||_{W^{m,p}(V_i)} \le \delta,$$

where

$$V_i = A \cap U(x_i^0, \frac{r_i}{2}).$$

Let $V_0 \subset\subset U$. Then

$$A \subset \cup_{i=0}^N V_i.$$

There exist functions $\zeta_i, i = 0, 1, \dots, N$, such that $\zeta_i \in \mathscr{C}_0^\infty(V_i), 0 \le \zeta_i \le 1$ and $\sum_{i=0}^N \zeta_i = 1$. Now, we define the function

$$v = \sum_{i=0}^N \zeta_i v_i.$$

Then

$$u = \sum_{i=0}^N \zeta_i u$$

and

$$||D^\alpha v - D^\alpha u||_{L^p(A)} = ||\sum_{i=0}^N D^\alpha(\zeta_i(v_i - u))||_{L^p(A)}$$

$$\le \sum_{i=0}^N ||D^\alpha(\zeta_i(v_i - u))||_{L^p(A)}$$

$$= \sum_{i=0}^{N} ||D^\alpha v_i - D^\alpha u||_{L^p(V_i)}$$

$$\leq \sum_{i=0}^{N} ||v_i - u||_{W^{m,p}(V_i)}$$

$$\leq \delta(N+1)$$

for $0 \leq |\alpha| \leq m$.

10.4 Extensions

Theorem 10.8 (Extension Theorem) *Let $A \subset\subset V$ be bounded subsets of $\mathbb{R}^n$ and assume the boundary ∂A is $\mathscr{C}^1$. There exists a linear operator $E : W^{1,p}(A) \to W^{1,p}(\mathbb{R}^n)$ such that*

1. *$Eu = u$ for any $u \in W^{1,p}(A)$,*
2. *$\mathrm{supp}E \subset V$,*
3. *$||Eu||_{W^{1,p}(\mathbb{R}^n)} \leq C||u||_{W^{1,p}(A)}$.*

Proof Fix $x^0 \in \partial A$.

Case 1. $u \in \mathscr{C}^\infty(\overline{A})$.

First of all, we suppose that locally, around x^0, the boundary ∂A belongs to $\{x_n = 0\}$. Since ∂A is $\mathscr{C}^1$, there is a ball U such that

$$U^+ = U \cap \{x_n \geq 0\} \subset \overline{A},$$
$$U^- = U \cap \{x_n \leq 0\} \subset \mathbb{R}^n \backslash A.$$

Now, define

$$\bar{u}(x) = \begin{cases} u(x) & \text{for} \quad x \in U^+, \\ -3u(x_1, x_2, \ldots, x_{n-1}, -x_n) + 4u(x_1, x_2, \ldots, x_{n-1}, -\dfrac{x_n}{2}) & \text{for} \quad x \in U^-. \end{cases}$$

We will show that $\bar{u} \in \mathscr{C}^1(U)$. In fact, let us define

$$u^+ = \bar{u}\big|_{x_n \geq 0, x \in U^+}, \quad u^- = \bar{u}\big|_{x_n \leq 0, x \in U^-}.$$

We have

$$u^+\big|_{x_n=0} = u(x_1, x_2, \ldots, x_{n-1}, 0),$$
$$u^-\big|_{x_n=0} = -3u(x_1, x_2, \ldots, x_{n-1}, 0) + 4u(x_1, x_2, \ldots, x_{n-1}, 0) = u(x_1, x_2, \ldots, x_{n-1}, 0).$$

Consequently

$$u^+_{\,|x_n=0} = u^-_{\,|x_n=0},$$

and then

$$\frac{\partial u^+}{\partial x_i}\bigg|_{x_n=0} = \frac{\partial u^-}{\partial x_i}\bigg|_{x_n=0}, \quad i = 1, 2, \ldots, n-1.$$

On the other hand,

$$\frac{\partial u^+}{\partial x_n}\bigg|_{x_n=0} = \frac{\partial u}{\partial x_n}\bigg|_{x_n=0},$$

$$\frac{\partial u^-}{\partial x_n}\bigg|_{x_n=0} = 3\frac{\partial u}{\partial x_n}\bigg|_{x_n=0} - 2\frac{\partial u}{\partial x_n}\bigg|_{x_n=0} = \frac{\partial u}{\partial x_n}\bigg|_{x_n=0},$$

so overall,

$$\frac{\partial u^+}{\partial x_n}\bigg|_{x_n=0} = \frac{\partial u^-}{\partial x_n}\bigg|_{x_n=0}$$

and $D^\alpha u^-_{\,|x_n=0} = D^\alpha u^+_{\,|x_n=0}$ is well defined for $0 \le |\alpha| \le 1$. Hence, $\bar{u} \in \mathscr{C}^1(U)$. Additionally,

$$\|\bar{u}\|_{W^{1,p}(U)} \le C\|u\|_{W^{1,p}(U^+)} \le C\|u\|_{W^{1,p}(A)}.$$

If ∂A does not belong on such hyperplane locally (in a neighbourhood of x^0), since the boundary is $\mathscr{C}^1$ there exists a function Φ with inverse Ψ, say $x = \Phi(y)$, $y = \Psi(x)$, that maps a neighbourhood of x^0 to a neighbourhood of $\Psi(x^0)$ in such a way that, locally, $\Psi(\partial A)$ lies on $\{y_n = 0\}$. Let $u^1(y) = u(x) = u(\Phi(y))$. As above, we construct a function $\bar{u}^1$ such that

$$\|\bar{u}^1\|_{W^{1,p}(U)} \le C\|u^1\|_{W^{1,p}(U^+)} \le C\|u^1\|_{W^{1,p}(A)},$$

and $\bar{u}^1 = u^1$ in U^+. If $W = \Psi(U)$, then

$$\|\bar{u}\|_{W^{1,p}(W)} \le C\|u\|_{W^{1,p}(A)}.$$

Now, we define the operator

$$Eu = \bar{u}.$$

Since $\bar{u}$ is bounded in $W^{1,p}(\bar{A})$, the map $u \mapsto Eu$ is linear and bounded.

Because ∂A is compact, there exist finitely many points $x_1^0, x_2^0, \ldots, x_N^0$, open sets W_i, extensions $\bar{u}_i$ of u on W_i such that if we take $W_0 \subset\subset A$, then we have

$$\partial A \subset \cup_{i=1}^N W_i, \quad A \subset \cup_{i=0}^N W_i.$$

Let $\{\zeta_i\}_{i=0}^N$ be the partition of unity corresponding to the system W_0, $W_1, \ldots, W_N$. Now, we put $\bar{u}_0 = u$. Then

$$\bar{u} = \sum_{i=0}^N \zeta_i \bar{u}_i$$

and

$$
\begin{aligned}
||\bar{u}||_{W^{1,p}(\mathbb{R}^n)} &= \left|\left| \sum_{i=0}^N \zeta_i \bar{u}_i \right|\right|_{W^{1,p}(\mathbb{R}^n)} \\
&\leq \sum_{i=0}^N ||\zeta_i \bar{u}_i||_{W^{1,p}(\mathbb{R}^n)} \\
&= \sum_{i=0}^N ||\zeta_i \bar{u}_i||_{W^{1,p}(W_i)} \\
&= \sum_{i=0}^N ||\bar{u}_i||_{W^{1,p}(W_i)} \\
&\leq \sum_{i=0}^N C||u||_{W^{1,p}(A)} \\
&= C(N+1)||u||_{W^{1,p}(A)}.
\end{aligned}
$$

Case 2. Let $u \in W^{1,p}(A)$. Then there exists a sequence $\{u_l\}_{l=1}^\infty$ in $\mathscr{C}^\infty(\overline{A}) \cap W^{1,p}(A)$ such that $u_l \to u, l \to \infty$, in $W^{1,p}(A)$. For any $u_l, l = 1, 2, \ldots$, we apply case 1. We also have

$$||Eu_m - Eu_l||_{W^{1,p}(\mathbb{R}^n)} = ||E(u_m - u_l)||_{W^{1,p}(\mathbb{R}^n)} \leq C||u_m - u_l||_{W^{1,p}(A)} \to_{l,m\to\infty} 0.$$

Consequently $\{Eu_m\}_{m=1}^\infty$ is a fundamental sequence in the Banach space $W^{1,p}(A)$, so it converges to some $\bar{u} \in W^{1,p}(A)$. But as $Eu_m = \bar{u}$ on $\overline{A}$, we have $\bar{u} = u$ on $\overline{A}$. This completes the proof.

Definition 10.5 We call Eu an extension of u to $\mathbb{R}^n$.

Exercise 10.10 Let $A \subset\subset V \subset \mathbb{R}^n$ be bounded sets with ∂A of class $\mathscr{C}^2$. Then there is a linear operator $E : W^{2,p}(A) \mapsto W^{2,p}(\mathbb{R}^n)$ such that

1. $Eu = u$ for $u \in W^{2,p}(A)$,
2. $\operatorname{supp} E \subset V$,
3. $\|Eu\|_{W^{2,p}(\mathbb{R}^n)} \leq C\|u\|_{W^{2,p}(A)}$.

Exercise 10.11 Extend $u \in W^{1,p}((0, \infty))$ onto $(-\infty, 0)$ by setting $\overline{u}(x) = u(-x)$. Prove that this extension $\overline{u}$ is an element of $W^{1,p}(\mathbb{R})$.

10.5   Traces

Theorem 10.9 (Trace Theorem) *Let A be a bounded set in $\mathbb{R}^n$ with $\mathscr{C}^1$ boundary ∂A. There exists a linear bounded operator $T : W^{1,p}(A) \mapsto W^{1,p}(\partial A)$ such that*

$$Tu = u_{|\partial A} \quad \text{if} \quad u \in W^{1,p}(A) \cap \mathscr{C}(\overline{A})$$

and

$$\|Tu\|_{L^p(\partial U)} \leq C\|u\|_{W^{1,p}(A)}.$$

Proof We assume $u \in \mathscr{C}^1(\overline{A})$ and take $x^0 \in \partial A$. We also suppose that ∂A intersected with some neighbourhood of x^0 lies on the plane $\{x_n = 0\}$. Let $r > 0$ be such that $A \cap U(x^0, r) \subset \{x_n = 0\}$. We consider $U(x^0, \frac{r}{2})$ and call $\Gamma = \partial(A \cap U(x^0, \frac{r}{2}))$. We choose $\zeta \in \mathscr{C}_0^\infty(U(x^0, r))$ so that $0 \leq \zeta \leq 1$ on $U(x^0, r), \zeta \equiv 1$ on $U(x^0, \frac{r}{2})$.

By denoting

$$x' = (x_1, x_2, \ldots, x_{n-1}) \in \mathbb{R}^{n-1} = \{x_n = 0\},$$

we have

$$\int_\Gamma |u|^p dx' \leq \int_{\{x_n=0\}} \zeta|u|^p dx = -\int_{U^+} (\zeta|u|^p)_{x_n} dx$$

$$= -\int_{U^+} \zeta_{x_n}|u|^p dx - p\int_{U^+} \zeta|u|^{p-1}(\operatorname{sign} u)u_{x_n} dx. \tag{10.5}$$

Young's inequality, with $\dfrac{1}{p} + \dfrac{1}{q} = 1$, gives

$$|u|^{p-1}|u_{x_n}| \leq \frac{(|u|^{p-1})^q}{q} + \frac{|u_{x_n}|^p}{p} \leq C(|u|^p + |Du|^p).$$

From this and (10.5), we deduce

$$\int_{\Gamma} |u|^p dx \leq C \int_{U^+} (|u|^p + |Du|^p) dx,$$

so

$$\|u\|_{L^p(\Gamma)} \leq C\|u\|_{W^{1,p}(U^+)} \leq C\|u\|_{W^{1,p}(A)}. \tag{10.6}$$

If we cannot find a neighbourhood of x^0 the restriction of ∂A to which belongs in $\{x_n = 0\}$, there exist a $\mathscr{C}^1$ map Φ, with inverse Ψ, mapping a neighbourhood of x^0 to a neighbourhood of $y^0 = \Psi(x^0)$ so that, locally, $\Psi(\partial A)$ lies in $\{y_n = \Psi(x_n) = 0\}$.

Since ∂A is compact, there exists a finite number of points $x_1^0, x_2^0, \ldots, x_N^0$ and balls $U(x_i^0, r_i) = V_i$ such that $\partial A \subset \cup_{i=1}^N V_i$, $V_0 \subset\subset A$ and

$$A \subset \cup_{i=0}^N V_i.$$

From (10.6),

$$\|u\|_{L^p(\Gamma_i)} \leq C\|u\|_{W^{1,p}(U)}.$$

Let $\{\zeta_i\}_{i=0}^N$ be the partition of unity of the system $\{V_i\}_{i=0}^N$. Then

$$\begin{aligned}
\|u\|_{L^p(\partial A)} &= \left\|\sum_{i=0}^N \zeta_i u\right\|_{L^p(\partial A)} \\
&\leq \sum_{i=0}^N \|\zeta_i u\|_{L^p(\partial A)} \\
&= \sum_{i=0}^N \|u\|_{L^p(\Gamma_i)} \\
&\leq C\|u\|_{W^{1,p}(A)}.
\end{aligned} \tag{10.7}$$

Now, define the operator $T : W^{1,p}(A) \mapsto W^{1,p}(\partial A)$ by

$$Tu = u_{|\partial A}.$$

Using (10.7), we see that

$$\|Tu\|_{L^p(\partial A)} \le C\|u\|_{W^{1,p}(A)}.$$

Let $u \in W^{1,p}(A) \cap \mathscr{C}(\overline{A})$. There exists a sequence $\{u_m\}_{m=1}^{\infty}$ in $\mathscr{C}^{\infty}(\overline{A})$ such that $u_m \to u$ in $W^{1,p}(A)$, as $m \to \infty$. From (10.7), we have

$$\|Tu_m - Tu_l\|_{L^p(\partial A)} \le C\|u_m - u_l\|_{W^{1,p}(A)} \to_{m,l\to\infty} 0.$$

Therefore the sequence $\{Tu_m\}_{m=1}^{\infty}$ is fundamental in the Banach space $L^p(\partial A)$, and as such it converges in $L^p(\partial A)$:

$$\lim_{m\to\infty} Tu_m = Tu.$$

As $Tu_m = u_{|\partial A}$, we infer that $Tu = u_{|\partial A}$ and T is a bounded operator.

Definition 10.6 We call Tu the trace of u on ∂U.

Exercise 10.12 Let A be a bounded set in $\mathbb{R}^n$ and assume ∂A is $\mathscr{C}^1$. Prove that Tu vanishes on ∂A, provided $u \in W_0^{1,p}(A)$.

Hint Use the fact that there exists a sequence $\{u_m\}_{m=1}^{\infty}$ in $\mathscr{C}_0^{\infty}(A)$ such that $u_m \to u$ in $W^{1,p}(A)$, as $m \to \infty$. Since Tu_m is zero on ∂A, we also have $Tu = 0$ on ∂A.

Exercise 10.13 Let A be a bounded set in $\mathbb{R}^n$ and let ∂A be $\mathscr{C}^1$. Take $u \in W^{1,p}(A)$ with $Tu = 0$ on ∂A. Prove that

$$\int_{\mathbb{R}^{n-1}} |u(x', x_n)|^p dx' \le Cx_n^{p-1} \int_0^{x_n} \int_{\mathbb{R}^{n-1}} |Du|^p dx' dt$$

for a.e. $x_n > 0$. Here $x' = (x_1, x_2, \ldots, x_{n-1})$.

Hint Use the extension theorem, then choose $\{u_m\}_{m=1}^{\infty}$ in $\mathscr{C}^1(\overline{R}_+^n)$ such that $u_m \to u$ in $W^{1,p}(\overline{R}_+^n)$, as $m \to \infty$. Then use the identity

$$u_m(x', x_n) - u_m(x', 0) = \int_0^{x_n} u_{mx_n}(x', s) ds.$$

Here $\overline{R}_+^n$ is the closure of $\{x = (x_1, x_2, \ldots, x_n) \in \mathbb{R}^n : x_n > 0\}$.

Exercise 10.14 Let A be a bounded set in $\mathbb{R}^n$ with ∂A of class $\mathscr{C}^1$. Take $u \in W^{1,p}(A)$ with $Tu = 0$ on ∂A. Prove that $u \in W_0^{1,p}(A)$.

Hint Use the extension theorem. Consider the function $\zeta \in \mathscr{C}^\infty(\mathbb{R})$ such that $\zeta \equiv 1$ on $[0,1]$, $0 \leq \zeta \leq 1$ on $\mathbb{R}$, $\zeta \equiv 0$ on $\mathbb{R}\backslash[0,2]$, and the two sequences $\zeta_m(x) = \zeta(mx_n)$, $w_m(x) = u(x)(1 - \zeta_m(x))$, $x \in \mathbb{R}^n_+$. Prove that $w_m \to_{m\to\infty} u$ in $W^{1,p}(\mathbb{R}^n_+)$. Mollify w_m to produce functions $u_m \in \mathscr{C}^\infty_0(\mathbb{R}^n_+)$ such that $u_m \to_{m\to\infty} u$ in $W^{1,p}(\mathbb{R}^n_+)$.

10.6 Sobolev Inequalities

Definition 10.7 Let $1 \leq p < n$. The Sobolev conjugate p^* of p is defined by

$$\frac{1}{p^*} = \frac{1}{p} - \frac{1}{n}.$$

Theorem 10.10 (Gagliardo-Nirenberg-Sobolev Inequality) *Let* $u \in \mathscr{C}^1_0(\mathbb{R}^n)$. *Then*

$$||u||_{L^{p^*}(\mathbb{R}^n)} \leq C||Du||_{L^p(\mathbb{R}^n)}$$

for some constant $C > 0$, $1 \leq p < n$.

Proof

Case 1.   Let $p = 1$. Then $p^* = \dfrac{n}{n-1}$ and we have to prove that

$$||u||_{L^{\frac{n}{n-1}}(\mathbb{R}^n)} \leq C||Du||_{L^1(\mathbb{R}^n)}.$$

Our first observation is

$$|u(x)| = \left| \int_{-\infty}^{x_i} u_{x_i}(x_1, \ldots, x_{i-1}, y_i, x_{i+1}, \ldots, x_n)dy_i \right|$$

$$\leq \int_{-\infty}^{x_i} |u_{x_i}(x_1, \ldots, x_{i-1}, y_i, x_{i+1}, \ldots, x_n)|dy_i$$

$$\leq \int_{-\infty}^{\infty} |Du(y_i)|dy_i, \quad i = 1, 2, \ldots, n.$$

Then

$$|u(x)|^{\frac{1}{n-1}} \leq \left(\int_{-\infty}^{\infty} |Du(y_i)|dy_i \right)^{\frac{1}{n-1}}, \quad i = 1, 2, \ldots, n,$$

i.e.,

$$|u(x)|^{\frac{1}{n-1}} \le \left(\int_{-\infty}^{\infty} |Du(y_1)| dy_1 \right)^{\frac{1}{n-1}},$$

$$|u(x)|^{\frac{1}{n-1}} \le \left(\int_{-\infty}^{\infty} |Du(y_2)| dy_2 \right)^{\frac{1}{n-1}},$$

$$\dots$$

$$|u(x)|^{\frac{1}{n-1}} \le \left(\int_{-\infty}^{\infty} |Du(y_n)| dy_n \right)^{\frac{1}{n-1}}.$$

We multiply the above inequalities and get

$$|u(x)|^{\frac{n}{n-1}} \le \prod_{i=1}^{n} \left(\int_{-\infty}^{\infty} |Du(y_i)| dy_i \right)^{\frac{1}{n-1}}.$$

Now, we integrate in the variable x_1 and obtain

$$\int_{-\infty}^{\infty} |u(x)|^{\frac{n}{n-1}} dx_1 \le \int_{-\infty}^{\infty} \prod_{i=1}^{n} \left(\int_{-\infty}^{\infty} |Du(y_i)| dy_i \right)^{\frac{1}{n-1}} dx_1$$

$$\le \left(\int_{-\infty}^{\infty} |Du(y_1)| dy_1 \right)^{\frac{1}{n-1}} \prod_{i=2}^{n} \left(\int_{-\infty}^{\infty} \int_{-\infty}^{\infty} |Du(y_i)| dy_i dx_1 \right)^{\frac{1}{n-1}}.$$

Integrating now in x_2 gives

$$\int_{-\infty}^{\infty} \int_{-\infty}^{\infty} |u(x)|^{\frac{n}{n-1}} dx_1 dx_2$$

$$\le \int_{-\infty}^{\infty} \left(\int_{-\infty}^{\infty} |Du(y_1)| dy_1 \right)^{\frac{1}{n-1}} \prod_{i=2}^{n} \left(\int_{-\infty}^{\infty} \int_{-\infty}^{\infty} |Du(y_i)| dy_i dx_1 \right)^{\frac{1}{n-1}} dx_2$$

$$\le \left(\int_{-\infty}^{\infty} \int_{-\infty}^{\infty} |Du(y_2)| dy_2 dx_1 \right)^{\frac{1}{n-1}} \left(\int_{-\infty}^{\infty} \int_{-\infty}^{\infty} |Du(y_1)| dy_1 dx_2 \right)^{\frac{1}{n-1}}$$

$$\times \prod_{i=3}^{n} \left(\int_{-\infty}^{\infty} \int_{-\infty}^{\infty} \int_{-\infty}^{\infty} |Du(y_i)| dy_i dx_1 dx_2 \right)^{\frac{1}{n-1}}.$$

Iterating,

$$\int_{\mathbb{R}^n} |u(x)|^{\frac{n}{n-1}} dx \le C\left(\int_{\mathbb{R}^n} |Du(x)| dx\right)^{\frac{n}{n-1}}, \tag{10.8}$$

and hence,

$$\|u\|_{L^{\frac{n}{n-1}}(\mathbb{R}^n)} \le C\|Du\|_{L^1(\mathbb{R}^n)}.$$

Case 2. Let $p > 1$. We put $v = |u|^\gamma$, where γ will be determined subsequently. We apply the inequality (10.8) to v and get

$$\left(\int_{\mathbb{R}^n} |u(x)|^{\frac{\gamma n}{n-1}} dx\right)^{\frac{n-1}{n}} \le C\int_{\mathbb{R}^n} |u(x)|^{\gamma-1}|Du(x)| dx.$$

By Hölder's inequality

$$\left(\int_{\mathbb{R}^n} |u(x)|^{\frac{\gamma n}{n-1}} dx\right)^{\frac{n-1}{n}} \le C\left(\int_{\mathbb{R}^n} |u(x)|^{\frac{(\gamma-1)p}{p-1}} dx\right)^{\frac{p-1}{p}} \|Du\|_{L^p(\mathbb{R}^n)}. \tag{10.9}$$

Now, we take $\gamma > 0$ such that

$$\frac{\gamma n}{n-1} = \frac{(\gamma-1)p}{p-1},$$

so

$$\gamma = \frac{p(n-1)}{n-p}, \quad \frac{\gamma n}{n-1} = \frac{pn}{n-p} = p^*.$$

From (10.9), we find

$$\left(\int_{\mathbb{R}^n} |u(x)|^{p^*} dx\right)^{\frac{n-1}{n}} \le C\left(\int_{\mathbb{R}^n} |u(x)|^{p^*} dx\right)^{\frac{p-1}{p}} \|Du\|_{L^p(\mathbb{R}^n)},$$

and then

$$\|u\|_{L^{p^*}(\mathbb{R}^n)} \le C\|Du\|_{L^p(\mathbb{R}^n)}.$$

This completes the proof.

Exercise 10.15 Let $1 \le p < n$ and consider A bounded, open in $\mathbb{R}^n$ with $\mathscr{C}^1$ boundary. For $u \in W^{1,p}(A)$ prove

$$\|u\|_{L^{p^*}(A)} \le \|u\|_{W^{1,p}(A)}.$$

Hint Apply the Extension theorem so to ensure the existence of the extension $\bar{u} \in W^{1,p}(\mathbb{R}^n)$ of u. Then choose a sequence $\{u_m\}_{m=1}^{\infty}$ in $\mathscr{C}_0^{\infty}(\mathbb{R}^n)$ such that $u_m \to \bar{u}$ in $W^{1,p}(\mathbb{R}^n)$, as $m \to \infty$. Apply the Gagliardo-Nirenberg-Sobolev inequality to $u_m - u_l$ and conclude that $\{u_m\}_{m=1}^{\infty}$ converges to $\bar{u}$ in $L^{p^*}(\mathbb{R}^n)$. Eventually, the Gagliardo-Nirenberg-Sobolev inequality on u_m gives the desired result.

Exercise 10.16 Let A be a bounded set in $\mathbb{R}^n$ with ∂A of class $\mathscr{C}^1$, $u \in W_0^{1,p}(A)$ for $1 \le p < n$. Prove

$$\|u\|_{L^q(A)} \le C\|Du\|_{L^p(A)}$$

for some constant $C > 0$ and for every $q \in [1, p^*]$.

Hint Use approximation and the Gagliardo-Nirenberg-Sobolev inequality.

Theorem 10.11 (Morrey Inequality) *Let* $n < p \le \infty$. *Then for every* $u \in \mathscr{C}^1(\mathbb{R}^n)$ *there exists a constant* $C = C(n, p) > 0$ *such that*

$$\|u\|_{\mathscr{C}^{0,\gamma}(\mathbb{R}^n)} \le C\|u\|_{W^{1,p}(\mathbb{R}^n)},$$

where $\gamma = 1 - \dfrac{n}{p}$.

Proof Let $U(x, r) \subset \mathbb{R}^n$ be arbitrary ball, $r > 0$, $s \in (0, r]$. Fix $w \in \partial U(0, 1)$, so

$$|u(x + sw) - u(x)| = \left| \int_0^s \frac{d}{dt} u(x + tw)dt \right| = \left| \int_0^s Du(x + tw) \cdot wdt \right|$$

$$\le \int_0^s |Du(x + tw) \cdot w|dt \le \int_0^s |Du(x + tw)|dt,$$

which we integrate on $\partial U(0, 1)$ and get

$$\int_{\partial U(0,1)} |u(x + sw) - u(x)|dS \le \int_{\partial U(0,1)} \int_0^s |Du(x + tw)|dtdS$$

$$= \int_0^s \int_{\partial U(0,1)} |Du(x + tw)|dSdt = \int_0^s \int_{\partial U(0,1)} |Du(x + tw)|\frac{t^{n-1}}{t^{n-1}}dSdt$$

$$(x + tw = y)$$

$$\le \int_0^s \int_{\partial U(x,t)} |Du(y)|\frac{1}{|x - y|^{n-1}}dS_ydt = \int_{U(x,s)} |Du(y)|\frac{1}{|x - y|^{n-1}}dy$$

$$\le \int_{U(x,r)} |Du(y)|\frac{1}{|x - y|^{n-1}}dy.$$

From here,

$$s^{n-1} \int_{\partial U(0,1)} |u(x+sw) - u(x)| dS \le s^{n-1} \int_{U(x,r)} |Du(y)| \frac{1}{|x-y|^{n-1}} dy.$$

Therefore

$$\int_0^r s^{n-1} \int_{\partial U(0,1)} |u(x+sw) - u(x)| dS ds \le \int_0^r s^{n-1} \int_{U(x,r)} |Du(y)| \frac{1}{|x-y|^{n-1}} dy ds,$$

so

$$\int_0^r s^{n-1} \int_{\partial U(0,1)} |u(x+sw) - u(x)| dS ds \le \frac{r^n}{n} \int_{U(x,r)} |Du(y)| \frac{1}{|x-y|^{n-1}} dy,$$

and substituting $x + sw = y$ finally

$$\int_0^r \int_{\partial U(x,s)} |u(y) - u(x)| dS_y ds \le \frac{r^n}{n} \int_{U(x,r)} |Du(y)| \frac{1}{|x-y|^{n-1}} dy$$

and

$$\frac{1}{r^n} \int_{U(x,r)} |u(y) - u(x)| dy \le \frac{1}{n} \int_{U(x,r)} |Du(y)| \frac{1}{|x-y|^{n-1}} dy.$$

We set $\dfrac{1}{r^n} \displaystyle\int_{U(x,r)} (\cdot) dy = \fint_{U(x,r)} (\cdot) dy$. Then there exists a constant $C > 0$ such that

$$\fint_{U(x,r)} |u(y) - u(x)| dy \le C \int_{U(x,r)} |Du(y)| \frac{1}{|x-y|^{n-1}} dy. \tag{10.10}$$

On the other hand,

$$|u(x)| = \fint_{U(x,1)} |u(y)| dy = \fint_{U(x,1)} |u(x) - u(y) + u(y)| dy$$

$$\le \fint_{U(x,1)} |u(x) - u(y)| dy + \fint_{U(x,1)} |u(y)| dy$$

$$\leq \fint_{U(x,1)} |u(x) - u(y)| \, dy + C\|u\|_{L^p(U(x,1))}$$

$$\leq \fint_{U(x,1)} |u(x) - u(y)| \, dy + C\|u\|_{L^p(\mathbb{R}^n)} \quad \text{(by (10.10))}$$

$$\leq \fint_{U(x,1)} \frac{|Du(y)|}{|x-y|^{n-1}} \, dy + C\|u\|_{L^p(\mathbb{R}^n)}$$

$$\leq C \int_{U(x,1)} \frac{|Du(y)|}{|x-y|^{n-1}} \, dy + C\|u\|_{L^p(\mathbb{R}^n)}$$

(Hölder's inequality)

$$\leq C \left(\int_{U(x,1)} |Du(y)|^p \, dy \right)^{\frac{1}{p}} \left(\int_{U(x,1)} \frac{1}{|x-y|^{\frac{(n-1)p}{p-1}}} \, dy \right)^{\frac{p-1}{p}} + C\|u\|_{L^p(\mathbb{R}^n)}$$

$$\leq C(\|u\|_{L^p(\mathbb{R}^n)} + \|Du\|_{L^p(\mathbb{R}^n)}) \leq C\|u\|_{W^{1,p}(\mathbb{R}^n)}.$$

Therefore

$$\sup_{x \in \mathbb{R}^n} |u(x)| \leq C\|u\|_{W^{1,p}(\mathbb{R}^n)}.$$

Let $x, y \in \mathbb{R}^n$ be arbitrary points and $W = U(x,r) \cap U(y,r)$. Then

$$|u(x) - u(y)| = \fint_W |u(x) - u(y)| \, dz = \fint_W |u(x) - u(z) + u(z) - u(y)| \, dz$$

$$\leq \fint_W |u(x) - u(z)| \, dz + \fint_W |u(z) - u(y)| \, dz.$$

$$(10.11)$$

The inequality (10.10) allows us to estimate

$$\fint_W |u(x) - u(z)| \, dz \leq \fint_{U(x,r)} |u(x) - u(z)| \, dz \leq C \int_{U(x,r)} \frac{|Du(z)|}{|x-z|^{n-1}} \, dz \text{(Hölder's inequality)}$$

$$\leq C \left(\int_{U(x,r)} |Du(z)|^p \, dz \right)^{\frac{1}{p}} \left(\int_{U(x,r)} \frac{1}{|x-z|^{\frac{(n-1)p}{p-1}}} \, dz \right)^{\frac{p-1}{p}}$$

$$\leq C \left(r^{n - \frac{(n-1)p}{p-1}} \right)^{\frac{p-1}{p}} \|Du\|_{L^p(U(x,r))} \leq C|x-y|^{1-\frac{n}{p}} \|Du\|_{L^p(\mathbb{R}^n)}$$

$$= C|x-y|^{\gamma} \|Du\|_{L^p(\mathbb{R}^n)} \leq C|x-y|^{\gamma} \|Du\|_{W^{1,p}(\mathbb{R}^n)}.$$

$$(10.12)$$

As above, we deduce

$$\overline{\int}_W |u(y) - u(z)|dz \le C|x - y|^\gamma \|Du\|_{W^{1,p}(\mathbb{R}^n)}.$$

From the latter, (10.11) and (10.12), we get

$$|u(x) - u(y)| \le C|x - y|^\gamma \|u\|_{W^{1,p}(\mathbb{R}^n)},$$

so

$$\sup_{x \ne y, x, y \in \mathbb{R}^n} \frac{|u(x) - u(y)|}{|x - y|^\gamma} \le C\|u\|_{W^{1,p}(\mathbb{R}^n)}$$

or

$$\|u\|_{\mathscr{C}^{0,\gamma}(\mathbb{R}^n)} \le C\|u\|_{W^{1,p}(\mathbb{R}^n)}.$$

This completes the proof.

To prove the Poincaré inequality we have a need of the following Rellich-Kondrachov compactness theorem. For its proof we refer the reader to [6].

Theorem 10.12 (Rellich-Kondrachov Compactness Theorem) *Let A be a bounded open set in $\mathbb{R}^n$ with $\mathscr{C}^1$ boundary and $1 \le p < n$. Then*

$$W^{1,p}(A) \hookrightarrow L^q(A)$$

for every $1 \le q < p^*$, where p^* is the Sobolev conjugate of p.

Exercise 10.17 Let A be a bounded open set in $\mathbb{R}^n$ with $\mathscr{C}^1$ boundary. Prove that $W^{1,p}(A) \hookrightarrow L^p(A)$, $1 \le p \le \infty$.

Definition 10.8 One calls

$$(u)_A = \frac{1}{|A|} \int_A u(y)dy$$

the average of u over A.

Theorem 10.13 (Poincaré Inequality) *Let A be a bounded, connected, open set in $\mathbb{R}^n$, $1 \le p \le \infty$. Then for every $u \in W^{1,p}(A)$ there exists a constant $C = C(u, p, A)$ such that*

$$\|u - (u)_A\|_{L^p(A)} \le C\|Du\|_{L^p(A)}.$$

Proof Let us suppose that there exists a function $u_k \in W^{1,p}(A)$ such that

$$||u_k - (u_k)_A||_{L^p(A)} > k||Du_k||_{L^p(A)}. \tag{10.13}$$

Let

$$v_k = \frac{u_k - (u_k)_A}{||u_k - (u_k)_A||_{L^p(A)}}.$$

Then

$$(v_k)_A = \frac{1}{||u_k - (u_k)_A||_{L^p(A)}} \overline{\int}_A (u_k(y) - (u_k)_A) dy$$

$$= \frac{1}{||u_k - (u_k)_A||_{L^p(A)}} \left(\overline{\int}_A u_k(y) dy - (u_k)_A \overline{\int}_A dy \right)$$

$$= \frac{1}{||u_k - (u_k)_A||_{L^p(A)}} ((u_k)_A - (u_k)_A) = 0$$

and

$$||v_k||_{L^p(A)} = \left\| \frac{u_k - (u_k)_A}{||u_k - (u_k)_A||_{L^p(A)}} \right\|_{L^p(A)} = \frac{||u_k - (u_k)_A||_{L^p(A)}}{||u_k - (u_k)_A||_{L^p(A)}} = 1.$$

From (10.13), we infer

$$\frac{1}{k} > \frac{||Du_k||_{L^p(A)}}{||u_k - (u_k)_A||_{L^p(A)}}$$
$$= \frac{||D(u_k - (u_k)_A)||_{L^p(A)}}{||u_k - (u_k)_A||_{L^p(A)}} \tag{10.14}$$
$$= \left\| D\frac{u_k - (u_k)_A}{||u_k - (u_k)_A||_{L^p(A)}} \right\|_{L^p(A)}$$
$$= ||Dv_k||_{L^p(A)}.$$

Consequently $\{v_k\}_{k=1}^{\infty}$ is a bounded sequence in $L^p(A)$ and $W^{1,p}(A)$. By the Rellich-Kondrachov compactness theorem $W^{1,p}(A) \hookrightarrow L^p(A)$. Therefore there is a subsequence $\{v_{k_j}\}_{j=1}^{\infty}$ that converges to some $v \in L^p(A)$. From (10.14), we have

$$\lim_{j \to \infty} ||Dv_{k_j}||_{L^p(A)} = 0.$$

Fix an arbitrary $\phi \in \mathscr{C}_0^{\infty}(A)$. Then

$$\int_A v\phi_{x_i} dx = \lim_{j \to \infty} \int_A v_{k_j} \phi_{x_i} dx = -\lim_{j \to \infty} \int_A v_{k_j x_i} \phi dx = 0$$

and $Dv = 0$. Consequently $v \in W^{1,p}(A)$. Since A is connected and $Dv = 0$, it follows that $v = $ const. Let $v = l$. Because $\|v_k\|_{L^p(A)} = 1$, we have $\|v\|_{L^p(A)} = 1$. From $(v_k)_A = 0$, we conclude that $(v)_A = 0$, so $l = 0$, $\|v\|_{L^p(A)} = 0$, which is a contradiction.

Exercise 10.18 (Poincaré Inequality on the Ball) Given $1 \le p \le \infty$, there exists a constant $C = C(u, p, A)$ such that

$$\|u - (u)_{U(x,r)}\|_{L^p(U(x,r))} \le Cr \|Du\|_{L^p(U(x,r))}.$$

Hint Consider the function $v(y) = u(x + ry)$, $y \in U_1$. Use the Poincaré inequality for U_1 and change variables $y = \dfrac{z - x}{r}$.

10.7 The Space H^{-s}

Definition 10.9 We denote by $H^{-s}(\mathbb{R}^n)$, for any $0 < s < \infty$, the dual space to $H_0^s(\mathbb{R}^n)$. In other words, $f \in H^{-s}(\mathbb{R}^n)$ is a bounded linear functional on $H_0^s(\mathbb{R}^n)$.
We also set

$$H^{-\infty}(\mathbb{R}^n) = \cup_{s \in R} H^s(\mathbb{R}^n).$$

Theorem 10.14 (Characterization of H^{-s}) *For any given $0 < s < \infty$, any element $u \in \mathscr{S}'(\mathbb{R}^n) \cap H^{-s}(\mathbb{R}^n)$ can be written as*

$$u = \sum_{|\alpha| \le s} D^\alpha h_\alpha, \quad with \quad h_\alpha \in L^2(\mathbb{R}^n).$$

Proof We have that

$$f = (1 + |\xi|^2)^{-\frac{s}{2}} \mathscr{F}(u) \in L^2(\mathbb{R}^n).$$

Then

$$\mathscr{F}(u) = (1 + |\xi|^2)^{\frac{s}{2}} f = \left(1 + \sum_{j=1}^n |\xi_j|^s\right) \frac{(1 + |\xi|^2)^{\frac{s}{2}} f}{1 + \sum_{j=1}^n |\xi_j|^s}$$

$$= \left(1 + \sum_{j=1}^n |\xi_j|^s\right) g = g + \sum_{j=1}^n \xi_j^s \left(\frac{|\xi_j|^s}{\xi_j^s} g\right),$$

where

$$g = \frac{(1 + |\xi|^2)^{\frac{s}{2}} f}{1 + \sum\limits_{j=1}^{n} |\xi_j|^s} \in L^2(\mathbb{R}^n).$$

We set

$$g_j = \frac{|\xi_j|^s}{(-i\xi_j)^s} g \in L^2(\mathbb{R}^n), \quad h = \mathscr{F}^{-1}(g), \quad h_j = \mathscr{F}^{-1}(g_j), \quad j = 1, 2, \ldots, n.$$

Then $h, h_j \in L^2(\mathbb{R}^n)$, $j = 1, 2, \ldots, n$, so

$$\mathscr{F}(u) = \mathscr{F}(h) + \sum_{j=1}^{n}(-i\xi_j)^s \mathscr{F}(h_j) = \mathscr{F}(h) + \sum_{j=1}^{n} \mathscr{F}(D_j^s h_j),$$

and therefore

$$u = h + \sum_{j=1}^{n} D_j^s h_j.$$

This completes the proof.

Exercise 10.19 Prove that $H^{-\infty}(\mathbb{R}^n) \subset \mathscr{S}'(\mathbb{R}^n)$.

10.8   Advanced Practical Problems

Problem 10.1 Let $k > 0$ be given, and set

$$U = \{(x, y) \in \mathbb{R}^2 : 0 < x < 1, x^k < y < 2x^k\},$$

$u(x, y) = y^\alpha$, $(x, y) \in U$. Find conditions on α so that $u \in H^m(U)$, $m = 1, 2, \ldots$.

Problem 10.2 Let $u(x) = |x|^{-\alpha}$, $x \in U_1$, $x \neq 0$. Find conditions on $\alpha > 0$, n and p so that $u \in W^{1,p}(U_1)$.

Problem 10.3 Find conditions for s so that $e^{-|x|} \in H^s(\mathbb{R}^n)$.

Problem 10.4 Find conditions for s so that $\delta \in H^s(\mathbb{R}^n)$.

Problem 10.5 Let $n > 1$. Prove that $u = \log\log\left(1 + \frac{1}{|x|}\right)$ belongs to $W^{1,p}(U_1)$.

Problem 10.6 Prove the following interpolation inequality

$$\int_A |Du|^2 dx \le \left(\int_A |u|^2 dx\right)^{\frac{1}{2}}\left(\int_A |D^2 u|^2 dx\right)^{\frac{1}{2}},$$

for every $u \in \mathscr{C}_0^\infty(A)$, $A \subset \mathbb{R}^n$ open and bounded. Using approximation, prove it for $u \in H^1(A) \cap H_0^1(A)$.

Problem 10.7 Prove the interpolation inequality

$$\int_A |Du|^p dx \le \left(\int_A |u|^p dx\right)^{\frac{1}{2}}\left(\int_A |D^2 u|^p dx\right)^{\frac{1}{2}}$$

for $2 \le p < \infty$ and every $u \in W^{2,p}(A) \cap W_0^{1,p}(A)$, where $A \subset \mathbb{R}^n$ is open and bounded.

Problem 10.8 Prove that any $u \in W^{1,p}((a,b))$ can be extended to $W^{1,p}(\mathbb{R})$.

Problem 10.9 Let A be an open bounded set in $\mathbb{R}^n$ with $\mathscr{C}^1$ boundary, take $u \in W^{m,p}(A)$, $m < \dfrac{n}{p}$. Prove that $u \in L^q(A)$, where $\dfrac{1}{q} = \dfrac{1}{p} - \dfrac{m}{n}$, and

$$||u||_{L^q(A)} \le C||u||_{W^{m,p}(A)}.$$

Hint Use the fact that $D^\alpha u \in L^p(A)$ for any $u \in W^{m,p}(A)$ and every α such that $|\alpha| \le m$. Applying the Gagliardo-Nirenberg-Sobolev inequality deduce

$$||D^\beta u||_{L^{p^*}(A)} \le C||D^\alpha u||_{L^p(A)} \le C||u||_{W^{m,p}(A)}$$

for some constant $C > 0$, $|\alpha| = m$ and $|\beta| = m - 1$, $\dfrac{1}{p^*} = \dfrac{1}{p} - \dfrac{1}{n}$. Conclude $u \in W^{m-1,p^*}(A)$. Then using again the Gagliardo-Nirenberg-Sobolev inequality, prove that

$$||D^\gamma u||_{L^{p^{**}}(A)} \le C||D^\beta u||_{L^{p^*}(A)} \le C||u||_{W^{m,p}(A)}.$$

Deduce that $u \in W^{m-2,p^{**}}$, where $\dfrac{1}{p^{**}} = \dfrac{1}{p^*} - \dfrac{1}{n} = \dfrac{1}{p} - \dfrac{2}{n}$, and so forth. Eventually conclude that $u \in W^{0,q}(A)$ and that the given inequality holds.

Problem 10.10 Let A be an open bounded set in $\mathbb{R}^n$ with $\mathscr{C}^1$ boundary. Take $u \in W^{m,p}(A)$, $m > \dfrac{n}{p}$ and prove

$$u \in \mathscr{C}^{m-\left[\frac{n}{p}\right]-1,\gamma}(\overline{A}),$$

where

$$\gamma = \begin{cases} \left[\dfrac{n}{p}\right] + 1 - \dfrac{n}{p}, & \text{if } \dfrac{n}{p} \text{ is not an integer} \\ \text{any positive number } < 1, & \text{otherwise.} \end{cases}$$

Also show that u satisfies the inequality

$$\|u\|_{\mathscr{C}^{m-\left[\frac{n}{p}\right]-1,\gamma}(A)} \leq C\|u\|_{W^{m,p}(A)}$$

for some positive constant C.

Hint Use the Extension theorem, Morrey's inequality and the Gagliardo-Nirenberg-Sobolev inequality.

Problem 10.11 Prove $H^{+\infty} = \cap_{s \in \mathbb{R}} H^s(\mathbb{R}^n)$.

Problem 10.12 Show $H^{+\infty}(\mathbb{R}^n) \subset \mathscr{C}^{\infty}(\mathbb{R}^n)$.

10.9   Notes and References

In this chapter we introduce Sobolev spaces and we deduct some of their basic properties. They are investigated approximations by smooth functions and it is proved the main extension theorem. It is defined a trace of an element of a Sobolev space and it is proved the trace theorem. In the chapter are deducted the Gagliardo-Nirenberg-Sobolev inequality, the Morrey inequality and the Poincaré inequality. They are introduced the dual spaces H^{-s} of the spaces H^s. Additional materials can be found in [1, 2, 6, 29, 30].

References

1. R. Adams, *Sobolev Spaces* (Academic Press, New York, 1975)
2. R. Adams, J.J.F. Fournier, *Sobolev Spaces*, 2nd edn. (Elsevier, Oxford, 2003)
3. N. Bourbaki, *Elements of Mathematics. General Topology. Part 1* (Hermann, Paris, 1966)
4. R. Courant, D. Hilbert, *Methods of Mathematical Physics*, vol. I (Interscience, New York, 1953)
5. R. Courant, D. Hilbert, *Methods of Mathematical Physics. Partial Differential Equations*, vol. II (Interscience, New York, 1962)
6. L. Evans, *Partial Differential Equations*, vol. 19 (AMS, Providence, 1997)
7. G. Friedlander, M. Joshi, *Introduction to the Theory of Distributions*, 2nd edn. (Cambridge University Press, New York, 1998)
8. B. Fuchssteiner, D. Laugwitz, *Funktional Analysis* (BI Wissenschaftsverlag, Zürich, 1974)
9. G.B. Folland, *Introduction to Partial Differential Equations*, 2nd edn. (Princeton University Press, Princeton, 1995)
10. G.B. Folland, *Real Analysis* (Wiley, New York, 1999)
11. O. Forster, *Analysis 1*, 8 Auflage. (Vieweg Verlag, Wiesbaden, 2006)
12. O. Forster, *Analysis 2*, 6 Auflage. (Vieweg Verlag, Wiesbaden, 2005)
13. O. Forster, *Analysis 3*, 3 Auflage. (Vieweg Verlag, Wiesbaden, 1984)
14. D. Gillbarg, N. Trudinger, *Elliptic Partial Differential Equations of Second Order*, 2nd edn. (Springer, Berlin, 1987)
15. M. Grosser, M. Kunzinger, M. Oberguggenberger, R. Steinbauer, *Geometric Theory of Generalized Functions* (Kluwer, Dordrecht, 2001)
16. D. Haroske, H. Triebel, *Distributions, Sobolev Spaces, Elliptic Equations* (European Math. Society, Zürich, 2008)
17. L. Hörmander, *The Analysis of Linear Partial Differential Operators*, vol. I, 2nd edn. (Springer, Berlin, 1990)
18. G. Hörmander, Analysis (Lecture notes, University of Vienna). Available electronically at http://www.mat.univie.ac.at/gue/material.html, 2008–09
19. R.V. Kadison, J.R. Ringrose, *Fundamentals of the Theory of Operator Algebras. Advanced Theory*, vol. II (Academic, New York, 1986)
20. M. Kunzinger, Distributionen theorie II (Lecture notes, University of Vienna, spring term 1998). Available from the author, 1998
21. E.H. Lieb, M. Loss, *Analysis. Graduate Studies in Mathematics*, vol. 14, 2nd edn. (American Mathematical Society, Providence, 2001)
22. F. Riesz, B.Sz. Nagy, *Vorlesungen über funktional analysis* (VEB Deutscher Verlag der Wissenschaften, Berlin, 1982)

© The Author(s), under exclusive license to Springer Nature Switzerland AG 2021
S. G. Georgiev, *Theory of Distributions*,
https://doi.org/10.1007/978-3-030-81265-2

23. X. Saint Raymond, *Elementary Introduction to the Theory of Pseudodifferential Operators* (CRC Press, Boca Raton, 1991)
24. H.H. Schaefer, *Topological Vector Spaces* (Springer, New York, 1966)
25. W. Schempp, B. Dreseler, *Einführung in die harmonische analyse* (B. G. Teubner, Stuttgart, 1980)
26. L.A. Steen, J.A. Seebach, *Counterexamples in Topology* (Dover, Mineola, 1995). Reprint of the second edition (1978)
27. E. Stein, R. Shakarchi, *Complex Analysis. Princeton Lectures in Analysis II* (Princeton University Press, Princeton, 2003)
28. R. Steinbauer, Locally convex vector spaces (Lecture notes, University of Vienna, fall term 2008). Available from the author, 2009
29. L. Tartar, *An Introduction to Sobolev Spaces and Interpolation Spaces* (Springer, Berlin, 2007)
30. R. Temam, *Navier-Stokes Equations* (North Holland, Amsterdam, 1977)
31. H. Triebel, *Theory of Function Spaces*. Monogr. Math., vol. 78 (Birkhäuser, Basel, 1983)
32. H. Triebel, *Theory of Function Spaces*. Monogr. Math., vol. 84 (Birkhäuser II, Basel, 1992)
33. H. Triebel, *Theory of Function Spaces*. Monogr. Math., vol. 91 (Birkhäuser III, Basel, 2006)
34. V. Vladimirov, *Elements of Mathematical Physics* (Dekker, New York, 1971)
35. V. Vladimirov, *Methods of the Theory of Generalized Functions* (Dekker, New York, 1984)
36. D. Werner, *Funktional Analysis*, fünfte Auflage. (Springer, Berlin, 2005)

Index

© The Author(s), under exclusive license to Springer Nature Switzerland AG 2021
S. G. Georgiev, *Theory of Distributions*,
https://doi.org/10.1007/978-3-030-81265-2

Printed in the United States
by Baker & Taylor Publisher Services